Practical
Digital
Electronics

Nigel P. Cook

Upper Saddle River, New Jersey
Columbus, Ohio

To Dawn, Candy, and Jon

Library of Congress Cataloging-in-Publication Data
Cook, Nigel P.,
    Practical digital electronics / Nigel P. Cook.
      p. cm.
    ISBN 0-13-111060-8
    1. Digital electronics—Textbooks. I. Title.

TK7868.D5C98 2004
621.381—dc21   2003056302

Editor in Chief: Stephen Helba
Acquisitions Editor: Dennis Williams
Development Editor: Kate Linsner
Production Editor: Rex Davidson
Production Coordinator: Holly Henjum, Carlisle Publishers Services
Design Coordinator: Diane Ernsberger
Cover Designer: Ali Mohrman
Production Manager: Pat Tonneman
Marketing Manager: Ben Leonard

This book was set in Times by Carlisle Communications, Ltd. It was printed and bound by
R. R. Donnelley & Sons Company. The cover was printed by Phoenix Color Corp.

Pearson Education Ltd.   Pearson Education Australia Pty. Limited
Pearson Education Singapore Pte. Ltd.   Pearson Education North Asia Ltd.
Pearson Education Canada, Ltd.   Pearson Educación de Mexico, S. A. de C.V.
Pearson Education—Japan   Pearson Education Malaysia Pte. Ltd.

10 9 8 7 6 5 4 3 2 1
ISBN: 0-13-111060-8

Books by Nigel P. Cook

Combination Books

Electronics: A Complete Course

Introductory DC/AC Electronics

Series Books

Practical Electricity

Practical Electronics

Practical Digital Electronics

Math

Introductory Mathematics, Fourth Edition
8 Math Chapters

Introductory Computer Mathematics, Second Edition
7 Math Chapters, 6 Computer Math Chapters

Mathematics for Electronics and Computers
7 Math Chapters, 10 Electronics Math Chapters,
6 Computer Math Chapters

Mathematics for Technical Trades
6 Math Chapters

DC/AC

Electronics: A Complete Course, Second Edition
9 DC/AC Chapters, 6 Devices Chapters,
14 Digital Chapters

Practical Electricity, Second Edition
14 DC/AC Chapters

Introductory DC/AC Circuits, Fifth Edition
14 DC/AC Chapters

Introductory DC/AC Electronics, Fifth Edition
14 DC/AC Chapters, 6 Devices Chapters

Devices

Practical Electronics, Second Edition
14 Devices Chapters

Introductory Semiconductor Electronics
19 Devices Chapters

Microwave Principles and Systems
7 Microwave Communication Chapters

Digital

Practical Digital Electronics
14 Digital Chapters

Introductory Digital Electronics
17 Digital Chapters

Digital Electronics with PLD Integration
17 Digital Chapters

For more information and a desk copy of any of these textbooks by Nigel P. Cook, call 1-800-228-7854, visit the Prentice Hall website at www.prenhall.com, or ask your local Prentice Hall representative.

Preface

INTRODUCTION

Since World War II, no branch of science has contributed more to the development of the modern world than *electronics.* It has stimulated dramatic advances in the fields of communication, computing, consumer products, industrial automation, test and measurement, and healthcare. It has now become the largest single industry in the world, exceeding the automobile and oil industries, with annual sales of electronic systems greater than $2 trillion. One of the most important trends in this huge industry has been a gradual shift from *analog electronics* to *digital electronics.* This movement began in the 1960s and is almost complete today. In fact, a recent statistic stated that, on average, 90% of the circuitry within electronic systems is now digital and only 10% is analog. This digitization of the electronics industry is merging sectors that were once separate. For example, two of the largest sectors or branches of electronics are *computing* and *communications.* Being able to communicate with each other using the common language of digital has enabled computers and communications to interlink so that computers can now function within communication-based networks, and communications networks can now function through computer-based systems. Industry experts call this merging *convergence,* and they predict that digital electronics will continue to unite the industry and stimulate progress in practically every field of human endeavor.

CHAPTER OUTLINE

SUPPLEMENTS

The following ancillaries accompanying this text provide extensive opportunity for further study and support:

- **Laboratory Manual.** This important supplement is co-authored by Nigel Cook and Gary Lancaster and offers numerous experiments designed to translate all of the theory from the text into practical experimentation.
- **Instructor's Solutions Manual.** Solutions for text problems and the lab experiments are provided.
- **PowerPoint™ Transparencies.** All of the figures and tables appearing in the text are available as PowerPoint slides on CD.
- **TestGen,** a computerized test bank, is available to instructors who wish to create customized tests.
- **Companion Website.** Located at http://www.prenhall.com/cook, this site is an online study guide enabling students to review text material and test their understanding.

Contents

Accessible Writing Style,
Supported by . . .

FIGURE 13-9 A Successive Approximation Analog-to-Digital Converter (ADC).

Successive Approximation ADC

An ADC circuit tests and sets each of the digital output word's bits in

The slow conversion time of the staircase ADC can be overcome by using the **successive approximation ADC,** shown in Figure 13-9(b). Before we discuss the operation of this circuit, let us first examine the successive approximation (SA) method. As an example, Figure 13-9(a) shows how the SA method can be used to determine the digital equivalent of the analog input by making only four comparisons. In the example shown highlighted in Figure 13-9(a), an analog input voltage of 11.2 V has been applied. Following the com- [...] shown in binary (and decimal in the inset), you can see that the SA method first sets [...] HIGH (1000, or decimal 8) and then tests to see whether this value is greater than [...] an the analog input voltage. Since the analog input voltage of 11.2 V is greater than

13 / ANALOG AND DIGITAL SIGNAL CONVERTERS

2-2-5 *Converting Information Signals*

Figure 2-7(a) shows how analog data can be converted into digital data, and Figure 2-7(b) shows how digital data can be converted into analog data. To give this process purpose, let us consider how the music information stored on a compact disc (CD) is first recorded and then played back. Like all sound, music is made up of waves of compressed air. When these

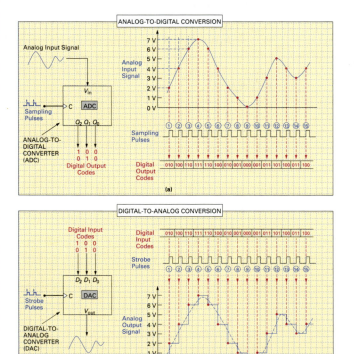

FIGURE 2-7 Converting Information Signals.

. . . Descriptive Illustrations

DEVICE: SN7400, SN5400 (Quad Two-Input NAND)

FIGURE 5-8 A Standard TTL Data Sheet.

Data Sheet
Breakdown Notes

HIGH Output Voltage

LOW Output Voltage

Highest rate at which a signal can change when applied to an input

Minimum time a pulse must be present at the input to cause a reliable change at the output

Minimum amount of time data must be available at an input

Minimum amount of time a data input must be maintained

Not a recommended condition

Current supply to IC when all outputs are HIGH or LOW

Schottky TTL line of products (labeled 74ALS00) which have power dissipation and propagation time figures of 1 mW and 4 ns. Some manufacturers, such as Fairchild, use the letter "F" for fast in their AS series of ICs; for example, 74F10, 74F14, and so on.

As you have seen, an IC contains quite a complex circuit, and this circuit has a detailed list of circuit characteristics. These characteristics or specifications are listed in a manufacturer's data sheet or spec sheet. As an example, Figure 5-8 shows a typical data sheet for a 7400 (quad two-input NAND) TTL IC.

Most data sheets can be broken down into five basic blocks:

a. pin-out and logic symbols
b. maximum ratings
c. recommended ratings

SECTION 5-1 / THE BIPOLAR FAMILY OF DIGITAL INTEGRATED CIRCUITS

Working Application
Circuits with IC Pinouts

FIGURE 8-3   Application—Sixteen-LED Back-and-Forth Flasher Circuit.

Light-Emitting Diode (LED) Display Decoders

The 7447 IC is one example of a display decoder/driver. Figure 8-4(a) shows how the seven segments of an LED display are labeled—*a, b, c, d, e, f,* and *g*—and Figure 8-4(b) shows how certain ON/OFF segment combinations can be used to display the decimal digits numbers 0 through 9.

Application—Seven-Segment LED Display Decoder/Driver

Figure 8-4(c) shows how a 7447 **BCD-to-seven-segment decoder** IC can be connected to drive a common-anode seven-segment LED display. The function table for this digital IC has been repeated in the upper inset in Figure 8-4(c). Looking at the first line of this table, for example, you can see that if a binary input of 0000 (0) is applied to the *ABCD* inputs, the 7447 will make LOW outputs *abcdef* and make HIGH output *g*. These outputs are applied to the cathodes of the LEDs within the common-anode display (TIL 312), as seen in the lower inset, and since all of the anodes are connected to +5 V, a LOW output will turn ON a segment, and a HIGH output will turn OFF a segment. The 7447, therefore, generates active-LOW outputs, which means that a "LOW" output signal will "activate" its associated output LED. In this example, the active-LOW outputs from the 7447 will turn ON segments *abcdef*, while the HIGH output will turn OFF segment *g*, causing

BCD-to-Seven-Segment Decoder
A decoder circuit that converts a BCD input code into an equivalent seven-segment output code.

SECTION 8-1 / DECODERS

167

Standard Test Equipment Testing

FIGURE 7-3   Testing with the Oscilloscope.

time and *fall time,* as seen in Figure 7-2(d). The rise time is defined as the time needed for the pulse to rise from 10% to 90% of the peak amplitude, while the fall time is the time needed for the pulse to fall from 90% to 10% of the peak amplitude.

When carefully studying pulse waveforms on the oscilloscope, you will be able to measure pulse width, PRT, rise time, and fall time. You will also see a few other conditions known as *overshoot, undershoot,* and *ringing.* These unwanted conditions tend to always accompany high-frequency pulse waveforms due to imperfections in the circuit, as seen in Figure 7-2(e). As you can see, the positive rising edge of the waveform "overshoots" the peak, and then a series of gradually decreasing sine wave oscillations occur, called ringing. At the end of the pulse negative falling edge "undershoots" the negative peak value.
cope's display can be used to measure amplitude and time,
e/amplitude measuring ability of the oscilloscope enables

FIGURE 7-9   An Internal Open Gate Input.

Digital Test Equipment Testing

Figure 7-9(b), the output remains permanently LOW. These conditions could only occur if the pin 10 input was permanently HIGH. As discussed previously, an open or floating input is equivalent to HIGH input, since both conditions have next-to-no input current. The cause, therefore, must be either an internal open at pin 10 (since this is equivalent to a HIGH input) or a $+V_{CC}$ short to the pin 10 input of this gate (which is less common than an open). In either case, however, the logic gate IC has an internal failure and will have to be replaced.

Figure 7-10 illustrates how a logic pulser and logic probe can be used to isolate an internal open gate output. In Figure 7-10(a), the logic gate's input at pin 10 has been jumper

FIGURE 7-10   An Internal Open Gate Output.

FIGURE 6-1   Constructing a Circuit Using Standard Logic Devices (SLDs).

Standard Logic Device (SLD) Coverage

design is then **downloaded** through the parallel port of the PC into the PLD. Switches are then connected to the inputs and LEDs are connected to the outputs to make a final test of the prototype.

In summary, the five-step process for creating a prototype using a PLD is

PLD Prototyping Process

Step 1:   Create the new circuit using the software's schematic editor.

Step 2:   Compile the circuit into a bitstream file that when loaded into the CPLD instruct it to act like the entered schematic.

Programmable Logic Device (PLD) Coverage

FIGURE 6-2   Constructing a Circuit Using Programmable Logic Devices (PLDs).

Step 3:   Verify the operation of your circuit using the software's functional and timing simulator.

Step 4:   Download the circuit file from the PC to the PLD.

Step 5:   Physically test the PLD by activating its inputs and monitoring its outputs.

This programmable logic prototyping method has the following advantages:

- With manual wiring reduced to a minimum, prototypes can be constructed, tested, and modified at a much faster rate.
- Wiring errors can be avoided.

Integrated Application Circuits Link Digital Device Building Blocks

FIGURE 10-31   Application—A Frequency Counter Circuit.

The BCD outputs of the four-stage counter are applied to a set of registers, and the output of these registers drives a four-digit seven-segment display via BCD-to-seven-segment display decoders. Referring to the divide-by-two and monostable block, you can see that the 7476 Q output is used to trigger a 74121 one-shot, whose RC network has been chosen to ‾‾a 1 µs pulse. Referring again to the timing waveforms in the inset in Figure 10-31, ‾‾e that the positive edge of the 7476 $\overline{Q}$ output triggers the 74121, which generates ‾‾ 1 µs pulse at its Q output and a negative 1 µs pulse at its $\overline{Q}$ output. The positive

‾0 / SEQUENTIAL LOGIC CIRCUITS

Calculator Sequences

■ **EXAMPLE:**

Convert hexadecimal 2BF9 to its binary equivalent.

■ *Solution:*

Hexadecimal number	2	B	F	9
Binary equivalent	0010	1011	1111	1001

CALCULATOR KEYS

Name: Entering and converting number bases

Function and Examples: Some calculators allow you to enter and convert number bases.

To enter a binary number, use the form:

0b*binaryNumber*   (for example: 0b11100110)
— Binary number with up to 32 digits
— Zero, not the letter *O,* and the letter *b*

To enter a hexadecimal number, use the form:

0h*hexadecimalNumber*   (for example: 0h89F2C)
— Hexadecimal number with up to 8 digits
— Zero, not the letter *O,* and letter *h*

If you enter without the 0b or 0h prefix, such as 11, it is always treated as a decimal number. If you omit the 0h prefix on a hexadecimal number containing A–F, all or part of the entry is treated as a variable.

Use the ▶ conversion operator.

integerExpression ▶ Bin
integerExpression ▶ Dec
integerExpression ▶ Hex

— For ▶, press 2nd [▶]. Also, you can select base conversions from the MATH/Base menu.

— For a binary or hex entry, you must use the 0b or 0h prefix.

For example, to convert 256 from decimal to binary:

256 ▶ Bin

n 256 ▶ Bin	0b100000000
n 0b101110 ▶ Hex	0h2E
0b101110 ▶ hex	
MATH RAD AUTO FUNC	2738

To convert 101110 from binary to hexadecimal:

0b101110 ▶ Hex

— Result use the 0b or 0h prefix to identify the base.

Final Chapter Combines All Coverage in a Component-Level, Full Schematic . . .

FIGURE 14-10   The Fetch-Execute Cycle for an MVI A Instruction.

input/output ports. For example, the contents of one register may be incremented or decremented. The fourth operation type includes instructions such as JMP, which causes the program counter and therefore the program to shift to point to a new memory location.

To summarize the operation of the microprocessor, we can say that the microprocessor keeps reading sequentially through the memory, one location after another, performing the indicated operations. Exceptions to this occur when a jump instruction is executed or at the occurrence of an **interrupt.** The 8085 has a few interrupt input pins, which are used to tell the microprocessor to stop its current task and perform another task immediately. Jump instructions or interrupts will cause the microprocessor to stop its sequential flow through a ... executing instructions from another address.

... at opcodes and data are intermixed in memory. One address might con-... next two addresses a jump address, the next address an opcode, and the

Interrupt
Involves suspension of the normal program that the microprocessor is executing in order to handle a sudden request for service (interrupt). The processor then jumps from the program it was executing to the interrupt service routine. When the interrupt service routine is completed, control returns to the interrupted program.

SECTION 14-2 / A MICROCOMPUTER SYSTEM   **403**

Simple, All-Purpose Microcomputer (SAM)

421

Analog to Digital

Moon Walk

In July of 1969 almost everyone throughout the world was caught at one stage or another gazing up to the stars and wondering. For U.S. astronaut Neil Armstrong, the age-old human dream of walking on the moon was close to becoming a reality. On earth, millions of people and thousands of newspapers and magazines waited anxiously to celebrate a successful moon landing. Then, the rather brief but eloquent message from Armstrong was transmitted 240,000 miles to Houston, Texas, where it was immediately retransmitted to a waiting world. This message was: "That's one small step for man, one giant leap for mankind." These words were received by many via television, but for magazines and newspapers this entire mission, including the speech from Armstrong, was converted into a special code made up of ON–OFF pulses that traveled from computer to computer. Every letter of every word was converted into a code that used the two symbols of the binary number system, 0 and 1. This code is still used extensively by all modern computers and is called *American Standard Code for Information Interchange,* or ASCII (pronounced "askey").

It is only fitting that these codes made up of zeros and ones conveyed the finale to this historic mission, since they played such an important role throughout. Commands encoded into zeros and ones were used to control almost everything, from triggering the takeoff to keeping the spacecraft at the proper angle for reentry into the earth's atmosphere.

No matter what its size or application, a digital electronic computer is quite simply a system that manages the flow of information in the form of zeros and ones. Referring to the ASCII code table in Figure 1-2 on p. 6, see if you can decode the following famous Armstrong message that reads from left to right:

```
0100010  1010100  1101000  1100001  1110100  0100111  1110011  0100000  1101111  1101110
1100101  0100000  1110011  1101101  1100001  1101100  1101100  0100000  1110011  1110100
1100101  1110000  0100000  1100110  1101111  1110010  0100000  1100001  0100000  1101101
1100001  1101110  0101100  0100000  1101111  1101110  1100101  0100000  1100111  1101101
1100001  1101110  1110100  0100000  1101100  1100101  1100001  1110000  0100000  1100110
1101111  1110010  0100000  1101101  1100001  1101110  1101011  1101001  1101110  1100100
0101110  0100010  0100000  0100000  0101101  1001110  1100101  1101001  1101100  0100000
1000001  1110010  1101101  1110011  1110100  1110010  1101111  1101110  1100111  0101100
0100000  1000001  1110000  1101111  1101100  1101100  1101111  0100000  0110001  0110001
```

Introduction

Since World War II, no branch of science has contributed more to the development of the modern world than electronics. It has stimulated dramatic advances in the fields of communication, computing, consumer products, industrial automation, test and measurement, and health care. It has now become the largest single industry in the world, exceeding the automobile and oil industries, with annual sales of electronic systems exceeding $2 trillion.

One of the most important trends in this huge industry has been a gradual shift from *analog electronics* to *digital electronics.* This movement began in the 1960s and is almost complete today. In fact, a recent statistic stated that, on average, 90% of the circuitry within electronic systems is now digital and only 10% is analog. This digitalization of the electronics industry is bringing sectors that were once separate closer together. For example, two of the largest sectors or branches of electronics are *computing* and *communications.* Being able to communicate with each other using the common language of digital has enabled computers and communications to interlink so that computers can now function within communication-based networks, and communications networks can now function through computer-based systems. Industry experts call this merging *convergence* and predict that digital electronics will continue to unite the industry and stimulate progress in practically every field of human endeavor.

Needless to say, this part of the book you are about to read involving digital electronic mathematics and concepts is an essential element in your study of electronics.

Analog

The representation of physical properties by a proportionally varying signal.

AC Analog Signal

An analog signal that alternates positive and negative.

AC Amplifier Circuit

An amplifier designed to increase the magnitude of an ac signal.

Analog Device or Component

A device or component that makes up a circuit designed to manage analog signals.

Analog Circuit

A circuit designed to manage analog signals.

DC Analog Signal

An analog signal that is always either positive or negative.

DC Amplifier Circuit

An amplifier designed to increase the magnitude of a dc signal.

Linear Circuit

A circuit in which the output varies in direct proportion to the input.

Before we begin, let's first review analog electronics and then compare it to digital, so that we can clearly see the differences between the two.

1-1-1 *Analog Data and Devices*

Figure 1-1(a) shows an electronic circuit designed to amplify speech information detected by a microphone. One of the easiest ways to represent data or information is to have a voltage change in direct proportion to the information it is representing. In the example in Figure 1-1(a), the *pitch and loudness* of the sound waves applied to the microphone should control the *frequency and amplitude* of the voltage signal from the microphone. The output voltage signal from the microphone is said to be an analog of the input speech signal. The word **analog** means "similar to," and in Figure 1-1(a) the electronic signal produced by the microphone is an analog (or similar) to the speech signal, since a change in speech "loudness or pitch" will cause a corresponding change in signal voltage "amplitude or frequency."

In Figure 1-1(b), a light detector or solar cell converts light energy into an electronic signal. This information signal represents the amount of light present, since changes in voltage amplitude result in a change in light-level intensity. Once again, the output electronic signal is an analog (or similar) to the sensed light level at the input.

Figure 1-1, therefore, indicates two analog circuits. The microphone in Figure 1-1(a) generates an **ac analog signal** that is amplified by an **ac amplifier circuit.** The microphone is considered an **analog device** or **component** and the amplifier an **analog circuit.** The light detector in Figure 1-1(b) would also be an analog component; however, in this example it generates a **dc analog signal** that is amplified by a **dc amplifier circuit.**

Both of the information signals in Figure 1-1 vary smoothly and continuously, in accordance with the natural quantities they represent (sound and light). Analog circuits are often called **linear circuits,** since linear by definition is *the variation of an output in direct*

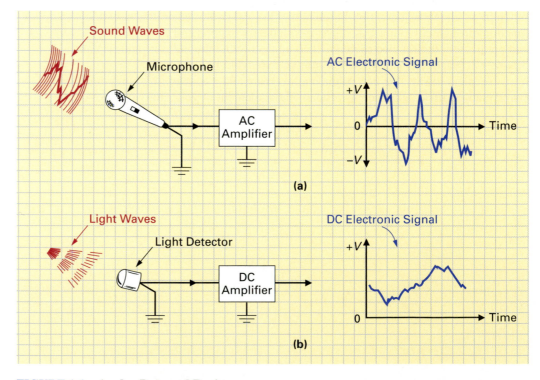

FIGURE 1-1 **Analog Data and Devices.**

proportion to the input. This linear circuit response is evident in Figure 1-1, where you can see that in both the dc and ac circuit, the output signal voltage varies in direct proportion to the sound or light signal input.

1-1-2   *Digital Data and Devices*

In **digital electronic circuits,** information is first converted into a coded group of pulses. To explain this concept, let's take a closer look at the example shown in Figure 1-2(a). This code consists of a series of HIGH and LOW voltages, in which the HIGH voltages are called "1s" (ones) and the LOW voltages are called "0s" (zeros). Figure 1-2(b) lists the ASCII code, which is one example of a digital code. Referring to Figure 1-2(a), you will notice that the "1101001" information or data stream code corresponds to lowercase *i* in the ASCII table shown highlighted in Figure 1-2(b). Computer keyboards are one of many devices that make use of the digital ASCII code. In Figure 1-2(c), you can see how the lowercase *i* ASCII code is generated whenever the *i* key is pressed, encoding the information *i* into a group of pulses (1101001).

The next question you may have is: Why do we go to all this trouble to encode our data or information into these two-state codes? The answer can best be explained by examining history. The early digital systems constructed in the 1950s made use of a decimal code that used 10 levels or voltages, with each of these voltages corresponding to one of the 10 digits in the decimal number system ($0 = 0$ V, $1 = 1$ V, $2 = 2$ V, $3 = 3$ V, up to $9 = 9$ V). The circuits that had to manage these decimal codes, however, were very complex since they had to generate one of 10 voltages and sense the difference among all 10 voltage levels. This complexity led to inaccuracy, since some circuits would periodically confuse one voltage level for a different voltage level. *The solution to the problem of circuit complexity and inaccuracy was solved by adopting a two-state system instead of a 10-state system.* Using a two-state or two-digit system, you can generate codes for any number, letter, or symbol, as we have seen in the ASCII table in Figure 1-2. The electronic circuits that manage these two-state codes are less complex, since they have to generate and sense only either a HIGH or LOW voltage. In addition, two-state circuits are much more accurate, since there is little room for error between the two extremes of ON and OFF, or HIGH voltage and LOW voltage.

Abandoning the 10-state system and adopting the two-state system for the advantages of circuit simplicity and accuracy meant that we were no longer dealing with the decimal number system. As a result, having only two digits (0 and 1) means that we are now operating in the two-state number system, which is called **binary.** Figure 1-3(a) shows how the familiar decimal system has 10 digits or levels labeled 0 through 9, and Figure 1-3(b) shows how we could electronically represent each decimal digit. With the base 2, or binary number system, we have only two digits in the number scale, and therefore only the digits 0 (zero) and 1 (one) exist in binary, as shown in Figure 1-3(c). These two states are typically represented in an electronic circuit as two different values of voltage (binary $0 =$ LOW voltage; binary $1 =$ HIGH voltage), as shown in Figure 1-3(d).

Using combinations of **binary digits** (abbreviated as **bits**) we can represent information as a binary code. This code is called a **digital signal** because it is an *information signal* that makes use of *binary digits.* Today, almost all information, from your telephone conversations to the music on your compact discs, is **digitized** or converted to binary data form.

You will probably hear the terms *analog* and *digital* used to describe the difference between the two types of meters shown in Figure 1-4. On the **analog readout multimeter** shown in Figure 1-4(a), the amount of pointer deflection across the scale is an analog (or similar) to the magnitude of the electrical property being measured. On the other hand, with the **digital readout multimeter** shown in Figure 1-4(b), the magnitude of the electrical property being measured is displayed using digits, which in this case are decimal digits.

Digital Electronic Circuit
A circuit designed to manage digital information signals.

Binary
Having only two alternatives, two-state.

Binary Digit
Abbreviated bit; it is either a 0 or 1 from the binary number system.

Bit
Binary digit.

Digital Signal
An electronic signal made up of binary digits.

Digitize
To convert an analog signal to a digital signal.

Analog Readout Multimeter
A multimeter that uses a movement on a calibrated scale to indicate a value.

Digital Readout Multimeter
A multimeter that uses digits to indicate a value.

FIGURE 1-2 Two-State Digital Information. (a) Information or Data Code for *i*. (b) ASCII Code. (c) Keyboard ASCII Generator.

FIGURE 1-3   Electronically Representing the Digits of a Number System.

FIGURE 1-4   Analog and Digital Readouts.

7

Use the following questions to test your understanding of Section 1-1.

1. What does the word *analog* mean?
2. What is an analog signal?
3. What is the ASCII code?
4. What were the two key reasons for adopting the two-state binary system in place of the 10-state decimal system?
5. How many digits does the binary number system have?
6. Describe the difference between an analog readout watch and a digital readout watch.

1-2   ANALOG AND DIGITAL SIGNAL CONVERSION

Analog-to-Digital Converter

A circuit that converts analog input signals into equivalent digital output signals.

Whenever you have two different forms, like analog and digital, there is always a need to convert back and forth between the two. Computers are used at the heart of almost every electronic system today because of their ability to quickly process and store large amounts of data, make systems more versatile, and perform many functions. Many of the input signals applied to a computer for storage or processing are analog in nature, and therefore a data conversion circuit is needed to interface these analog inputs with a digital system. Similarly, a data conversion circuit may also be needed to interface the digital computer system with an analog output device.

Figure 1-5 shows a simplified block diagram of a computer. No matter what the application, a computer has three basic blocks: a microprocessor unit, a memory unit, and an input/output unit.

Getting information into the computer is the job of input devices, and as you can see in Figure 1-5, there are two input paths into a computer: one path from analog input devices,

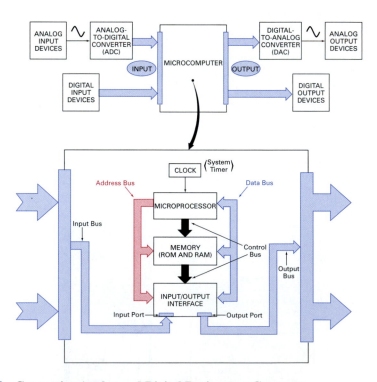

FIGURE 1-5   **Connecting Analog and Digital Devices to a Computer.**

and the other path from digital input devices. Information in the form of light, sound, heat, pressure, or any other real-world quantity is analog in nature and has an infinite number of input levels. Sensors or transducers of this type, such as photoelectric cells for light or microphones for sound, will generate analog signals. Since the computer operates on information in only digital or binary form, a translation or conversion is needed if the computer is going to be able to understand or interpret an analog signal. This signal processing is achieved by an **analog-to-digital converter (ADC),** which transforms the varying analog input voltage into equivalent digital codes that the computer can understand. Once the analog information has been encoded into digital form, the information can then enter the computer through an electronic doorway called the **input port.** Some input devices, such as a computer keyboard, automatically generate a digital code for each key that is pressed (ASCII codes). The codes generated by these digital input devices can be connected directly to the computer, without having to pass through a converter.

In contrast, the digital output information from a computer's **output port** can be used to drive either an analog device or a digital device. If an analog output is desired, as in the case of a voice signal for a speaker, the digital computer's output will have to be converted into a corresponding equivalent analog signal by a **digital-to-analog converter (DAC).** If only a digital output is desired, as in the case of a printer, which translates the binary codes into printed characters, the digital computer output can be connected directly to the output device without the need for a converter.

Input Port

An electronic doorway used for receiving information into a computer.

Output Port

An electronic doorway used for transmitting information out of a computer.

Digital-to-Analog Converter

A circuit that converts digital input signals into equivalent analog output signals.

SELF-TEST EVALUATION POINT FOR SECTION 1-2

Use the following questions to test your understanding of Section 1-2.

1. What are the three basic blocks of a computer?
2. Give the full names of the following acronyms:
   a. ADC   **b.** DAC
3. An ADC will convert a/an _____ input into an equivalent _____ output.
4. A DAC will convert a/an _____ input into an equivalent_____ output.

REVIEW QUESTIONS

Multiple Choice Questions

1. Which of the following is an example of a digital electronic circuit?
   a. Amplifier   **c.** Power supply
   b. MPU   **d.** ON/OFF lighting circuit

2. Which of the following is an example of an analog electronic circuit?
   a. Amplifier   **c.** ASCII keyboard
   b. CPU   **d.** ON/OFF lighting circuit

3. Digital circuits based on the binary number system are often referred to as _____ circuits. On the other hand, the output of analog circuits varies in direct proportion to the input, which is why analog circuits are often called _____ circuits.
   a. Ten-state, linear   **c.** Digital, 10-state
   b. Linear, two-state   **d.** Two-state, linear

4. A digital multimeter displays the measured quantity using
   a. Binary digits   **c.** Decimal digits
   b. A pointer and scale   **d.** An abacus

5. Why do digital electronic systems use the two-state system instead of the 10-state system?
   a. Circuit simplicity   **c.** Both (a) and (b)
   b. Accuracy   **d.** None of the above

6. A _____ converts an analog input voltage into a proportional digital output code.
   a. ROM   **c.** ADC
   b. DAC   **d.** RWM

7. A _____ converts a digital input code into a proportional analog output voltage.
   a. ROM   **b.** DAC   **c.** ADC   **d.** RWM

8. What are the three basic blocks of a computer?
   a. ADC, I/O, memory   **c.** MPU, I/O, memory
   b. DAC, ADC, I/O   **d.** ROM, MPU, I/O

Web Site Questions

Go to the web site http://www.prenhall.com/cook, select the textbook *Practical Digital Electronics,* select this chapter, and then follow the instructions when answering the multiple-choice practice problems.

Number Systems and Codes

Werner Von Siemens

Conducting Achievement

Werner Von Siemens (1816–1892) was born in Lenthe, Germany. He became familiar with the recently developed electric telegraph during military service. Later in 1847, together with skilled mechanic J. G. Halske, he founded the electrical firm of Siemens and Halske. This firm, under Siemens's guidance, became one of the most important electrical undertakings in the world. Siemens invented cable insulation, an armature for large generators, and the dynamo, or electric generator, which converts mechanical energy to electric energy.

Karl Wilhelm Siemens (1823–1883), Werner's brother, is also well known for his work in the fields of electricity and heat. At the age of nineteen he visited England expressly to patent his electroplating invention and never left, making England his home. From 1848 onward, Wilhelm represented his brother's firm in London and became an acknowledged authority. The company installed much of the overland telegraph cable then in existence, as well as an underwater telegraph cable, using submarines. One month before he died, Wilhelm was knighted as Sir William Siemens in acknowledgment of his achievements.

In the family tradition, Alexander Siemens (1847–1928), a nephew of William, went to England in 1867 and worked his way up, beginning in Siemens's workshops. In 1878 he became manager of the electric department and was responsible for the installation of electric light at Godalming, Surrey, the first English town to be lit with electricity. Like many other members of the family, he patented several inventions and, after the death of Sir William, he became company director.

In honor of the Siemens family's achievements, conductance (G) is measured in the unit siemens (S), with a resistance or impedance of 1 Ω being equal to a conductance or admittance of 1 S.

Introduction

One of the best ways to understand anything new is to compare it with something that you are already familiar with so that the differences are highlighted. In this chapter we will be examining the *binary number system,* which is the language used by digital computer electronic circuits. To best understand this new number system, we will compare it with the number system most familiar to you, the *decimal* (base 10) *number system.* Although this system is universally used to represent quantities, many people do not fully understand its weighted structure. This review of decimal has been included so that you can compare the base 10 number system with the system used internally by digital electronics, the binary (base 2) number system.

In addition to binary, two other number systems are widely used in conjunction with digital electronics: the octal (base 8) number system, and the hexadecimal (base 16) number system. These two number systems are used as a type of shorthand for large binary numbers and enable us to represent a large group of binary digits with only a few octal or hexadecimal digits.

The last section in this chapter covers many of the different binary codes, which use a combination of 1s and 0s to represent letters, numbers, symbols, and other information.

Keep in mind throughout this chapter that different number systems are just another way to count, and as with any new process, you will become more proficient and more at ease with practice as we proceed through the chapters that follow.

2-1 THE DECIMAL NUMBER SYSTEM

The decimal system of counting and keeping track of items was first created by Hindu mathematicians in India in A.D. 400. Since it involved the use of fingers and thumbs, it was natural that this system would have 10 digits. The system found its way to all the Arab countries by A.D. 800, where it was named the Arabic number system, and from there it was eventually adopted by nearly all the European countries by A.D. 1200, where it was called the **decimal number system.**

The key feature that distinguishes one number system from another is the number system's **base** or *radix*. This base indicates the number of digits that will be used. The decimal number system, for example, is a base 10 number system, which means that it uses 10 digits (0 through 9) to communicate information about an amount. A subscript is sometimes included after a number when different number systems are being used, to indicate the base of the number. For example, $12,567_{10}$ is a base 10 number, whereas 10110_2 is a base 2 number.

2-1-1 *Positional Weight*

The position of each digit of a decimal number determines the weight of that digit. A 1 by itself, for instance, is worth only 1, whereas a 1 to the left of three 0s makes the 1 worth 1000.

In decimal notation, each position to the left of the decimal point indicates an increased positive power of 10, as seen in Figure 2-1(a). The total quantity or amount of the number is therefore determined by the size and the weighted position of each digit. For example, the value shown in Figure 2-1(a) has six thousands, zero hundreds, one ten, and nine ones, which combined makes a total of 6019_{10}.

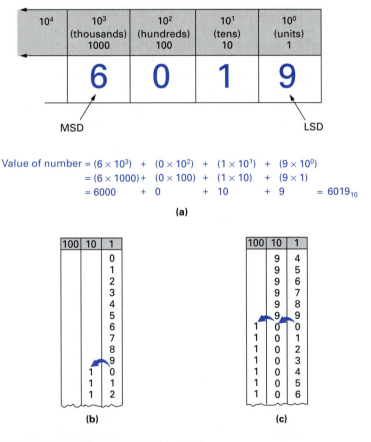

FIGURE 2-1   **The Decimal (Base 10) Number System.**

Decimal Number System
A base 10 number system.

Base
With number systems, it describes the number of digits used.

In the decimal number system, the leftmost digit is called the **most significant digit (MSD),** and the rightmost digit is called the **least significant digit (LSD).** Applying these definitions to the example in Figure 2-1(a), we see that the 6 is the MSD, since its position carries the most weight, and the 9 is the LSD, since its position carries the least weight.

> **Most Significant Digit (MSD)**
> The leftmost, largest-weight digit in a number.

> **Least Significant Digit (LSD)**
> The rightmost, smallest-weight digit in a number.

2-1-2 *Reset and Carry*

Before proceeding to the binary number system, let us review one other action that occurs when counting in decimal. This action, which is familiar to us all, is called **reset and carry.** Referring to Figure 2-1(b), you can see that a reset and carry operation occurs after a count of 9. The units column, which has reached its maximum count, resets to 0 and carries a 1 to the tens column, resulting in a final count of 10.

> **Reset and Carry**
> An action that occurs when a column has reached its maximum count.

This reset and carry action will occur in any column that reaches its maximum count. For example, Figure 2-1(c) shows how two reset and carry operations will take place after a count of 99. The units column resets and carries a 1 to the tens column, which in turn resets and carries a 1 to the hundreds column, resulting in a final count of 100.

SELF-TEST EVALUATION POINT FOR SECTION 2-1

Use the following questions to test your understanding of Section 2-1.

1. What is the difference between the Arabic number system and the decimal number system?
2. What is the base or radix of the decimal number system?
3. Describe the positional weight of each of the digits in the decimal number 2639.
4. What action occurs when a decimal column advances beyond its maximum count?

2-2 THE BINARY NUMBER SYSTEM

As in the decimal system, the value of a binary digit is determined by its position relative to the other digits. In the decimal system, each position to the left of the decimal point increases by a power of 10. Similarly, in the binary number system, since it is a base 2 (bi) number system, each place to the left of the **binary point** increases by a power of 2. Figure 2-2 shows how columns of the binary and decimal number systems have different weights. For example, with binary, the columns are weighted so that 2^0 is one, 2^1 is two, 2^2 is four, 2^3 is eight $(2 \times 2 \times 2 = 8)$, and so on.

> **Binary Point**
> A symbol used to separate the whole from the fraction in a binary number.

As we know, the base or radix of a number system also indicates the maximum number of digits used by the number system. The base 2 binary number system uses only the

FIGURE 2-2   A Comparison of Decimal and Binary.

DECIMAL 10^2 (100)	10^1 (10)	10^0 (1)		BINARY 2^3 (8)	2^2 (4)	2^1 (2)	2^0 (1)
		0					0
		1					1
		2				1	0
		3				1	1
		4			1	0	0
		5			1	0	1
		6			1	1	0
		7			1	1	1
		8		1	0	0	0
		9		1	0	0	1
	1	0		1	0	1	0
	1	1		1	0	1	1
	1	2		1	1	0	0

first two digits on the number scale, 0 and 1. The 0s and 1s in binary are called binary digits, or *bits,* for short.

2-2-1   *Positional Weight*

As in the decimal system, each column in binary carries its own weight, as seen in Figure 2-3(a). With the decimal number system, each position to the left increases 10 times. With binary, the weight of each column to the left increases 2 times. The first column, therefore, has a weight of 1, the second column has a weight of 2, the third column 4, the fifth 8, and so on. The value or quantity of a binary number is determined by the digit in each column and the positional weight of the column. For example, in Figure 2-3(a), the binary number 101101_2 is equal in decimal to 45_{10}, since we have 1×32, 1×8, 1×4, and 1×1 ($32 + 8 + 4 + 1 = 45$). The 0s in this example are not multiplied by their weights (16 and 2), since they are "0" and of no value. The leftmost binary digit is called the most significant bit (MSB) since it carries the most weight, and the rightmost digit is called the least significant bit (LSB), since it carries the least weight. Applying these definitions to the example in Figure 2-3(a), we see that the 1 in the thirty-twos column is the MSB, and the 1 in the units column is the LSB.

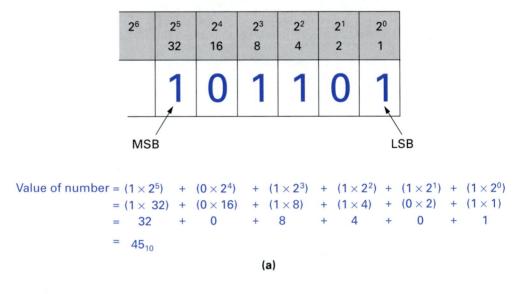

$$
\begin{aligned}
\text{Value of number} &= (1 \times 2^5) + (0 \times 2^4) + (1 \times 2^3) + (1 \times 2^2) + (1 \times 2^1) + (1 \times 2^0) \\
&= (1 \times 32) + (0 \times 16) + (1 \times 8) + (1 \times 4) + (0 \times 2) + (1 \times 1) \\
&= 32 + 0 + 8 + 4 + 0 + 1 \\
&= 45_{10}
\end{aligned}
$$

(a)

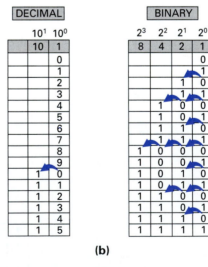

(b)

FIGURE 2-3   The Binary (Base 2) Number System.

2-2-2 Reset and Carry

The reset and carry action occurs in binary in exactly the same way as it does in decimal; however, since binary has only two digits, a column will reach its maximum digit much sooner, and therefore the reset and carry action in the binary number system will occur much more frequently.

Referring to Figure 2-3(b), you can see that the binary counter begins with 0 and advances to 1. At this stage, the units column has reached its maximum, and therefore the next count forces the units column to reset and carry into the next column, producing a count of 0010_2 (2_{10}). The units column then advances to a count of 0011_2 (3_{10}). At this stage, both the units column and twos column have reached their maximums. As the count advances by 1 it will cause the units column to reset and carry a 1 into the twos column, which will also have to reset and carry a 1 into the fours column. This will result in a final count of 0100_2 (4_{10}). The count will then continue to 0101_2 (5_{10}), 0110_2 (6_{10}), 0111_2 (7_{10}), and then 1000_2 (8_{10}), which, as you can see in Figure 2-3(b), is a result of three reset and carries.

Comparing the binary reset and carry with the decimal reset and carry in Figure 2-3(b), you can see that since the binary number system uses only two digits, binary numbers quickly turn into multidigit figures. For example, a decimal eight (8) uses only one digit, whereas a binary eight (1000) uses four digits.

2-2-3 Converting Binary Numbers to Decimal Numbers

Binary numbers can easily be converted to their decimal equivalent by simply adding together all the column weights that contain a binary 1, as we did previously in Figure 2-3(a).

EXAMPLE:

Convert the following binary numbers to their decimal equivalents.

  a. 1010

  b. 101101

Solution:

  a. Binary column weights 32 16 8 4 2 1
      Binary number   1 0 1 0
      Decimal equivalent $= (1 \times 8) + (0 \times 4) + (1 \times 2) + (0 \times 1)$
      $= 8 + 0 + 2 + 0 = 10_{10}$

  b. Binary column weights 32 16 8 4 2 1
      Binary number   1   0   1 1 0 1
      Decimal equivalent $= (1 \times 32) + (0 \times 16) + (1 \times 8) + (1 \times 4)$
      $+ (0 \times 2) + (1 \times 1)$
      $= 32 + 0 + 8 + 4 + 0 + 1 = 45_{10}$

EXAMPLE:

The LEDs in Figure 2-4 are being used as a 4-bit (4-binary-digit) display. When the LED is OFF, it indicates a binary 0, and when the LED is ON, it indicates a binary 1. Determine the decimal equivalent of the binary displays shown in Figure 2-4(a), (b), and (c).

Solution:

  a. $0101_2 = 5_{10}$

  b. $1110_2 = 14_{10}$

  c. $1001_2 = 9_{10}$

(a)

○ ON
● OFF

(b)

(c)

FIGURE 2-4 **Using LEDs to Display Binary Numbers.**

2-2-4 *Converting Decimal Numbers to Binary Numbers*

To convert a decimal number to its binary equivalent, continually subtract the largest possible power of two until the decimal number is reduced to zero, placing a binary 1 in columns that are used and a binary 0 in the columns that are not used. To see how simple this process is, refer to Figure 2-5, which shows how decimal 53 can be converted to its binary equivalent. As you can see in this example, the first largest power of two that can be subtracted from decimal 53 is 32, and therefore a 1 is placed in the thirty-twos column, and 21 remains. The largest power of two that can be subtracted from the remainder 21 is 16, and therefore a 1 is placed in the sixteens column, and 5 remains. The next largest power of two that can be sub-

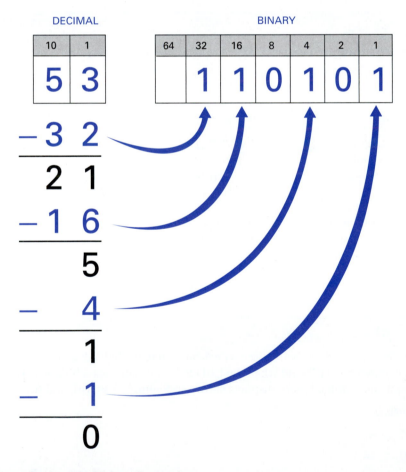

FIGURE 2-5 **Decimal-to-Binary Conversion.**

tracted from 5 is 4, and therefore a 1 is placed in the fours column, and 1 remains. The final 1 is placed in the units column, and therefore decimal 53 is represented in binary as 110101, which indicates that the value is the sum of 1×32, 1×16, 1×4, and 1×1.

EXAMPLE:

Convert the following decimal numbers to their binary equivalents.

  a. 25

  b. 55

Solution:

  a. Binary column weights   32 16 8 4 2 1
       Binary number   1   1 0 0 1
       Decimal equivalent $25 - 16 = 9, 9 - 8 = 1, 1 - 1 = 0$

  b. Binary column weights   32 16 8 4 2 1
       Binary number   1   1   0 1 1 1
       Decimal equivalent $55 - 32 = 23, 23 - 16 = 7, 7 - 4 = 3,$
                              $3 - 2 = 1, 1 - 1 = 0$

EXAMPLE:

Figure 2-6 shows a strip of magnetic tape containing binary data. The shaded dots indicate magnetized dots (which are equivalent to binary 1s), and the unshaded circles on the tape represent unmagnetized points (which are equivalent to binary 0s). What binary values should be stored in rows 4 and 5 so that decimal 19_{10} and 33_{10} are recorded?

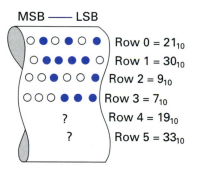

FIGURE 2-6   **Binary Numbers on Magnetic Tape.**

Solution:

  a. Binary column weights   32 16 8 4 2 1
       Binary number   1   0 0 1 1
       Decimal equivalent $19 - 16 = 3, 3 - 2 = 1, 1 - 1 = 0$
       Stored magnetic data   ○●○○●●

  b. Binary column weights   32 16 8 4 2 1
       Binary number   1   0   0 0 0 1
       Decimal equivalent $32 - 32 = 1, 1 - 1 = 0$
       Stored magnetic data   ●○○○○●

Figure 2-7(a) shows how analog data can be converted into digital data, and Figure 2-7(b) shows how digital data can be converted into analog data. To give this process purpose, let us consider how the music information stored on a compact disc (CD) is first recorded and then played back. Like all sound, music is made up of waves of compressed air. When these

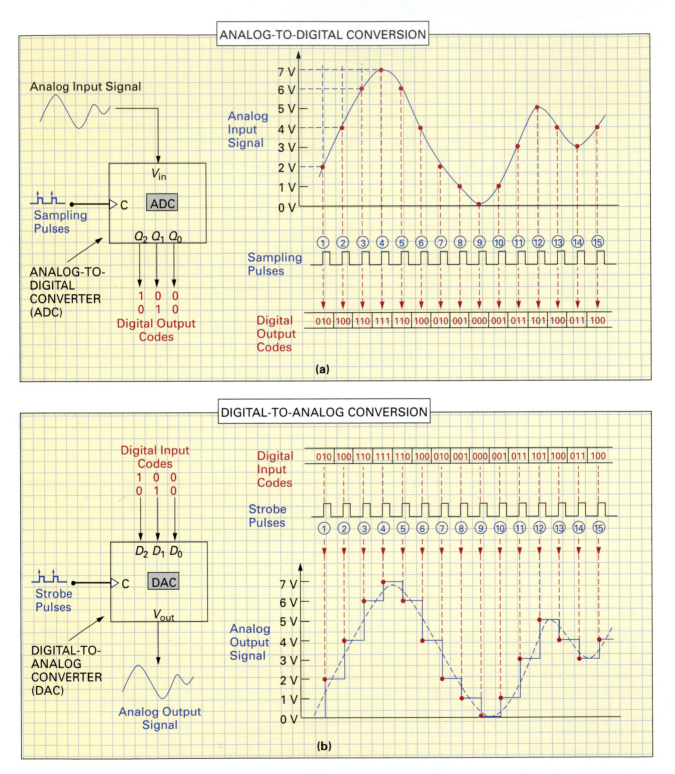

FIGURE 2-7   Converting Information Signals.

waves strike the diaphragm of a microphone, an analog voltage signal is generated. In the past, this analog data was recorded on magnetic tapes or as a grooved track on a record. These data storage devices were susceptible to wear and tear, temperature, noise, and age. Using digital recording techniques, binary codes are stored on a compact disc to achieve near-perfect fidelity to live sound.

Figure 2-7(a) shows how an analog-to-digital converter (ADC) is used during the recording process to convert the analog music input into a series of digital output codes. These digital codes are used to control the light beam of a recording laser so that it will engrave the binary 0s and 1s onto a compact disc (CD) in the form of pits or spaces. The ADC is triggered into operation by a sampling pulse that causes it to measure the input voltage of the analog signal at that particular time and generate an equivalent digital output code. For example, on the active edge of sampling pulse 1, the analog input voltage is at 2 V, so the binary code 010 (decimal 2) is generated at the output of the ADC. On the active edge of sampling pulse 2, the analog voltage has risen to 4 V, so the binary code 100 (decimal 4) is generated at the output of the ADC. On the active edge of sampling pulses 3, 4, and 5, the binary codes 110 (decimal 6), 111 (decimal 7), and 110 (decimal 6) are generated at the output, representing the analog voltages 6 V, 7 V, and 6 V, respectively, and so on.

Figure 2-7(b) shows how a digital-to-analog converter (DAC) is used during the music playback process to convert the digital codes stored on a CD to an analog output signal. Another laser is used during playback to read the pits and spaces on the compact disc as 0s and 1s. These codes are then applied to the DAC, which converts the digital input codes into discrete voltages. The DAC is triggered into operation by a strobe pulse that causes it to convert the code currently being applied at the input into an equivalent output voltage.

For example, on the active edge of strobe pulse 1, the digital code 010 (decimal 2) is being applied to the DAC, so it will generate 2 V at its output. On the active edge of strobe pulse 2, the digital code 100 (decimal 4) is being applied to the DAC, so it will generate 4 V at its output. On the active edge of strobe pulse 3, the digital code 110 (decimal 6) is being applied to the DAC, so it will generate 6 V at its output. On the active edge of strobe pulse 4, the digital code 111 (decimal 7) is being applied to the DAC, so it will generate 7 V at its output. On the active edge of strobe pulse 5, the digital code 110 (decimal 6) is being applied to the DAC, so it will generate 6 V at its output, and so on. If the output of a DAC is then applied to a low-pass filter, the discrete voltage steps can be blended into a smooth wave that closely approximates the original analog wave, as shown by the dashed line in Figure 2-7(b).

SELF-TEST EVALUATION POINT FOR SECTION 2-2

Use the following questions to test your understanding of Section 2-2.

1. What is the decimal equivalent of 11010_2?
2. Convert 23_{10} to its binary equivalent.
3. What are the full names for the acronyms LSB and MSB?
4. Convert 110_{10} to its binary equivalent.

2-3   THE HEXADECIMAL NUMBER SYSTEM

If digital electronic circuits operate using binary numbers and we operate using decimal numbers, why is there any need for us to have any other system? The hexadecimal, or hex, system is used as a sort of shorthand for large strings of binary numbers, as will be explained in this section. To begin, let us examine the basics of this number system and then look at its application.

Hexadecimal
A base 16 number system.

Hexadecimal means "sixteen," and this number system has 16 different digits, as shown in Figure 2-8(a), which shows a comparison between decimal, hexadecimal, and binary. Looking at the first 10 digits in the decimal and hexadecimal columns, you can see that there is no difference between the two columns; however, beyond 9, hexadecimal makes use of the letters A, B, C, D, E, and F. Digits 0 through 9 and letters A through F make up the 16 total digits of the hexadecimal number system. Comparing hexadecimal with decimal once again, you can see that $A_{16} = 10_{10}$, $B_{16} = 11_{10}$, $C_{16} = 12_{10}$, $D_{16} = 13_{10}$, $E_{16} = 14_{10}$ and $F_{16} = 15_{10}$. Having these extra digits means that a column reset and carry will not occur until the count has reached the last and largest digit, F. The hexadecimal column in Figure 2-8(a) shows how reset and carry will occur whenever a column reaches its maximum digit, F, and this action is further illustrated in the examples in Figure 2-8(b).

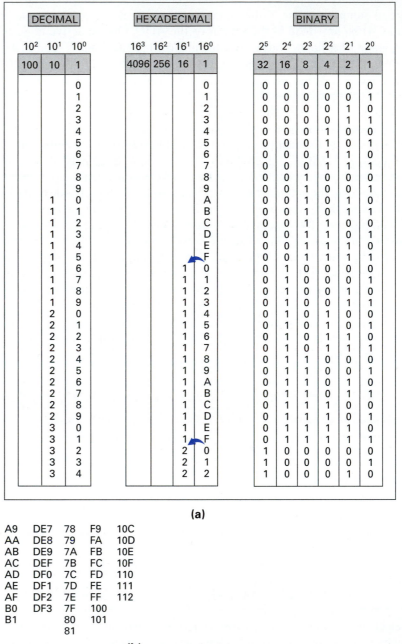

(a)

A9	DE7	78	F9	10C
AA	DE8	79	FA	10D
AB	DE9	7A	FB	10E
AC	DEF	7B	FC	10F
AD	DF0	7C	FD	110
AE	DF1	7D	FE	111
AF	DF2	7E	FF	112
B0	DF3	7F	100	
B1		80	101	
		81		

(b)

FIGURE 2-8 **Hexadecimal Rest and Carry. (a) Number System Comparison. (b) Hex Counting Examples.**

CHAPTER 2 / NUMBER SYSTEMS AND CODES

16^4	16^3	16^2	16^1	16^0
65,536	4096	256	16	1
			4	C

$$\text{Value of Number} = (4 \times 16) + (C \times 1)$$
$$= (4 \times 16) + (12 \times 1)$$
$$= \quad 64 \quad + \quad 12 \quad = 76_{10}$$

FIGURE 2-9 **Positional Weight of the Hexadecimal Number System.**

2-3-1 *Converting Hexadecimal Numbers to Decimal Numbers*

To find the decimal equivalent of a hexadecimal number, simply multiply each hexadecimal digit by its positional weight. For example, referring to Figure 2-9, you can see that the positional weight of each of the hexadecimal columns, as expected, is progressively 16 times larger as you move from right to left. The hexadecimal number 4C, therefore, indicates that the value is 4×16 ($4 \times 16 = 64$) and $C \times 1$. Since C is equal to 12 in decimal (12×1), the result of $12 + 64$ is 76, so hexadecimal 4C is equivalent to decimal 76.

EXAMPLE:

Convert hexadecimal 8BF to its decimal equivalent.

Solution:

Hexadecimal column weights   4096   256   16   1
                              8   B   F

$$\text{Decimal equivalent} = (8 \times 256) + (B \times 16) + (F \times 1)$$
$$= 2048 + (11 \times 16) + (15 \times 1)$$
$$= 2048 + 176 + 15$$
$$= 2239$$

2-3-2 *Converting Decimal Numbers to Hexadecimal Numbers*

Decimal-to-hexadecimal conversion is achieved in the same way as decimal to binary. First, subtract the largest possible power of 16, and then keep subtracting the largest possible power of 16 from the remainder. Each time a subtraction takes place, add a 1 to the respective column until the decimal value has been reduced to zero. Figure 2-10 illustrates this procedure with an example showing how decimal 425 is converted to its hexadecimal equivalent. To begin, the largest possible power of sixteen (256) is subtracted once from 425 and therefore a 1 is placed in the 256s column, leaving a remainder of 169. The next largest power of 16 is 16, which can be subtracted 10 times from 169, and therefore the hexadecimal equivalent of 10, which is A, is placed in the sixteens column, leaving a remainder of 9. Since nine 1s can be subtracted from the remainder of 9, the units column is advanced nine times, giving us our final hexadecimal result, 1A9.

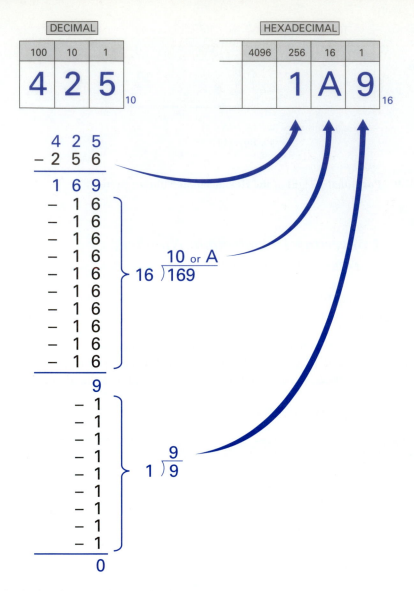

FIGURE 2-10 Decimal-to-Hexadecimal Conversion.

■ **EXAMPLE:**

Convert decimal 4525 to its hexadecimal equivalent.

■ *Solution:*

Hexadecimal column weights 4096   256   16   1
                              1   1   A   D

Decimal equivalent $= 4525 - 4096 = 429, 429 - 256 = 173,$
                      $173 - 16 - 16 - 16 - 16 - 16 - 16 -$
                      $16 - 16 - 16 - 16 = 13, 13 - 1 -$
                      $1 - 1 - 1 - 1 - 1 - 1 - 1 - 1 -$
                      $1 - 1 - 1 - 1 = 0$

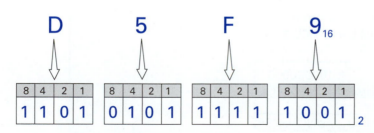

FIGURE 2-11   Representing Binary Numbers in Hexadecimal.

2-3-3   *Converting between Binary and Hexadecimal*

As mentioned at the beginning of this section, hexadecimal is used as a shorthand for representing large groups of binary digits. To illustrate this, Figure 2-11(a) shows how a 16-bit binary number, which is more commonly called a 16-bit binary **word,** can be represented by 4 hexadecimal digits. To explain this, Figure 2-11(b) shows how a 4-bit binary word can have any value from 0_{10} (0000_2) to 15_{10} (1111_2), and since hexadecimal has the same number of digits (0 through F), we can use one hexadecimal digit to represent 4 binary bits. As you can see in Figure 2-11(a), it is much easier to work with a number like $AE73_{16}$ than with 1010111001110011_2.

To convert from hexadecimal to binary, we simply do the opposite, as shown in Figure 2-12. Since each hexadecimal digit represents 4 binary digits, a 4-digit hexadecimal number will convert to a 16-bit binary word.

Word

An ordered set of characters that is treated as a unit.

FIGURE 2-12   Hexadecimal-to-Binary Conversion.

■ **EXAMPLE:**

Convert hexadecimal 2BF9 to its binary equivalent.

■ *Solution:*

Hexadecimal number	2	B	F	9
Binary equivalent	0010	1011	1111	1001

CALCULATOR KEYS

Name: Entering and converting number bases

Function and Examples: Some calculators allow you to enter and convert number bases.

To enter a binary number, use the form:

0b*binaryNumber*   (for example: 0b11100110)

— Binary number with up to 32 digits

— Zero, not the letter *O,* and the letter *b*

To enter a hexadecimal number, use the form:

0h*hexadecimalNumber*   (for example: 0h89F2C)

— Hexadecimal number with up to 8 digits

— Zero, not the letter *O,* and letter *h*

If you enter without the 0b or 0h prefix, such as 11, it is always treated as a decimal number. If you omit the 0h prefix on a hexadecimal number containing A–F, all or part of the entry is treated as a variable.

Use the ▶ conversion operator.

integerExpression ▶ Bin
integerExpression ▶ Dec
integerExpression ▶ Hex

— For ▶, press 2nd [▶]. Also, you can select base conversions from the MATH/Base menu.

— For a binary or hex entry, you must use the 0b or 0h prefix.

For example, to convert 256 from decimal to binary:

256 ▶ Bin

n 256 ▶ Bin	0b100000000
n 0b101110 ▶ Hex	0h2E
0b101110 ▶ hex	
MATH RAD AUTO FUNC	2738

To convert 101110 from binary to hexidecimal:

— Result use the 0b or 0h prefix to identify the base.

0b101110 ▶ Hex

Convert binary 110011100001 to its hexadecimal equivalent.

Solution:

Binary number	1100	1110	0001
Hexadecimal equivalent	C	E	1

SELF-TEST EVALUATION POINT FOR SECTION 2-3

Use the following questions to test your understanding of Section 2-3.

1. What is the base of each?
   a. Decimal number system
   b. Binary number system
   c. Hexadecimal number system
2. What are the decimal and hexadecimal equivalents of 10101_2?
3. Convert $1011\ 1111\ 0111\ 1010_2$ to its hexadecimal equivalent.
4. What are the binary and hexadecimal equivalents of 33_{10}?

2-4 BINARY CODES

The process of converting a decimal number to its binary equivalent is called *binary coding.* The result of the conversion is a binary number or code that is called **pure binary.** There are, however, binary codes used in digital circuits other than pure binary, and in this section we will examine some of the most frequently used.

Pure Binary
Uncoded binary.

2-4-1 *The Binary Coded Decimal (BCD) Code*

No matter how familiar you become with binary, it will always be less convenient to work with than the decimal number system. For example, it will always take a short time to convert 1111000_2 to 120_{10}. Designers realized this disadvantage early on and developed a binary code that had decimal characteristics and that was appropriately named **binary coded decimal (BCD).** Being a binary code, it has the advantages of a two-state system, and since it has a decimal format, it is also much easier for an operator to interface via a decimal keypad or decimal display with systems such as pocket calculators and wristwatches.

Binary-Coded Decimal (BCD)

A code in which each decimal digit is represented by a group of 4 binary bits.

The BCD code expresses each decimal digit as a 4-bit word, as shown in Figure 2-13(a). In this example, decimal 1753 converts to a BCD code of 0001 0111 0101 0011, with the first 4-bit code ($0001_2 = 1_{10}$) representing the 1 in the thousands column, the second 4-bit code ($0111_2 = 7_{10}$) representing the 7 in the hundreds column, the third 4-bit code ($0101_2 = 5_{10}$) representing the 5 in the tens column, and the fourth 4-bit code ($0011_2 = 3_{10}$) representing the 3 in the units column. As can be seen in Figure 2-13(a), the subscript BCD is often used after a BCD code to distinguish it from a pure binary number ($1753_{10} = 0001\ 0111\ 0101\ 0011_{BCD}$).

Figure 2-13(b) compares decimal, binary, and BCD. As you can see, the reset and carry action occurs in BCD at the same time it does in decimal. This is because BCD was designed to have only ten 4-bit binary codes, 0000, 0001, 0010, 0011, 0100, 0101, 0110, 0111, 1000, and 1001 (0 through 9), to make it easy to convert between this binary code and decimal. Binary codes 1010, 1011, 1100, 1101, 1110, and 1111 (A_{16} through F_{16}), are invalid codes that are not used in BCD because they are not used in decimal.

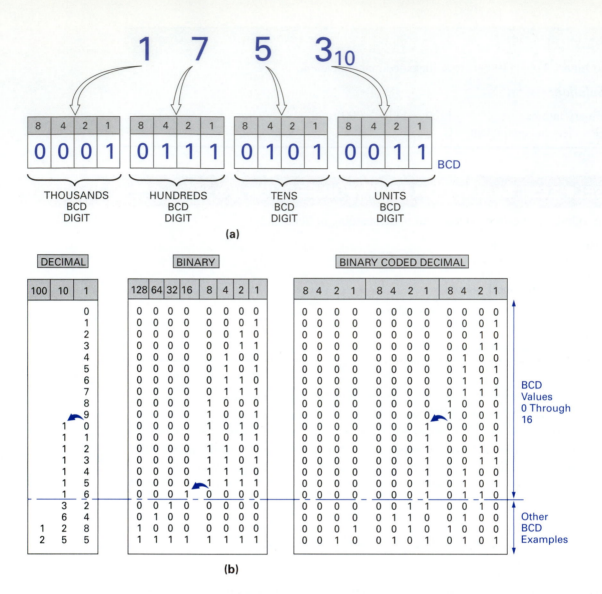

FIGURE 2-13 Binary Coded Decimal (BCD). (a) Decimal-to-BCD Conversion Example. (b) A Comparison of Decimal, Binary, and BCD.

■ **EXAMPLE:**

Convert the BCD code 0101 1000 0111 0000 to decimal, and the decimal number 369 to BCD.

■ *Solution:*

BCD code	0101	1000	0111	0000
Decimal equivalent	5	8	7	0
Decimal number	3	6	9	
BCD code	0011	0110	1001	

2-4-2   *The Excess-3 Code*

Excess-3 Code

A code in which the decimal digit *n* is represented by the 4-bit binary code *n* + 3.

The **excess-3 code** is similar to BCD and is often used in some applications because of certain arithmetic operation advantages. Referring to Figure 2-14(a), you can see that, like BCD, each decimal digit is converted into a 4-bit binary code. The difference with the excess-3 code, however, is that a value of 3 is added to each decimal digit before it is converted into

DECIMAL	BCD		EXCESS-3	
0	0000	0000	0011	0011
1	0000	0001	0011	0100
2	0000	0010	0011	0101
3	0000	0011	0011	0110
4	0000	0100	0011	0111
5	0000	0101	0011	1000
6	0000	0110	0011	1001
7	0000	0111	0011	1010
8	0000	1000	0011	1011
9	0000	1001	0011	1100
10	0001	0000	0100	0011
11	0001	0001	0100	0100
12	0001	0010	0100	0101
13	0001	0011	0100	0110
14	0001	0100	0100	0111
15	0001	0101	0100	1000

(b)

FIGURE 2-14   The Excess-3 Code. (a) Example. (b) Code Comparison.

a 4-bit binary code. Although this code has an offset or excess of 3, it still uses only ten 4-bit binary codes (0011 through 1100) like BCD. The invalid 4-bit values in this code are therefore 0000, 0001, 0010 (decimal 0 through 2), and 1101, 1110, 1111 (decimal 13 through 15). Figure 2-14(b) shows a comparison between decimal, BCD, and excess-3.

EXAMPLE:

Convert decimal 408 to its excess-3 code equivalent.

Solution:

Decimal number	4	0	8
+3 =	7	3	11
Excess-3 binary	0111	0011	1011

2-4-3   *The Gray Code*

The **gray code,** shown in Figure 2-15, is a nonweighted binary code, which means that the position of each binary digit carries no specific weight or value. This code, named after its inventor, was developed so that only one of the binary digits changes as you step from one code group to the next code group in sequence. For example, in pure binary, a change from 3 (0011) to 4 (0100) results in a change in 3 bits, whereas in the gray code only 1 bit changes.

The minimum-change gray code is used to reduce errors in digital electronic circuitry. To explain this in more detail, when a binary count changes, it takes a very small amount of time for the bits to change from 0 to 1, or from 1 to 0. This transition time can produce an incorrect intermediate code. For example, if when changing from pure binary 0011 (3) to 0100 (4), the LSB were to switch slightly faster than the other bits, an incorrect code would be momentarily generated, as shown in the following example.

Binary	Decimal
0 0 1 1	3
0 0 1 0	Error
0 1 0 0	4

Gray Code

A nonweighted binary code in which sequential numbers are represented by codes that differ by one bit.

DECIMAL	BINARY	GRAY CODE
0	0000	0000
1	0001	0001
2	0010	0011
3	0011	0010
4	0100	0110
5	0101	0111
6	0110	0101
7	0111	0100
8	1000	1100
9	1001	1101
10	1010	1111
11	1011	1110
12	1100	1010
13	1101	1011
14	1110	1001
15	1111	1000

FIGURE 2-15   The Gray Code.

This momentary error could trigger an operation that should not occur, resulting in a system malfunction. Using the gray code eliminates these timing errors, since only 1 bit changes at a time. The disadvantage of the gray code is that the codes must be converted to pure binary if numbers need to be added, subtracted, or used in other computations.

2-4-4   *The American Standard Code for Information Interchange (ASCII)*

Alphanumeric Code

A code used to represent the letters of the alphabet and decimal numbers.

To this point we have discussed only how we can encode numbers into binary codes. Digital electronic computers must also be able to generate and recognize binary codes that represent letters of the alphabet and symbols. The ASCII code is the most widely used **alphanumeric code** (alphabet and numerical code). As I write this book the ASCII codes for each of these letters is being generated by my computer keyboard, and the computer is decoding these ASCII codes and then displaying the alphanumeric equivalent (letter, number, or symbol) on my computer screen.

Figure 2-16 lists all the 7-bit ASCII codes and the full names for the abbreviations used. This diagram also shows how each of the 7-bit ASCII codes is made up of a 4-bit group that indicates the row of the table and a 3-bit group that indicates the column of the table. For example, the uppercase letter K is in column 100 (4_{10}) and in row 1011 (11_{10}) and is therefore represented by the ASCII code 1001011.

EXAMPLE:

List the ASCII codes for the message "Digital."

Solution:

$$D = 1000100$$
$$i = 1101001$$
$$g = 1100111$$
$$i = 1101001$$
$$t = 1110100$$
$$a = 1100001$$
$$l = 1101100$$

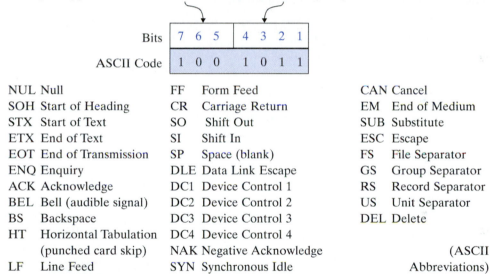

Column		0	1	2	3	4	5	6	7
Row Bits	4321 ‖ 765 ➡	000	001	010	011	**100**	101	110	111
0	0000	NUL	DLE	SP	0	@	P	\	p
1	0001	SOH	DC1	!	1	A	Q	a	q
2	0010	STX	DC2	"	2	B	R	b	r
3	0011	ETX	DC3	#	3	C	S	c	s
4	0100	EOT	DC4	$	4	D	T	d	t
5	0101	ENQ	NAK	%	5	E	U	e	u
6	0110	ACK	SYN	&	6	F	V	f	v
7	0111	BEL	ETB	'	7	G	W	g	w
8	1000	BS	CAN	(	8	H	X	h	x
9	1001	HT	EM	)	9	I	Y	i	y
10	1010	LF	SUB	*	:	J	Z	j	z
11	**1011**	VT	ESC	+	;	**K**	[	k	{
12	1100	FF	FS	,	<	L	\	l	¦
13	1101	CR	GS	-	=	M	]	m	}
14	1110	SO	RS	.	>	N	⌢	n	~
15	1111	SI	US	/	?	O	—	o	DEL

Example: Upper Case "K" = Column 4   (100), Row 11 (1011)

Bits	7 6 5	4 3 2 1
ASCII Code	1 0 0	1 0 1 1

NUL Null
SOH Start of Heading
STX Start of Text
ETX End of Text
EOT End of Transmission
ENQ Enquiry
ACK Acknowledge
BEL Bell (audible signal)
BS Backspace
HT Horizontal Tabulation (punched card skip)
LF Line Feed
VT Vertical Tabulation

FF Form Feed
CR Carriage Return
SO Shift Out
SI Shift In
SP Space (blank)
DLE Data Link Escape
DC1 Device Control 1
DC2 Device Control 2
DC3 Device Control 3
DC4 Device Control 4
NAK Negative Acknowledge
SYN Synchronous Idle
ETB End of Transmission Block

CAN Cancel
EM End of Medium
SUB Substitute
ESC Escape
FS File Separator
GS Group Separator
RS Record Separator
US Unit Separator
DEL Delete

(ASCII Abbreviations)

FIGURE 2-16 The ASCII Code.

EXAMPLE:

Figure 2-17 illustrates a 7-bit register. A register is a circuit that is used to store or hold a group of binary bits. If a switch is closed, the associated "Q output" is grounded and 0 V (binary 0) is applied to the output. On the other hand, if a switch is open, the associated "Q output" is pulled HIGH due to the connection to + 5 V via the 10 kΩ resistor, and therefore + 5 V (binary 1) is applied to the output. What would be the pure binary value, hexadecimal value, and ASCII code stored in this register?

Solution:

$$\begin{array}{ccccccc} Q_6 & Q_5 & Q_4 & Q_3 & Q_2 & Q_1 & Q_0 \end{array}$$

Pure binary   1   0   0   0   0   1 1_2
Hexadecimal 43_{16}
ASCII   C   (uppercase C)

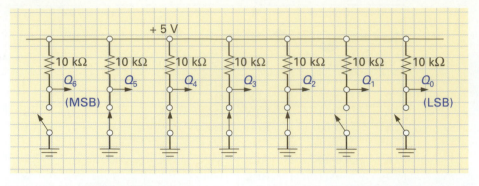

FIGURE 2-17 A 7-Bit Switch Register.

Use the following questions to test your understanding of Section 2-4.

1. Convert 7629_{10} to BCD.
2. What is the decimal equivalent of the excess-3 code 0100 0100?
3. What is a minimum-change code?
4. What does the following string of ASCII codes mean?

   1000010 1100001 1100010 1100010 1100001 1100111 1100101

REVIEW QUESTIONS

Multiple Choice Questions

1. Binary is a base _____ number system, decimal
   a base _____ number system, and hexadecimal
   a base _____ number system.
   a. 1, 2, 3   **c.** 10, 2, 3
   b. 2, 10, 16   **d.** 8, 4, 2

2. What is the binary equivalent of decimal 11?
   a. 1010   **c.** 1011
   b. 1100   **d.** 0111

3. Which of the following number systems is used as a shorthand for binary?
   a. Decimal   **c.** Binary
   b. Hexadecimal   **d.** Both a and b

4. What would be displayed on a hexadecimal counter if its count was advanced by 1 from $39FF_{16}$?
   a. 4000   **c.** 3900
   b. 3A00   **d.** 4A00

5. What is the base of 101?
   a. 10   **c.** 16
   b. 2   **d.** Could be any base

6. Which of the following is an invalid BCD code?
   a. 1010   **c.** 0111
   b. 1000   **d.** 0000

7. What is the decimal equivalent of 1001 0001 1000 0111$_{BCD}$?
   a. 9187   **c.** 8659
   b. 9A56   **d.** 254,345

8. What is the pure binary equivalent of 15_{16}?
   a. 0000 1111   **c.** 0101 0101
   b. 0001 0101   **d.** 0000 1101

9. Which of the following is an invalid excess-3 code?
   a. 1010   **c.** 0111
   b. 1000   **d.** 0000

10. Since the lower 4 bits of the ASCII code for the numbers 0 through 9 are the same as their pure binary equivalent, what is the 7-bit ASCII code for the number 6?
    a. 0111001   **c.** 0110011
    b. 0110111   **d.** 0110110

Practice Problems

11. What is the pure binary output of the switch register shown in Figure 2-18 and what is its decimal equivalent?

12. Convert the following into their decimal equivalents:
    a. 110111_2
    b. $2F_{16}$
    c. 10110_{10}

13. What would be the decimal equivalent of the LED display in Figure 2-19 if it were displaying each of the following:
    a. Pure binary   **c.** Excess-3
    b. BCD

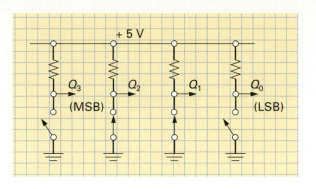

FIGURE 2-18 Four-Bit Switch Register.

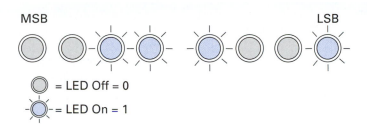

= LED Off = 0

= LED On = 1

FIGURE 2-19 8-Bit LED Display.

14. Convert the following binary numbers to octal and hexadecimal:
- **a.** 111101101001
- **b.** 1011
- **c.** 111000
- **d.** 1111111

15. Convert the following decimal numbers to BCD:
- **a.** 2,365
- **b.** 24

16. Give the ASCII codes for the following:
- **a.** ?
- **b.** $
- **c.** 6

17. Identify the following codes.

DECIMAL	(A)		(B)		(C)		(D)
0	0000	0000	0000	0011	0011	011	0000
1	0000	0001	0001	0011	0100	011	0001
2	0000	0010	0011	0011	0101	011	0010
3	0000	0011	0010	0011	0110	011	0011
4	0000	0100	0110	0011	0111	011	0100
5	0000	0101	0111	0011	1000	011	0101

18. Convert the decimal number 23 into each of the following:
- **a.** ASCII
- **b.** Pure binary
- **c.** BCD
- **d.** Hexadecimal
- **e.** Octal
- **f.** Excess-3

19. If a flashlight is used to transmit ASCII codes, with a dot (short flash) representing a 0 and a dash (long flash) representing a 1, what is the following message?

0000010	1010000	1110010	1100001	1100011	1110100	1101001
1100011	1100101	0100000	1001101	1100001	1101011	1100101
1110011	0100000	1010000	1100101	1110010	1100110	1100101
1100011	1110100	0000011				

20. Convert the following:
- **a.** $1110110_2 = $ _____ $_{16}$
- **b.** $175_{10} = $ _____ $_2$
- **c.** $ABC_{16} = $ _____ $_2$
- **d.** $00110110_2 = $ _____ $_8$

Web Site Questions

Go to the web site http://prenhall.com/cook, select the textbook *Practical Digital Electronics*, select this chapter, and then follow the instructions when answering the multiple-choice practice problems.

Logic Gates

Wireless

Lee De Forest was born in Council Bluffs, Iowa on August 26, 1873. He obtained his Ph.D. from Yale Sheffield Scientific School in 1899 with a thesis that is recognized as the first paper on radio communications in the United States.

De Forest was an astounding practical inventor; however, he wasn't a good businessman and many of his inventions were stolen and exploited by business partners.

In 1902, he invented an electrolytic radio detector and an alternating-current transmitter while working for the Western Electric Company. He also developed an optical sound track for movies which was rejected by film producers. Later another system was used that was based on De Forest's principles.

In 1907, De Forest patented his *audion* detector or triode (three-electrode) vacuum tube. This tube was a more versatile vacuum tube than the then available diode (two-electrode) vacuum tube, or *Fleming Valve,* invented by English electrical engineer, Sir John Fleming. By adding a third electrode, called a control grid, De Forest made amplification possible because the tube could now be controlled. The triode's ability to amplify as well as rectify led to the development of radio communications, or wireless as it was called, and later to television. De Forest also discovered that by cascading (connecting end-to-end) amplifiers, a higher gain could be attained. This ability led to long distance communications because weak long-range input signals could be increased in magnitude to a usable level.

In 1905, he began experimenting with speech and music broadcasts. In 1910 De Forest transmitted the singing voice of Enrico Caruso and, in so doing, became one of the pioneers of radio broadcasting.

Slowly the usefulness of the triode as a generator, amplifier, and detector of radio waves became apparent. However, it was not until World War I that the audion became an invaluable electronic device and was manufactured in large quantities.

As well as working on the technical development of radio, De Forest gave many practical public demonstrations of wireless communications. He became internationally renowned and was often referred to as "the father of radio." Before his fame and fortune, however—just before everyone became aware of the profound effect that his audion tube would have on the world—an incident occurred. While trying to sell stock in his company, De Forest was arrested on charges of fraud and his device, which was to launch a new era in electronics and make possible radio and television, was at this time called "a strange device like an incandescent lamp . . . which has proved worthless."

3

Introduction

Within any digital electronic system you will find that diodes and transistors are used to construct **logic gate circuits.** Logic gates are in turn used to construct flip-flop circuits, and flip-flops are used to construct register, counter, and a variety of other circuits. These logic gates, therefore, are used as the basic building blocks for all digital circuits, and their purpose is to control the movement of binary data and binary instructions. Logic gates, and all other digital electronic circuits, are often referred to as **hardware** circuits. By definition, the hardware of a digital electronic system includes all of the electronic, magnetic, and mechanical devices of a digital system. In contrast, the **software** of a digital electronic system includes the binary data (like pure binary and ASCII codes) and binary instructions that are processed by the digital electronic system hardware. To use an analogy, we could say that a compact disk player is hardware, and the music information stored on a compact disk and processed by the player is software. Just as a CD player is useless without CDs, a digital system's hardware is useless without software. In other words, the information on the CD determines what music is played, and similarly the digital software determines the actions of the digital electronic hardware.

Every digital electronic circuit uses logic-gate circuits to manipulate the coded pulses of binary language. These logic-gate circuits are constructed using diodes and transistors, and they are the basic decision-making elements. Digital circuits are often referred to as switching circuits because their control devices (diodes and transistors) are switched between the two extremes of ON and OFF. They are also called **two-state circuits** because their control devices are driven into one of two states: either into the saturation state (fully ON), or cutoff state (fully OFF). These two modes of operation are used to represent the two binary digits of 1 and 0.

In summary, a logic gate accepts inputs in the form of HIGH or LOW voltages, judges these input combinations based on a predetermined set of rules, and then produces a single output in the form of a HIGH or LOW voltage. The term *logic* is used because the output is predictable or logical, and the term *gate* is used because only certain input combinations will "unlock the gate." For example, any HIGH input to an OR gate will unlock the gate and allow the HIGH at the input to pass through to the output.

Logic Gate Circuits

Circuits containing predictable gate functions that either open or close their outputs.

Hardware

A term used to describe the electronic, magnetic, and mechanical devices of a digital system.

Software

A term used to describe the binary instructions and data processed by digital electronic system hardware.

Two-state Circuits

Switching circuits that are controlled to be either ON or OFF.

3-1 BASIC LOGIC GATES

The two basic types of logic gates used in digital circuits are the OR gate and the AND gate.

3-1-1 *The OR Gate*

The **OR gate** can have two or more inputs, but will always have a single output. Figure 3-1(a) shows how an OR gate can be constructed using two diodes and a resistor. Figure 3-1(b) shows a table listing all of the input possibilities for this two-input OR gate. This table is often referred to as a **truth table** or **function table** since it details the "truth" or the way in which this circuit will function. The insets in Figure 3-1(b) simplify how the circuit will basically produce one of two outputs. If both inputs are LOW, both diode switches are open and the output will be LOW. On the other hand, any HIGH input will cause the associated diode switch to close and connect the HIGH input through to the output. We could summarize the operation of this gate, therefore, by saying that *if either* A *OR* B *is HIGH, the output* Y *will be HIGH, and only when both inputs are LOW will the output be LOW.*

Figure 3-1(c) shows how an OR gate can be constructed using transistors as switches instead of diodes. In this OR gate circuit, three two-state transistor switches are included to construct a three-input OR gate. The truth table for this OR gate circuit is shown in Figure 3-1(d). Although another input has been added to this circuit, the action of the gate still remains the same since any HIGH input will still produce a HIGH output, and only when all of the inputs are LOW will the output be LOW. To explain the operation of this transistor logic gate in more detail, the insets in Figure 3-1(d) show how the transistor switches operate to produce one of two output voltage levels. When all inputs are LOW, the transistor's base-emitter junctions are reverse-biased, all transistors are OFF, and the pull-down resistor will take the *Y* output LOW. On the other hand, any HIGH input to the base of an NPN transistor will forward-bias the associated transistor and cause the transistor's collector-to-emitter junction to be equivalent to a closed switch, and therefore switch the HIGH ($+V_{CC}$) collector supply voltage through to the *Y* output.

Figure 3-1(e) shows the logic schematic symbols for a two- and three-input OR gate and the abbreviated operation expressions for the OR gate. Logic gates with a large number of inputs will have a large number of input combinations; however, the gate's operation will still be the same. To calculate the number of possible input combinations, you can use the following formula:

$$n = 2^x$$

n = number of input combinations
2 is included because we are dealing with a base-2 number system
x = number of inputs

◼ EXAMPLE:

Construct a truth table for a four-input OR gate and show the output logic level for each and every input combination.

◼ *Solution:*

Figure 3-2(a) shows the truth table for a four-input OR gate. The number of possible input combinations will be

$$n = 2^x$$
$$n = 2^4 = 16$$

(Calculator Sequence: press keys, 2 y^x 4 =)

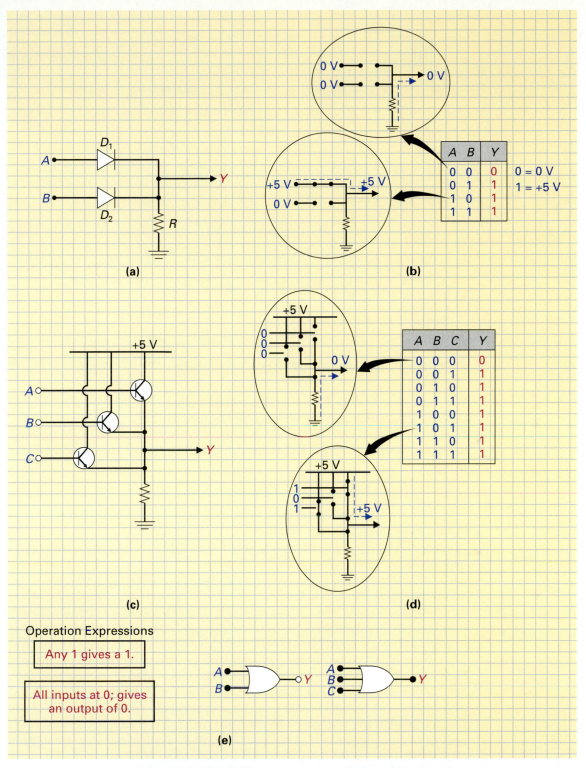

FIGURE 3-1 The OR Gate. (a) Two-Input Diode Circuit. (b) Two-Input Truth Table. (c) Three-Input Transistor Circuit. (d) Three-Input Truth Table. (e) OR Gate Symbols and Operation Expressions.

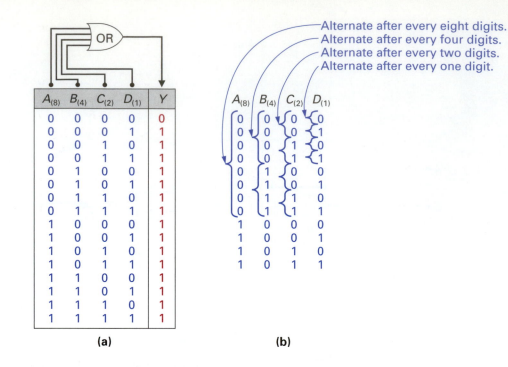

$A_{(8)}$	$B_{(4)}$	$C_{(2)}$	$D_{(1)}$	Y
0	0	0	0	0
0	0	0	1	1
0	0	1	0	1
0	0	1	1	1
0	1	0	0	1
0	1	0	1	1
0	1	1	0	1
0	1	1	1	1
1	0	0	0	1
1	0	0	1	1
1	0	1	0	1
1	0	1	1	1
1	1	0	0	1
1	1	0	1	1
1	1	1	0	1
1	1	1	1	1

(a)

(b)

FIGURE 3-2 **A Four-Input OR Gate.**

With only two possible digits (0 and 1) and four inputs, we can have a maximum of sixteen different input combinations. Whether the OR gate has 2, 3, 4, or 444 inputs, it will still operate in the same predictable or logical way: When all inputs are LOW, the output will be LOW, and when any input is HIGH, the output will be HIGH. This is shown in the Y column in Figure 3-2(a).

An easy way to construct these truth tables is to first calculate the maximum number of input combinations and then start in the units (1s) column (D column) and move down, alternating the binary digits after every single digit (0101010101, and so on) up to the maximum count, as shown in Figure 3-2(b). Then go to the twos (2s) column (C column) and move down, alternating the binary digits after every two digits (001100110011, and so on) up to the maximum count. Then go to the fours (4s) column (B column) and move down, alternating the binary digits after every four digits (000011110000, and so on) up to the maximum count. Finally, go to the eights (8s) column (A column) and move down, alternating the binary digits after every eight digits (0000000011111111, and so on) up to the maximum count.

These input combinations—or binary words—will start at the top of the truth table with binary 0 and then count up to a maximum value that is always one less than the maximum number of combinations. For example, with the four-input OR gate in Figure 3-2(a), the truth table begins with a count of 0000_2 (0_{10}) and then counts up to a maximum of 1111_2 (15_{10}). The maximum count within a truth table (1111_2 or 15_{10}) is always one less than the maximum number of combinations (16_{10}) because 0000_2 (0_{10}) is one of the input combinations (0000, 0001, 0010, 0011, 0100, 0101, 0110, 0111, 1000, 1001, 1010, 1011, 1100, 1101, 1110, 1111 is a total of sixteen different combinations, with fifteen being the maximum count). Stated with a formula:

$$Count_{max} = 2^x - 1$$

For example, in Figure 3-2: $Count_{max} = 2^4 - 1 = 16 - 1 = 15$ (Calculator Sequence: press keys, $2\ y^x\ 4 - 1 =$)

CHAPTER 3 / LOGIC GATES

Referring to the circuit in Figure 3-3(a), list the 4-bit binary words generated at the outputs of the four OR gates when each of the 0 through 9 push buttons is pressed.

■ *Solution:*

An encoder circuit is a code generator, and the circuit in Figure 3-3(a) is a decimal-to-binary code-generator circuit. Whenever a decimal push button is pressed, a 4-bit binary word equivalent to the decimal value is generated at the outputs of OR gates *D, C, B,* and *A*. The table in Figure 3-3(b) lists the pure binary codes generated for each of the decimal push buttons. For example, when push button 6 is pressed, a HIGH input is applied to OR gates *C* and *B*, and therefore their outputs will go HIGH, while the outputs of OR gates *C* and *A* will stay LOW because all of their inputs are LOW due to the pull-down resistors. In this instance, the 4-bit binary code 0110_2 (6_{10}) will be generated.

Key	Bit 4	Bit 3	Bit 2	Bit 1
0	0	0	0	0
1	0	0	0	1
2	?	?	?	?
3	?	?	?	?
4	?	?	?	?
5	?	?	?	?
6	?	?	?	?
7	?	?	?	?
8	?	?	?	?
9	?	?	?	?

(a)

	D	C	B	A
Key	8	4	2	1
0	0	0	0	0
1	0	0	0	1
2	0	0	1	0
3	0	0	1	1
4	0	1	0	0
5	0	1	0	1
6	0	1	1	0
7	0	1	1	1
8	1	0	0	0
9	1	0	0	1

Binary Words

(b)

FIGURE 3-3 Application—A Decimal-to-Binary Encoder Circuit Using OR Gates.

Rather than construct OR gates using discrete components such as the diode and transistor circuits shown in Figure 3-1(a) and (c), you can experiment with some of the digital integrated circuits (ICs) available from semiconductor manufacturers. These digital ICs contain complete logic-gate circuits. Figure 3-4(a) introduces the "7432 digital IC," which contains four (quad) two-input OR gates within a single 14-pin package. All four OR gates within this IC will operate if power is connected to the IC by using a dc power supply to apply +5 V to pin 14 and ground to pin 7. To test the OR gate that has its inputs connected to pins 1 and 2 and its output at pin 3, we will apply a square wave from a function generator to the OR gate's pin 1 input and a HIGH/LOW control input to the OR gate's pin 2 input, and monitor the output of the OR gate at pin 3 with an oscilloscope.

FIGURE 3-4 **Testing the Operation of Digital IC Containing Four Two-Input OR Gates.**

Referring to the timing diagram in Figure 3-4(b), you can see that whenever the square wave input (pin 1) or the control input (pin 2) are HIGH (shown shaded), the output will also be HIGH. Only when both inputs to the OR gate are LOW will the output be LOW. You may want to try constructing this test circuit in a lab and test each one of the four OR gates in a 7432 IC.

3-1-2   *The AND Gate*

Like the OR gate, an **AND gate** can have two or more inputs but will always have a single output. Figure 3-5(a) shows how an AND gate can be constructed using two diodes

AND Gate

A logic gate that will give a HIGH output only if all inputs are HIGH.

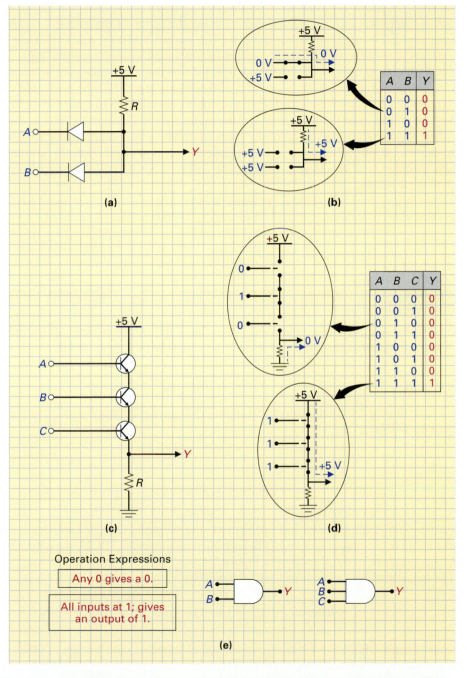

FIGURE 3-5   The AND Gate. (a) Two-Input Diode Circuit. (b) Two-Input Truth Table. (c) Three-Input Transistor Circuit. (d) Three-Input Truth Table. (e) AND Gate Symbols.

and a resistor. Figure 3-5(b) shows the truth table for this two-input AND gate. The insets in Figure 3-5(b) simplify how the circuit will basically produce one of two outputs. Since both of the diode anodes are connected to +5 V via *R,* any LOW input to the AND gate will turn ON the associated diode, as shown in the first inset in Figure 3-5(b). In this instance, the diode or diodes will be equivalent to a closed switch and therefore switch the LOW input through to the output. On the other hand, when both *A* AND *B* inputs are HIGH, both diodes will be reverse-biased and equivalent to open switches, and the output *Y* will be pulled HIGH, as shown in the second inset in Figure 3-5(b). We could summarize the operation of this gate, therefore, by saying that *any LOW input will cause a LOW output, and only when both* A *AND* B *inputs are HIGH will the output* Y *be HIGH*.

Figure 3-5(c) shows how a three-input AND gate can be constructed using three two-state transistor switches. The truth table for this AND gate circuit is shown in Figure 3-5(d). Although another input has been added to this circuit, the action of the gate still remains the same since any LOW input will still produce a LOW output, and only when all of the inputs are HIGH will the output be HIGH. To explain the operation of this transistor logic gate in more detail, the insets in Figure 3-5(d) show how the transistor switches operate to produce one of two output voltage levels. When any of the inputs are LOW, the transistor's base-emitter junction is reverse-biased, the OFF transistor is equivalent to an open switch between collector and emitter, and the pull-down resistor will take the *Y* output LOW. On the other hand, when all inputs are HIGH, all of the NPN transistor's base-emitter junctions will be forward-biased, causing all of the transistor's collector-to-emitter junctions to be equivalent to closed switches and therefore switching the HIGH ($+V_{CC}$) collector supply voltage through to the *Y* output.

Figure 3-5(e) shows the logic schematic symbols for a two- and three-input AND gate, and the abbreviated operation expressions for the AND gate.

EXAMPLE:

Develop a truth table for a five-input AND gate. Show the output logic level for each and every input combination, and indicate the range of values within the truth table.

Solution:

Figure 3-6 shows the truth table for a five-input AND gate. The number of possible input combinations will be

$$n = 2^x$$
$$n = 2^5 = 32$$

A five-input AND gate will still operate in the same predictable or logical way: When any input is LOW, the output will be LOW, and when all inputs are HIGH, the output will be HIGH. This is shown in the *Y* column in Figure 3-6.

The range of values within the truth table will be

$$Count_{max} = 2^5 - 1 = 32 - 1 = 31$$

With the five-input OR gate in Figure 3-6, the truth table begins with a count of 00000_2 (0_{10}) and then counts up to a maximum of 11111_2 (31_{10}).

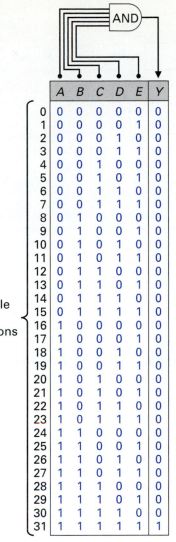

		A	B	C	D	E	Y
	0	0	0	0	0	0	0
	1	0	0	0	0	1	0
	2	0	0	0	1	0	0
	3	0	0	0	1	1	0
	4	0	0	1	0	0	0
	5	0	0	1	0	1	0
	6	0	0	1	1	0	0
	7	0	0	1	1	1	0
	8	0	1	0	0	0	0
	9	0	1	0	0	1	0
	10	0	1	0	1	0	0
	11	0	1	0	1	1	0
	12	0	1	1	0	0	0
	13	0	1	1	0	1	0
32 Possible	14	0	1	1	1	0	0
Input	15	0	1	1	1	1	0
Combinations	16	1	0	0	0	0	0
	17	1	0	0	0	1	0
	18	1	0	0	1	0	0
	19	1	0	0	1	1	0
	20	1	0	1	0	0	0
	21	1	0	1	0	1	0
	22	1	0	1	1	0	0
	23	1	0	1	1	1	0
	24	1	1	0	0	0	0
	25	1	1	0	0	1	0
	26	1	1	0	1	0	0
	27	1	1	0	1	1	0
	28	1	1	1	0	0	0
	29	1	1	1	0	1	0
	30	1	1	1	1	0	0
	31	1	1	1	1	1	1

FIGURE 3-6 A Five-Input AND Gate.

■ EXAMPLE:

Referring to the circuit in Figure 3-7(a), describe what would be present at the output of the AND gate for each of the CONTROL switch positions.

■ *Solution:*

When the CONTROL switch is put in the ENABLE position, a HIGH is applied to the lower input of the AND gate, as shown in Figure 3-7(b). In this mode, the output *Y* will follow the square wave input, since when the square wave is HIGH, the AND gate inputs are both HIGH, and *Y* is HIGH. When the square wave is LOW, the AND gate will have a LOW input, and *Y* will be LOW. The *Y* output, therefore, follows the square wave input, and the AND gate is said to be equivalent to a closed switch, as shown in Figure 3-7(b). On the other hand, when the CONTROL switch is put in the DISABLE position, a

FIGURE 3-7 Application—An Enable/Disable Control Circuit Using an AND Gate.

LOW is applied to the lower input of the AND gate, as shown in Figure 3-7(c). In this mode, the output Y will always remain LOW, since any LOW input to an AND gate will always result in a LOW output. In this situation, the AND gate is said to be equivalent to an open switch, as shown in Figure 3-7(c).

When used in applications such as this, the AND gate is said to be acting as a **controlled switch.**

Like the OR gate digital IC, there are several digital ICs available containing AND gates. Figure 3-8(a) introduces the "7408 digital IC," which contains four (quad) two-input AND gates and shows how to test the AND gates to see if they operate as controlled switches. As before, all four AND gates within this IC will operate if power is connected to the IC by using a dc power supply to apply +5 V to pin 14 and ground to pin 7. To test the AND gate that has its inputs connected to pins 1 and 2 and its output at pin 3, we will apply a square wave from a function generator to the AND gate's pin 1 input and a HIGH/LOW control input to the AND gate's pin 2 input and monitor the output of the AND gate at pin 3 with an oscilloscope. Referring to the timing diagram in Figure 3-8(b), you can see that only when both inputs are HIGH (shown shaded), will the output be HIGH. You may want to try constructing this test circuit in a lab and testing each one of the four AND gates in a 7408 IC.

FIGURE 3-8 Testing the Operation of a Digital IC Containing Four Two-Input AND Gates.

Use the following questions to test your understanding of Section 3-1.

1. The two basic logic gates are the _____ gate and the _____ gate.
2. With an OR gate, any binary 1 input will give a binary _____ output.
3. Which logic gate can be used as a controlled switch?
4. Only when all inputs are LOW will the output of an _____ gate be LOW.
5. With an AND gate, any binary 0 input will give a binary _____ output.
6. Only when all inputs are HIGH will the output of an _____ gate be HIGH.

Although the basic OR and AND logic gates can be used to construct many digital circuits, in some circuit applications other logic operations are needed. For this reason, semiconductor manufacturers made other logic-gate types available. In this section, we will examine three inverting-type logic gates, called the NOT gate, the NOR gate, and the NAND gate. We will begin with the NOT or INVERTER gate, which was introduced earlier.

3-2-1   *The NOT Gate*

The NOT, or logic INVERTER gate, is the simplest of all the logic gates because it has only one input and one output. Figure 3-9(a) shows how a NOT gate can be constructed using two resistors and a transistor. Figure 3-9(b) shows the truth table for the NOT gate. The insets in Figure 3-9(b) simplify how the circuit will produce one of two outputs. Referring to the upper inset, you can see that when the input is LOW, the transistor's base-emitter diode will be reverse-biased and so the transistor will turn OFF. In this instance, the transistor is equivalent to an open switch between emitter and collector, and therefore the +5 V supply voltage will be connected to the output. A LOW input, therefore, will be converted or inverted to a HIGH output. On the other hand, referring to the lower inset in Figure 3-9(b), you can see that when the input is HIGH, the transistor's base-emitter diode will be forward-biased and the transistor will turn ON. In this circuit condition, the transistor is equivalent to a closed switch between emitter and collector, and the output will be switched LOW. A HIGH input, therefore, will be converted or inverted to a LOW output. We could summarize the operation of this gate by saying that *the output logic level is "NOT" the same as the input logic level* (a 1 input is NOT the same as the 0 output, or a 0 input is NOT the same as the 1 output), hence the name NOT gate. The output is, therefore, always the **complement,** or opposite, of the input.

Complement

Opposite.

FIGURE 3-9   The NOT Gate or Logic Inverter. (a) Transistor Circuit. (b) Truth Table. (c) Logic Symbols for a NOT Gate. (d) Timing Diagram Showing Input-to-Output Inversion.

Figure 3-9(c) shows the schematic symbols normally used to represent the NOT gate. The triangle is used to represent the logic circuit, while the **bubble,** or small circle either before or after the triangle, is used to indicate the complementary or inverting nature of this circuit. Figure 3-9(d) shows a typical example of a NOT gate's input/output waveforms. As you can see from these waveforms, the output waveform is an inverted version of the input waveform.

Bubble

A small circle symbol used to signify an invert function.

■ **EXAMPLE:**

What will be the outputs from the circuit in Figure 3-10(a), assuming the square wave input shown?

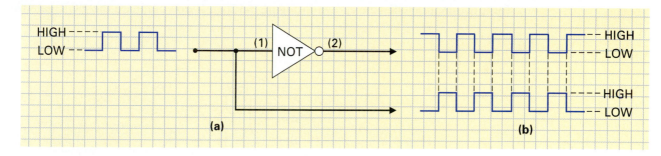

FIGURE 3-10   Application—A Two-Phase Signal Generator Circuit Using a NOT Gate.

■ *Solution:*

As can be seen in Figure 3-10(b), the two-phase signal generator circuit will supply two square wave output signals that are 180° out of phase due to the inverting action of the NOT gate.

■ **EXAMPLE:**

What will be the outputs from each of the six NOT gates within the 7404 digital IC shown in Figure 3-11?

■ *Solution:*

The NOT gate will always invert its input, and therefore:

	NOT GATE INPUT	NOT GATE OUTPUT
A	HIGH	LOW
B	LOW	HIGH
C	LOW	HIGH
D	HIGH	LOW
E	HIGH	LOW
F	LOW	HIGH

FIGURE 3-11   A Digital IC Containing Six NOT Gates.

3-2-2   *The NOR Gate*

The name "NOR" is a combination of the two words "NOT" and "OR" and is used to describe a type of logic gate that contains an OR gate followed by a NOT gate, as shown in Figure 3-12(a). The **NOR gate** could be constructed using the previously discussed transistor OR gate circuit followed by a transistor NOT gate circuit, as shown in the inset in Figure 3-12(a). The NOR gate should not be thought of as a new type of logic gate because it isn't. It is simply a combination of two previously discussed gates—an OR gate and a NOT gate.

Figure 3-12(b) shows the standard symbol used to represent the NOR gate. Studying the NOR gate symbol, you can see that the OR symbol is used to show how the two inputs are first "ORed," and that the bubble is used to indicate that the result from the OR operation is inverted before being applied to the output. As a result, the NOR gate's output is simply the opposite, or complement, of the OR gate, as can be seen in the truth table for a two-input NOR gate shown in Figure 3-12(c). If you compare the NOR gate truth

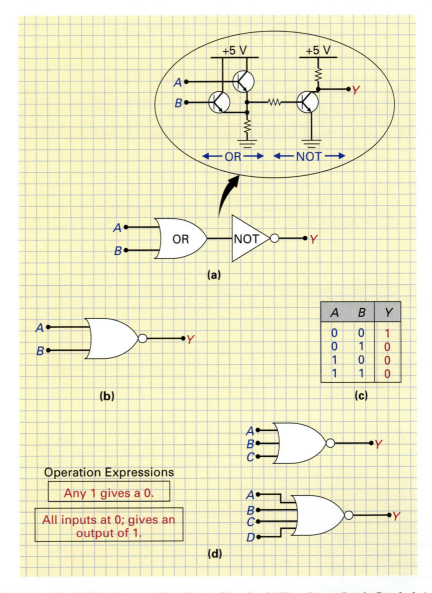

A	B	Y
0	0	1
0	1	0
1	0	0
1	1	0

(a)

(b)

(c)

Operation Expressions

Any 1 gives a 0.

All inputs at 0; gives an output of 1.

(d)

FIGURE 3-12   **The NOR Gate. (a) Two-Input Circuit. (b) Two-Input Logic Symbol. (c) Two-Input Truth Table. (d) NOR Gate Operation Expressions and Three- and Four-Input Logic Symbols.**

table to the OR gate truth table, you can see that the only difference is that the output is inverted:

> OR Gate: When all inputs are 0, the output is **0,** and any 1 input gives a **1** output.
> NOR Gate: When all inputs are 0, the output is **1,** and any 1 input gives a **0** output.

The operation expressions for the NOR gate are listed in Figure 3-12(d), along with the standard symbols for a three-input and four-input NOR gate.

EXAMPLE:

What will be the outputs from each of the four NOR gates within the 7402 digital IC shown in Figure 3-13?

Solution:

When all inputs are 0, the output is 1, and when any input is 1, the output is 0. Therefore:

	NOR GATE INPUTS		NOR GATE OUTPUT
A	HIGH	HIGH	LOW
B	HIGH	LOW	LOW
C	LOW	LOW	HIGH
D	LOW	HIGH	LOW

FIGURE 3-13   **A Digital IC Containing Four NOR gates.**

EXAMPLE:

Referring to the circuit shown in Figure 3-14, determine whether the LED will flash at a rate of 1 Hz or 2 Hz for the inputs and switch position shown.

Solution:

The circuit shown in Figure 3-14 is an example of a multiplexer circuit, which is a circuit that switches only one of its inputs through to a single output. In this circuit, two AND gates

FIGURE 3-14   **Application—A Data Selector or Multiplexer Circuit Using a NOR Gate.**

are operated as controlled switches, with the data select switch determining which of the AND gates is enabled. In this example, the upper AND gate is enabled, and so the *A*-data input (1 Hz signal) will be switched through to the NOR gate's upper input. The lower input to the NOR gate will be LOW, because the lower AND gate has been disabled by the data select switch. (AND: Any 0 input gives a 0 output.) Since any HIGH input to a NOR gate will give a LOW output, the NOR gate will basically act as an INVERTER and invert the 1 Hz square wave input. The anode of the LED is connected to +5 V, so when the output of the NOR gate goes LOW, the LED will turn ON, and when the output of the NOR gate goes HIGH, the LED will turn OFF. This multiplexer circuit will therefore switch the 1 Hz *A*-data input through to the output, and the LED will flash ON and OFF at a 1 Hz rate.

If the data select switch were placed in the *B* position, the 2 Hz *B*-data input would be switched through to the output and the LED would flash ON and OFF at a 2 Hz rate.

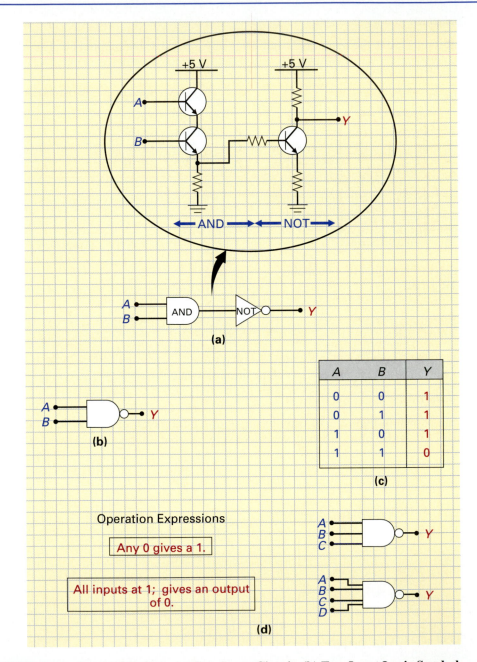

A	B	Y
0	0	1
0	1	1
1	0	1
1	1	0

(c)

Operation Expressions

Any 0 gives a 1.

All inputs at 1; gives an output of 0.

(d)

FIGURE 3-15 The NAND Gate. (a) Two-Input Circuit. (b) Two-Input Logic Symbol. (c) Two-Input Truth Table. (d) NAND Gate Operation Expressions and Three- and Four-Input Logic Symbols.

3-2-3 *The NAND Gate*

Like NOR, the name "NAND" is a combination of the two words "NOT" and "AND," and it describes another type of logic gate that contains an AND gate followed by a NOT gate, as shown in Figure 3-15(a). The **NAND gate** could be constructed using the previously discussed transistor AND gate circuit followed by a transistor NOT gate circuit, as shown in the inset in Figure 3-15(a). The NAND gate should also not be thought of as a new type of logic gate because it is simply an AND gate and a NOT gate combined to perform the NAND logic function.

Figure 3-15(b) shows the standard symbol used to represent the NAND gate. Studying the NAND gate symbol, you can see that the AND symbol is used to show how the two inputs are first "ANDed," and the bubble is used to indicate that the result from the AND operation is inverted before being applied to the output. As a result, the NAND gate's output is simply the opposite, or complement, of the AND gate, as can be seen in the truth table for a two-input NAND gate shown in Figure 3-15(c). If you compare the NAND gate truth table to the AND gate truth table, you can see that the only difference is that the output is inverted. To compare:

AND Gate: Any 0 input gives a **0** output, and only when both inputs are 1 will the output be **1.**

NAND Gate: Any 0 input gives a **1** output, and only when both inputs are 1 will the output be **0.**

The operation expressions for the NAND gate are listed in Figure 3-15(d), along with the standard symbols for a three-input and four-input NAND gate.

■ **EXAMPLE:**

What will be the outputs from each of the four NAND gates within the 7400 digital IC shown in Figure 3-16?

■ *Solution:*

With the NAND gate, any 0 input gives a 1 output, and only when both inputs are 1 will the output be 0. Therefore:

7400 - Quad Two-Input NAND IC

	NAND GATE INPUTS		NAND GATE OUTPUT
A	HIGH	HIGH	LOW
B	LOW	LOW	HIGH
C	HIGH	LOW	HIGH
D	LOW	HIGH	HIGH

FIGURE 3-16 A Digital IC Containing Four NAND Gates.

■ **EXAMPLE:**

Which of the LEDs in Figure 3-17 would turn ON if the binary input at *A* and *B* were both HIGH?

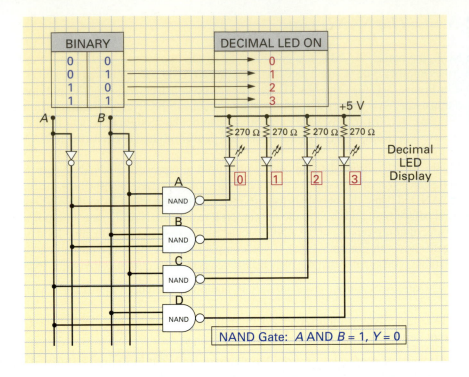

FIGURE 3-17 Application—A Binary-to-Decimal Decoder Circuit Using NAND Gates.

■ *Solution:*

The circuit shown in Figure 3-17 is an example of a decoder circuit, which is a circuit that translates coded characters into a more understandable form. In this example, the circuit is used to decode a 2-bit binary code and then activate a decimal display. A set of four NAND gates is used to drive the decimal display, which contains four LEDs numbered 0 through 3. Since the NAND gate outputs are connected to the cathodes of the LEDs, and the anodes are connected to +5 V, a LOW output from any NAND gate will turn ON its associated LED. As we now know, a NAND gate will only give a LOW output when both of its inputs are HIGH; and the inputs in this circuit are controlled by the A and B—and inverted A and B—binary input lines. For example, when A and B inputs are both LOW, the inverted A and B lines will both be HIGH. Only NAND gate D will have both of its inputs HIGH, so its output will go LOW, causing the number 3 LED to turn ON. As a result, when binary 11_2 is applied to the input of the decoder, the output will display decimal 3_{10}. Therefore, each of the four 2-bit binary input codes will activate one of the equivalent decimal output LEDs.

SELF-TEST EVALUATION POINT FOR SECTION 3-2

Use the following questions to test your understanding of Section 3-2.

1. Which of the logic gates has only one input?
2. With a NOT gate, any binary 1 input will give a binary _____ output.
3. Which logic gate is a combination of an AND gate followed by a logic INVERTER?
4. Only when all inputs are LOW will the output of a/an _____ gate be HIGH.
5. With a NAND gate, any binary 0 input will give a binary _____ output.
6. Only when all inputs are HIGH will the output of a/an _____ gate be LOW.

EXCLUSIVE LOGIC GATES

In this section, we will discuss the final two logic gate types: the exclusive-OR (XOR) gate and exclusive-NOR (XNOR) gate. Although these two logic gates are not used as frequently as the five basic OR, AND, NOT, NOR, and NAND gates, their function is ideal in some circuit applications.

3-3-1   *The XOR Gate*

The **exclusive-OR (XOR) gate** logic symbol is shown in Figure 3-18(a), and its truth table is shown in Figure 3-18(b). Like the basic OR gate, the operation of the XOR is dependent

Exclusive-OR Gate
A logic gate circuit that will give a HIGH output if any odd number of binary 1s are applied to the input.

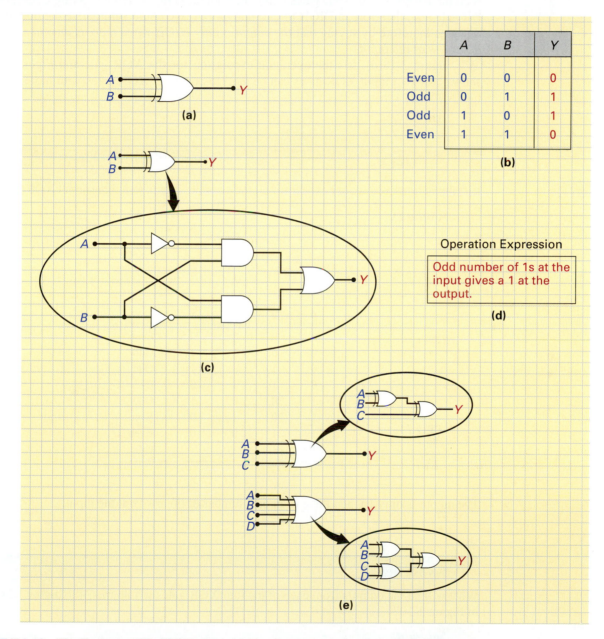

	A	B	Y
Even	0	0	0
Odd	0	1	1
Odd	1	0	1
Even	1	1	0

(b)

Operation Expression

Odd number of 1s at the input gives a 1 at the output.

(d)

FIGURE 3-18   The Exclusive-OR (XOR) Gate. (a) Two-Input Logic Symbol. (b) Two-Input Truth Table. (c) Basic Gate Circuit. (d) Operation Expression. (e) Three- and Four-Input XOR Gates.

on the HIGH or binary 1 inputs. With the basic OR, any 1 input would cause a 1 output. With the XOR, any odd number of binary 1s at the input will cause a binary 1 at the output. Looking at the truth table in Figure 3-18(b), you can see that when the binary inputs are 01 or 10, there is one binary 1 at the input (odd) and the output is 1. When the binary inputs are 00 or 11, however, there are two binary 0s or two binary 1s at the input (even) and the output is 0. To distinguish the XOR gate symbol from the basic OR gate symbol, you may have noticed that an additional curved line is included across the input.

Exclusive-OR logic gates are constructed by combining some of our previously discussed basic logic gates. Figure 3-18(c) shows how two NOT gates, two AND gates, and an OR gate can be connected to form an XOR gate. If inputs A and B are both HIGH or both LOW (even input), the AND gates will both end up with a LOW at one of their inputs, so the OR gate will have both of its inputs LOW, and therefore the final output will be LOW. On the other hand, if A is HIGH and B is LOW or if B is HIGH and A is LOW (odd input), one of the AND gates will have both of its inputs HIGH, giving a HIGH to the OR gate and, therefore, a HIGH to the final output. We could summarize the operation of this gate therefore by saying that *the output is only HIGH when there is an odd number of binary 1s at the input.*

Figure 3-18(d) lists the operation expression for the XOR gate and Figure 3-18(e) shows how two-input XOR gates can be used to construct XOR gates with more than two inputs.

EXAMPLE:

List the outputs you would expect for all the possible inputs to a three-input XOR gate.

Solution:

$$n = 2^x$$

$$n = 2^3 = 8$$

A	B	C	Y	
0	0	0	0	Even
0	0	1	1	Odd
0	1	0	1	Odd
0	1	1	0	Even
1	0	0	1	Odd
1	0	1	0	Even
1	1	0	0	Even
1	1	1	1	Odd

The output is only 1 when there is an odd number of binary 1s at the input.

EXAMPLE:

What will be the outputs from each of the four XOR gates within the 7486 digital IC shown in Figure 3-19?

Solution:

With the XOR gate, the output is only HIGH when there is an odd number of binary 1s at the input. Therefore:

CHAPTER 3 / LOGIC GATES

	XOR GATE INPUTS		XOR GATE OUTPUT
A	HIGH	HIGH	LOW
B	HIGH	LOW	HIGH
C	LOW	LOW	LOW
D	LOW	HIGH	HIGH

FIGURE 3-19 **A Digital IC Containing Four XOR Gates.**

3-3-2   *The XNOR Gate*

The **exclusive-NOR (XNOR) gate** is simply an XOR gate followed by a NOT gate, as shown in Figure 3-20(a). Its logic symbol, which is shown in Figure 3-20(b), is the same as the XOR symbol except for the bubble at the output, which is included to indicate that the result of the XOR operation is inverted before it appears at the output. Like the NAND and NOR gates, the XNOR is not a new gate, but simply a previously discussed gate with an inverted output.

The truth table for the XNOR gate is shown in Figure 3-20(c). Comparing the XOR gate to the XNOR gate, the XOR gate's output is only HIGH when there is an odd number of binary 1s at the input, whereas with the XNOR gate, the output is only HIGH when there is an even number of binary 1s at the input. The operation expression for the XNOR gate is listed in Figure 3-20(d).

Exclusive-NOR Gate
A NOT-exclusive OR gate that will give a HIGH output if any even number of binary 1s are applied to the input.

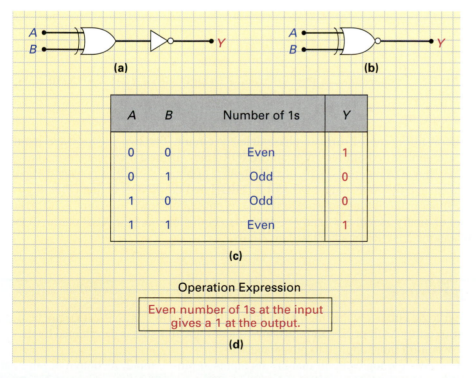

FIGURE 3-20   The Exclusive-NOR (XNOR) Gate. (a) Circuit. (b) Logic Symbol. (c) Truth Table. (d) Operation Expression.

What will be the *X* and *Y* outputs from the circuit in Figure 3-21(a) if the *A* and *B* inputs are either out-of-phase or in-phase?

FIGURE 3-21   **Application—A Phase Detector Circuit Using an XOR and XNOR Gate.**

Solution:

The circuit in Figure 3-21(a) is a phase detector circuit. The two square wave inputs at *A* and *B* are applied to both the XOR and XNOR gates. If these two waveforms are out-of-phase, as shown in Figure 3-21(b), an odd number of binary 1s will be constantly applied to both gates, and the XOR output (*X*) will go HIGH and the XNOR output (*Y*) will go LOW. On the other hand, if the two waveforms are in-phase, as shown in Figure 3-21(c), an even number of binary 1s will constantly be applied to both gates, and the XOR output (*X*) will go LOW and the XNOR output (*Y*) will go HIGH.

EXAMPLE:

The four logic circuits within the 74135 digital IC, shown in Figure 3-22(a), can be made to operate as either XOR or XNOR logic gates. Their logic function is determined by the HIGH or LOW control input applied to pins 4 and 12.

Consider the logic circuit that has its inputs at pins 1 and 2, its output at pin 3, and a control input at pin 4. How will this logic circuit function if its control input is first made LOW and then made HIGH?

FIGURE 3-22 A Digital IC Containing Four Logic Gates That Can Be Controlled to Act as XOR Gates or XNOR Gates.

■ *Solution:*

Figure 3-22(b) follows the logic levels throughout the logic circuit when the control input is LOW, and an odd and even number of binary 1s are applied to the inputs *A* and *B*. Studying the *Y* output for these odd and even inputs, you can see that an odd input results in a HIGH output, and an even input results in a LOW output. When the control input is LOW, therefore, the logic circuits function as XOR gates.

Figure 3-22(c) follows the logic levels throughout the logic circuit when the control input is HIGH, and an odd and even number of binary 1s are applied to the inputs *A* and *B*. Studying the *Y* output for these odd and even inputs, you can see that an odd input results in a LOW output, and an even input results in a HIGH output. When the control input is HIGH, therefore, the logic circuits function as XNOR gates.

SELF-TEST EVALUATION POINT FOR SECTION 3-3

Use the following questions to test your understanding of Section 3-3.

1. Which type of logic gate will always give a HIGH output when any of its inputs are HIGH?
2. Only when an odd number of 1s is applied to the input, will the output of an _____ gate be HIGH.
3. Only when an even number of 1s is applied to the input, will the output of an _____ gate be HIGH.
4. Which type of logic gate will always give a LOW output when an odd number of 1s is applied at the input?

Dependency Notation

A coding system on schematic diagrams that uses notations to indicate how an output is dependent on inputs.

The logic symbols presented so far in this chapter have been used for many years in the digital electronics industry. In 1984 the *Institute of Electrical and Electronic Engineers (IEEE)* and the *American National Standards Institute (ANSI)* introduced a new standard for logic symbols, which is slowly being accepted by more and more electronics companies. The advantage of this new standard is that instead of using distinctive shapes to represent logic gates, it uses a special **dependency notation system.** Simply stated, the new standard uses a *notation* (or note) within a rectangular or square block to indicate how the output is *dependent* on the input. Figure 3-23 compares traditional logic symbols with the newer IEEE/ANSI logic symbols and describes the meaning of the dependency notations.

You should be familiar with both logic symbol types since you will come across both in industry schematics.

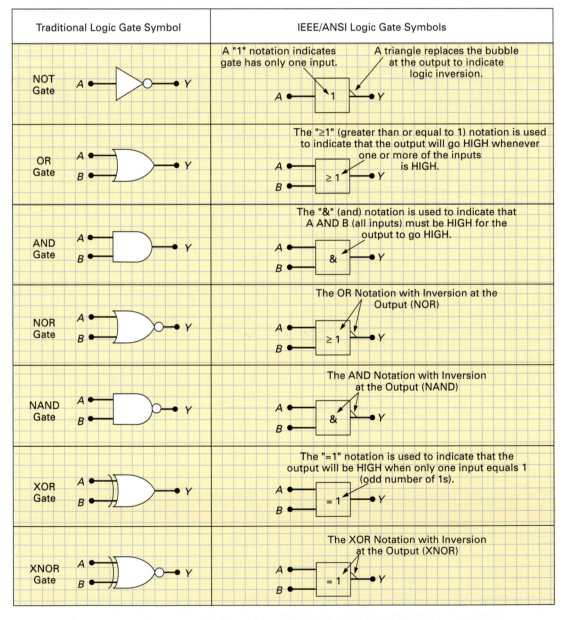

FIGURE 3-23   **Traditional and IEEE/ANSI Symbols for Logic Gates.**

■ EXAMPLE:

Identify which of the logic circuits shown in Figure 3-24 is traditional and which is using the IEEE/ANSI standard.

■ *Solution:*

Figure 3-24(a) shows the traditional logic symbol and Figure 3-24(b) shows the IEEE/ANSI logic symbol for a 7400 digital IC (quad two-input NAND).

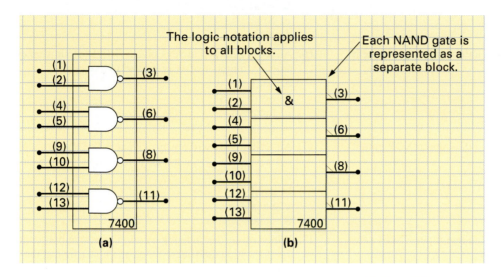

FIGURE 3-24 **Traditional and IEEE/ANSI Logic Symbols for a 7400 Digital IC (Quad Two-Input NAND).**

SELF-TEST EVALUATION POINT FOR SECTION 3-4

Use the following questions to test your understanding of Section 3-4.

1. Which type of logic-gate symbols makes use of rectangular blocks?
2. What advantage does the IEEE/ANSI standard have over the traditional symbol standard?

REVIEW QUESTIONS

Multiple-Choice Questions

1. Which device is currently being used extensively as a two-state switch in digital electronic circuits?
   a. Transistor
   b. Vacuum tube
   c. Relay
   d. Toggle switch

2. How many input combinations would a four-input logic gate have?
   a. Eight
   b. Sixteen
   c. Thirty-two
   d. Four

3. What would be the maximum count within the truth table for a four-input logic gate?
   a. 7
   b. 15
   c. 17
   d. 31

4. What would be the output from a three-input OR gate if its inputs were 101?
   a. 1
   b. 0
   c. Unknown
   d. None of the above

5. What would be the output from a three-input AND gate if its inputs were 101?
   a. 1
   b. 0
   c. Unknown
   d. None of the above

6. What would be the output from a three-input NAND gate if its inputs were 101?
   a. 1
   b. 0
   c. Unknown
   d. None of the above

7. What would be the output from a three-input NOR gate if its inputs were 101?
   a. 1
   b. 0
   c. Unknown
   d. None of the above

8. What would be the output from a three-input XOR gate if its inputs were 101?
   a. 1
   b. 0
   c. Unknown
   d. None of the above

9. What would be the output from a three-input XNOR gate if its inputs were 101?
   a. 1
   b. 0
   c. Unknown
   d. None of the above

10. What would be the output from a NOT gate if its input were 1?
    a. 1
    b. 0
    c. Unknown
    d. None of the above

11. Which of the following logic gates is ideal as a controlled switch?
    a. OR
    b. AND
    c. NOR
    d. NAND

12. Which of the following logic gates will always give a LOW output whenever a LOW input is applied?
    a. OR
    b. AND
    c. NOR
    d. NAND

13. Which of the following logic gates will always give a HIGH output whenever a LOW input is applied?
    a. OR
    b. AND
    c. NOR
    d. NAND

14. Which of the following logic gates will always give a HIGH output whenever a HIGH input is applied?
    a. OR
    b. AND
    c. NOR
    d. NAND

15. Which of the following logic gates will always give a LOW output whenever a HIGH input is applied?
    a. OR
    b. AND
    c. NOR
    d. NAND

16. An XOR gate will always give a HIGH output whenever an _____ number of 1s is applied to the input.
    a. odd
    b. even
    c. unknown
    d. none of the above

17. An XNOR gate will always give a HIGH output whenever an _____ number of 1s is applied to the input.
    a. odd
    b. even
    c. unknown
    d. none of the above

18. Which of the dependency notation logic symbols contains an "&" within the square block and no triangle at the output?
    a. OR
    b. AND
    c. NOR
    d. NAND

19. Which of the dependency notation logic symbols contains a "≥1" within the square block and no triangle at the output?
    a. OR
    b. AND
    c. NOR
    d. NAND

20. Which of the dependency notation logic symbols contains a "≥1" within the square block and a triangle at the output?
    a. OR
    b. AND
    c. NOR
    d. NAND

Practice Problems

21. In the home security system shown in Figure 3-25(a), a two-input logic gate is needed to actuate an alarm (output *Y*) if either the window is opened (input *A*) or the door is opened (input *B*). Which decision-making logic gate should be used in this application?

22. In the office temperature control system shown in Figure 3-25(b), a thermostat (input *A*) is used to turn ON and OFF a fan (output *Y*). A thermostat-enable switch is connected to the other input of the logic gate

WINDOW	DOOR	ALARM
Closed (0)	Closed (0)	OFF (0)
Closed (0)	Open (1)	ON (1)
Open (1)	Closed (0)	ON (1)
Open (1)	Open (1)	ON (1)

(a)

ENABLE	THERMOSTAT	FAN
OFF (0)	COLD (0)	OFF (0)
OFF (0)	HOT (1)	OFF (0)
ON (1)	COLD (0)	OFF (0)
ON (1)	HOT (1)	ON (1)

(b)

FIGURE 3-25 Application—Decision-Making Logic Gates.

(input *B*), and it will either enable thermostat control of the fan or disable thermostat control of the fan. Which decision-making logic gate should be used in this application?

23. Briefly describe the operation of the photodiode receiver circuit in Figure 3-26. What will be the output at *Y* when light is being applied to the photodiode and when light is not being applied?

24. Briefly describe the operation of the gated LED flasher circuit shown in Figure 3-27.

25. Which logic gate would produce the output waveform shown in Figure 3-28?

26. Choose one of the six logic gates shown in Figure 3-29(a) and then sketch the output waveform that would result for the input waveforms shown in Figure 3-29(b).

27. Develop truth tables for each of the logic circuits shown in Figure 3-30.

28. State the logic operation and sketch the IEEE/ANSI logic symbol for each of the logic circuits shown in Figure 3-30.

29. Referring to the circuit in Figure 3-31, what will be the outputs at Y_0, Y_1, Y_2, and Y_3 if the control line input is
   a. HIGH?
   b. LOW?

30. Referring to the circuit in Figure 3-31, what will the output logic levels be if the control line input is low?

Web Site Questions

Go to the web site http://www.prenhall.com/cook, select the textbook *Practical Digital Electronics,* select this chapter, and then follow the instructions when answering the multiple choice practice problems.

FIGURE 3-26 Application—Photodiode Receiver.

FIGURE 3-27   Application—Gated LED Flasher Circuit.

FIGURE 3-28   Timing Analysis of a Logic Gate.

FIGURE 3-29   Timing Waveforms.

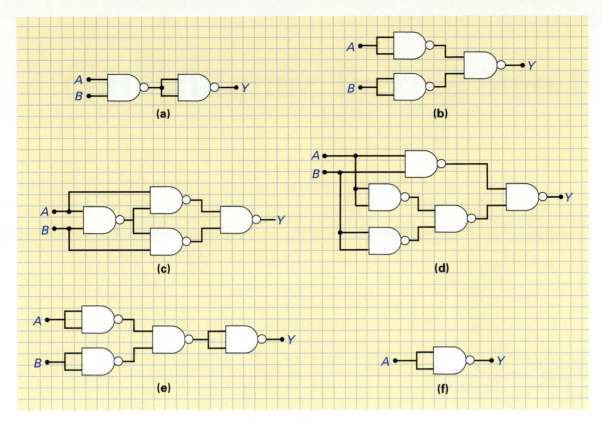

FIGURE 3-30   Using NAND Gates as Building Blocks for Other Gates.

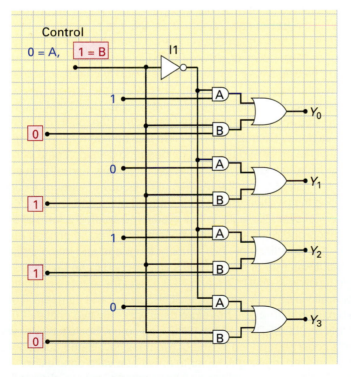

FIGURE 3-31   Application—Four-Bit Word Multiplexer.

61

Logic Circuit Simplification

From Folly to Foresight

George Boole was born in the industrial town of Lincoln in eastern England in 1815. His parents were poor tradespeople, and even though there was a school for boys in Lincoln, there is no record of him ever attending. In those hard times, children of the working class had no hope of receiving any form of education, and their lives generally followed the familiar pattern of their parents. George Boole, however, was to break the mold. He would rise up from these humble beginnings to become one of the most respected mathematicians of his day.

Boole's father had taught himself a small amount of mathematics, and since his son of six seemed to have a thirst for learning, he began to pass on all of his knowledge. At eight years old, George had surpassed his father's understanding and craved more. He quickly realized that his advancement was heavily dependent on his understanding Latin. Luckily, a family friend who owned a local book shop knew enough about the basics of Latin to get him on his way, and once he had taught Boole all he knew, Boole continued with the books at his disposal. By the age of twelve he had conquered Latin, and by fourteen he had added Greek, French, German, and Italian to his repertoire.

At the age of sixteen, however, poverty stood in his way. Since his parents could no longer support him, he was forced to take a job as a poorly paid teaching assistant. After studying the entire school system, he left four year laters and opened his own school in which he taught all subjects. It was in this role that he discovered that his mathematics was weak, so he began studying the mathematical journals at the local library in an attempt to stay ahead of his students. He quickly discovered that he had a talent for mathematics. As well as mastering all the present-day ideas, he began to develop some of his own, which were later accepted for publication. After a stream of articles, he became so highly regarded that he was asked to join the mathematics faculty at Queens College in 1849.

After accepting the position, Boole concentrated more on his ideas, one of which was to develop a system of symbolic logic. In this system he created a form of algebra that had its own set of symbols and rules. Using this system, Boole could encode any statement that had to be proven (a proposition) into his symbolic language and then manipulate it to determine whether it was true or false. Boole's algebra had three basic operations that are often called logic functions—AND, OR, and NOT. Using these three operations, Boole could perform such operations as add, subtract, multiply, divide, and compare. These logic functions were binary in nature and therefore dealt with only two entities—TRUE or FALSE, YES or NO, OPEN or CLOSED, ZERO or ONE, and so on. Boole's theory was that if all logical arguments could be reduced to one of two basic levels, the questionable middle ground would be removed, making it easier to arrive at a valid conclusion.

At the time, Boole's system, which was later called "Boolean algebra," was either ignored or criticized by colleagues who called it a folly with no practical purpose. Almost a century later, however, scientists would combine George Boole's Boolean algebra with binary numbers and make possible the digital electronic computer.

Introduction

In 1854 George Boole invented a symbolic logic that linked mathematics and logic. Boole's logical algebra, which today is known as **Boolean algebra,** states that each variable (input or output) can assume one of two values or states—true or false.

Up until 1938, Boolean algebra had no practical application until Claude Shannon used it to analyze telephone switching circuits in his MIT thesis. In his paper he described how the two variables of the Boolean algebra (true and false) could be used to represent the two states of the switching relay (open and closed).

Today, the mechanical relays have been replaced by semiconductor switches. However, Boolean algebra is still used to express both simple and complex two-state logic functions in a convenient mathematical format. These mathematical expressions of logic functions make it easier for technicians to analyze digital circuits and are a primary design tool for engineers. By using Boolean algebra, circuits can be made simpler, less expensive, and more efficient.

In this chapter we will discuss the basics of Boolean algebra and then see how it can be applied to logic gate circuit simplification.

Boolean Algebra

A form of algebra invented by George Boole that deals with classes, propositions, and ON/OFF circuit elements associated with operators such as NOT, OR, NOR, AND, NAND, XOR, and XNOR.

The operation of each of the seven basic logic gates can be described with a Boolean expression. To explain this in more detail, let us first consider the most basic of all the logic gates, the NOT gate.

4-1-1   *The NOT Expression*

In Figure 4-1 you can see the previously discussed NOT gate, and because of the inversion that occurs between input and output, Y is always equal to the opposite, or complement, of input A.

$$Y = \text{NOT } A$$

Therefore, if the input were 0, the output would be 1.

$$Y = \text{NOT } 0 = 1$$

If, on the other hand, the A input were 1, the output would be 0.

$$Y = \text{NOT } 1 = 0$$

In Boolean algebra, this inversion of the A input is indicated with a bar over the letter A as follows:

$$Y = \overline{A} \text{ (pronounced "Y equals not A")}$$

Using this Boolean expression, we can easily calculate the output Y for either of the two input A conditions. For example, if $A = 0$,

$$Y = \overline{A} = \overline{0} = 1$$

If, on the other hand, the A input is 1,

$$Y = \overline{A} = \overline{1} = 0$$

4-1-2   *The OR Expression*

The operation of the OR gate can also be described with a Boolean expression, as seen in Figure 4-2(a). In Boolean algebra, the "+" sign is used to indicate the OR operation.

$$Y = A + B \text{ (pronounced "Y equals A or B")}$$

The expression $Y = A + B$ (Boolean equation), is the same as $Y = A$ OR B (word equation). If you find yourself wanting to say "plus" instead of "OR," don't worry; it will require some

FIGURE 4-1   The Boolean Expression for a NOT Gate.

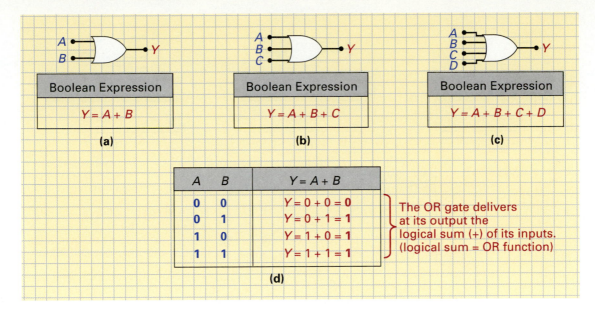

FIGURE 4-2 The Boolean Expression for an OR Gate.

practice to get used to this traditional symbol being used in a new application. Just as the word "wind" can be used in two different ways ("a gusty wind" or "to wind a clock"), then so can the "+" symbol be used to mean either addition or the OR operation. With practice you will get more comfortable with this symbol's dual role.

If OR gates have more than two inputs, as seen in Figure 4-2(b) or (c), the OR expression is simply extended, as follows:

For a three-input OR gate: $Y = A + B + C$
For a four-input OR gate: $Y = A + B + C + D$

The truth table for a two-input OR gate has been repeated in Figure 4-2(d). Using the Boolean expression for this gate, we can determine the output Y by substituting all of the possible 1 and 0 input combinations for the input variables A and B. Studying this truth table you can see that the output of the OR gate is a 1 when either A OR B is 1. George Boole described the OR function as *Boolean addition,* which is why the "plus" (addition) symbol was used. Boolean addition, or *logical addition,* is different from normal addition, and the OR gate is said to produce at its output the *logical sum* of the inputs.

4-1-3 *The AND Expression*

In Boolean algebra, the AND operation is indicated with a "·", and therefore the Boolean expression for the AND gate shown in Figure 4-3(a) would be

$Y = A \cdot B$ (pronounced "Y equals A and B")

This "·" is also used to indicate multiplication, and therefore like the "+" symbol, the "·" symbol has two meanings. In multiplication, the "·" is often dropped; however, the meaning is still known to be multiplication. For example:

$V = I \times R$ is the same as $V = I \cdot R$ or $V = IR$

In Boolean algebra the same situation occurs, in that

$Y = A \cdot B$ is the same as $Y = AB$

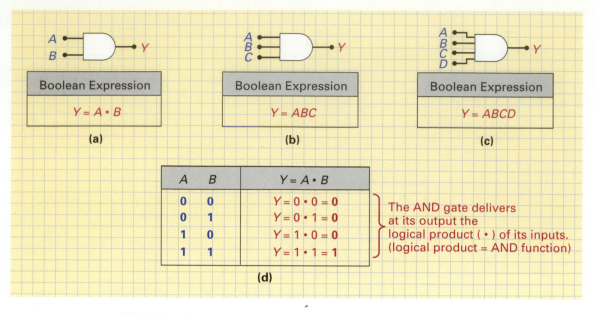

Boolean Expression

$Y = A \cdot B$

(a)

Boolean Expression

$Y = ABC$

(b)

Boolean Expression

$Y = ABCD$

(c)

A	B	$Y = A \cdot B$
0	0	$Y = 0 \cdot 0 = 0$
0	1	$Y = 0 \cdot 1 = 0$
1	0	$Y = 1 \cdot 0 = 0$
1	1	$Y = 1 \cdot 1 = 1$

The AND gate delivers at its output the logical product (•) of its inputs. (logical product = AND function)

(d)

FIGURE 4-3 The Boolean Expression for an AND Gate.

If AND gates have more than two inputs, as seen in Figure 4-3(b) and (c), the AND expression is simply extended.

    The truth table for a two-input AND gate is shown in Figure 4-3(d). Using the Boolean expression, we can calculate the output Y by substituting all of the possible 1 and 0 input combinations for the input variables A and B. Looking at the truth table, you can see that the output of the AND gate is a 1 only when both A and B are 1. George Boole described the AND function as *Boolean multiplication* or *logical multiplication,* which is why the period (multiplication) symbol was used. Boolean multiplication is different from standard multiplication as you have seen, and the AND gate is said to produce at its output the *logical product* of its inputs.

■ EXAMPLE:

What would be the Boolean expression for the circuit shown in Figure 4-4, and what would be the output if $A = 0$ and $B = 1$?

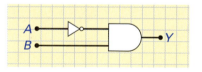

FIGURE 4-4 Single, Bubble Input AND Gate

■ *Solution:*

The A input in Figure 4-4 is first inverted before it is ANDed with the B input. The Boolean equation would be, therefore

$$Y = \overline{A} \cdot B$$

Using this expression, we can substitute the A and B for the 0 and 1 inputs, and then determine the output Y.

$$Y = \overline{A} \cdot B = \overline{0} \cdot 1 = 1 \cdot 1 = 1$$

Y equals NOT A and B = NOT 0 AND 1 = 1 AND 1 = 1)

EXAMPLE:

Give the Boolean expression and truth table for the logic circuit shown in Figure 4-5(a).

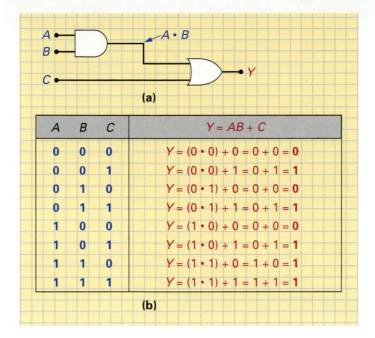

(a)

A	B	C	Y = AB + C
0	0	0	Y = (0 • 0) + 0 = 0 + 0 = 0
0	0	1	Y = (0 • 0) + 1 = 0 + 1 = 1
0	1	0	Y = (0 • 1) + 0 = 0 + 0 = 0
0	1	1	Y = (0 • 1) + 1 = 0 + 1 = 1
1	0	0	Y = (1 • 0) + 0 = 0 + 0 = 0
1	0	1	Y = (1 • 0) + 1 = 0 + 1 = 1
1	1	0	Y = (1 • 1) + 0 = 1 + 0 = 1
1	1	1	Y = (1 • 1) + 1 = 1 + 1 = 1

(b)

FIGURE 4-5 *(A* AND *B)* OR *(C)* **Logic Circuit.**

■ *Solution:*

To begin with, there are three inputs or variables to this circuit: *A, B,* and *C.* Inputs *A* and *B* are ANDed together ($A \cdot B$) and the result is then ORed with input *C* (+ *C*). The resulting equation will be

$$Y = (A \cdot B) + C \text{ or } Y = AB + C$$

As you can see in the truth table in Figure 4-5(b), the *Y* output is HIGH when *A* AND *B* are HIGH, OR if *C* is HIGH.

In the two previous examples, we have gone from a logic circuit to a Boolean equation. Now let us reverse the procedure and see how easy it is to go from a Boolean equation to a logic circuit.

EXAMPLE:

Sketch the logic gate circuit for the following Boolean equation.

$$Y = AB + CD$$

■ *Solution:*

Studying the equation, we can see that there are four variables: *A, B, C,* and *D.* Input *A* is ANDed with input *B*, giving $A \cdot B$ or *AB*, as seen in Figure 4-6(a). Input *C* is ANDed with input *D*, giving *CD*, as seen in Figure 4-6(b). The ANDed *AB* output and the ANDed *CD* output are then ORed to produce a final output *Y,* as seen in Figure 4-6(c).

$$Y = \overline{A + B} \text{ (pronounced ``Y equals not A or B'')}$$

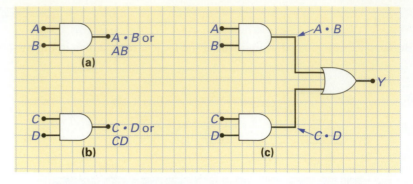

FIGURE 4-6 (*A* AND *B*) OR (*C* AND *D*) Logic Circuit.

4-1-4   *The NOR Expression*

The Boolean expression for the two-input NOR gate shown in Figure 4-7(a) will be

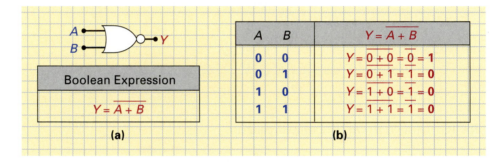

A	B	$Y = \overline{A + B}$
0	0	$Y = \overline{0 + 0} = \overline{0} = 1$
0	1	$Y = \overline{0 + 1} = \overline{1} = 0$
1	0	$Y = \overline{1 + 0} = \overline{1} = 0$
1	1	$Y = \overline{1 + 1} = \overline{1} = 0$

Boolean Expression

$Y = \overline{A + B}$

(a)   **(b)**

FIGURE 4-7   The Boolean Expression for a NOR Gate.

To explain why this expression describes a NOR gate, let us examine each part in detail. In this equation, the two input variables *A* and *B* are first ORed (indicated by the *A* + *B* part of the equation), and then the result is complemented, or inverted (indicated by the bar over the whole OR output expression). The truth table in Figure 4-7(b) tests the Boolean expression for a NOR gate for all possible input combinations.

While studying Boolean algebra, Augustus DeMorgan discovered two important theorems. The first theorem stated that a bubbled input AND gate was logically equivalent to a NOR gate. These two logic gates are shown in Figure 4-8(a) and (b), respectively, with their associ-

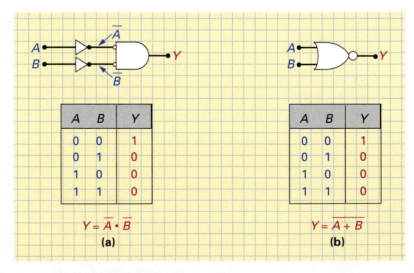

A	B	Y
0	0	1
0	1	0
1	0	0
1	1	0

$Y = \overline{A} \cdot \overline{B}$

(a)

A	B	Y
0	0	1
0	1	0
1	0	0
1	1	0

$Y = \overline{A + B}$

(b)

FIGURE 4-8   DeMorgan's First Theorem.

ated truth tables. Comparing the two truth tables you can see how any combination of inputs to either gate circuit will result in the same output, making the two circuits interchangeable.

■ EXAMPLE:

Give the two Boolean equations for the two logic gate circuits shown in Figure 4-8(a) and (b).

■ *Solution:*

With the bubbled input AND gate shown in Figure 4-8(a), you can see that input A is first inverted by a NOT gate, giving $\overline{A}$. Input B is also complemented by a NOT gate, giving $\overline{B}$. These two complemented inputs are then ANDed, giving the following final function:

$$Y = \overline{A} \cdot \overline{B}$$

With the NOR gate shown in Figure 4-8(b), inputs A and B are first ORed and then the result is complemented, giving

$$Y = \overline{A + B}$$

DeMorgan's first theorem, therefore, is as follows:

> **DeMorgan's First Theorem:** $\overline{A} \cdot \overline{B} = \overline{A + B}$

Studying the differences between the left equation and the right equation, you can see that there are actually two basic changes: The line is either broken or solid, and the logical sign is different. Therefore, if we wanted to convert the left equation into the right equation, all we would have to do is follow this simple procedure:

$$Y = \overline{A} \cdot \overline{B} \rightarrow Y = \overline{A + B}$$

"Mend the line, and change the sign."

To convert the right equation into the left equation, all we would have to do is follow this simple procedure:

"Break the line, and change the sign."

$$Y = \overline{A} \cdot \overline{B} \leftarrow Y = \overline{A + B}$$

This ability to interchange an AND function with an OR function, and to interchange a bubbled input with a bubbled output, will come in handy when we are trying to simplify logic circuits, as you will see in this chapter. For now, remember the rules on how to "DeMorganize" an equation.

> *Mend the line, and change the sign.*
> *Break the line, and change the sign.*

4-1-5   *The NAND Expression*

The Boolean expression for the two-input NAND gate shown in Figure 4-9(a) will be

> $Y = \overline{A \cdot B}$ or $Y = \overline{AB}$ (pronounced "Y equals not A and B")

A	B	$Y = \overline{A \cdot B}$
0	0	$Y = \overline{0 \cdot 0} = \overline{0} = 1$
0	1	$Y = \overline{0 \cdot 1} = \overline{0} = 1$
1	0	$Y = \overline{1 \cdot 0} = \overline{0} = 1$
1	1	$Y = \overline{1 \cdot 1} = \overline{1} = 0$

Boolean Expression

$Y = \overline{AB}$

(a)　　　　　(b)

FIGURE 4-9　**The Boolean Expression for a NAND Gate.**

This expression states that the two inputs A and B are first ANDed (indicated by the $A \cdot B$ part of the equation), and then the result is complemented (indicated by the bar over the AND expression).

The truth table shown in Figure 4-9(b) tests this Boolean equation for all possible input combinations.

■　EXAMPLE:

DeMorganize the following equation, and then sketch the logically equivalent circuits with their truth tables and Boolean expressions.

$$Y = \overline{A} + \overline{B}$$

■　*Solution:*

To apply DeMorgan's theorem to the equation $Y = \overline{A} + \overline{B}$, we simply apply the rule "mend the line, and change the sign," as follows:

$$\overline{A} + \overline{B} = \overline{A \cdot B}$$

The circuits for these two equations can be seen in Figure 4-10(a) and (b), and if you study the truth tables, you can see that the outputs of both logic gates are identical for all input combinations.

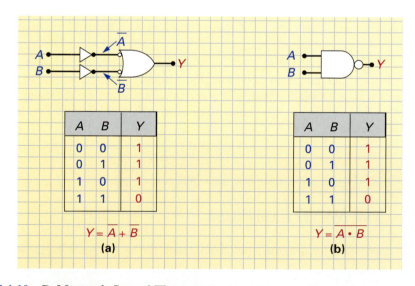

A	B	Y
0	0	1
0	1	1
1	0	1
1	1	0

$Y = \overline{A} + \overline{B}$

(a)

A	B	Y
0	0	1
0	1	1
1	0	1
1	1	0

$Y = \overline{A \cdot B}$

(b)

FIGURE 4-10　**DeMorgan's Second Theorem.**

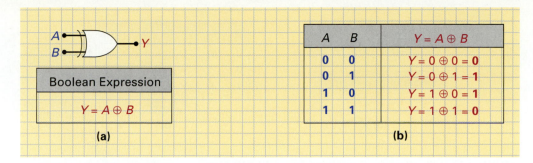

FIGURE 4-11   The Boolean Expression for an XOR Gate.

This observation—that a bubbled input OR gate is interchangeable with a bubbled output AND gate, or NAND gate—was first made by DeMorgan and is referred to as DeMorgan's second theorem.

DeMorgan's Second Theorem: $\overline{A} + \overline{B} = \overline{A \cdot B}$

4-1-6   *The XOR Expression*

In Boolean algebra, the "⊕" symbol is used to describe the exclusive-OR action. This means that the Boolean expression for the two-input XOR gate shown in Figure 4-11(a) will be

$Y = A \oplus B$ (pronounced "Y equals A exclusive-OR B")

The truth table in Figure 4-11(b) shows how this Boolean expression is applied to all possible input combinations.

EXAMPLE:

Give the Boolean expression and truth table for the circuit shown in Figure 4-12(a).

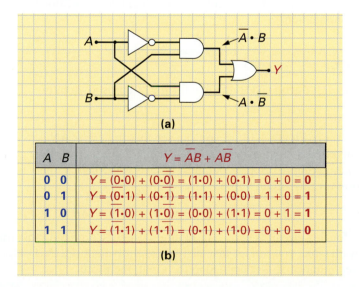

FIGURE 4-12   (NOT A AND B) OR (A AND NOT B).

FIGURE 4-13 The Boolean Expression for an XNOR Gate.

■ *Solution:*

Looking at the circuit in Figure 4-12(a), you can see that the upper AND gate has an inverted A input $(\overline{A})$ and a B input, giving a result of $\overline{A} \cdot B$. This $\overline{A} \cdot B$ output is then ORed with the input from the lower AND gate, which has an inverted B input and A input, giving a result of $A \cdot \overline{B}$. The final equation will therefore be

$$Y = (\overline{A} \cdot B) + (A \cdot \overline{B}) \ \text{ or } \ Y = \overline{A}B + A\overline{B}$$

The truth table for this logic circuit that combines five logic gates is shown in Figure 4-12(b). Looking at the input combinations and the output, you can see that the output is only 1 when an odd number of 1s appears at the input. This circuit is therefore acting as an XOR gate, and so

$$\overline{A}B + A\overline{B} = A \oplus B$$

4-1-7 *The XNOR Expression*

The Boolean expression for the two-input XNOR gate shown in Figure 4-13(a) will be

$$Y = \overline{A \oplus B} \text{ (pronounced "Y equals not A exclusive-OR B")}$$

From this equation, we can see that the two inputs A and B are first XORed and then the result is complemented. The truth table in Figure 4-13(b) tests this Boolean expression for all possible input combinations.

■ **EXAMPLE:**

Give the Boolean expression and truth table for the circuit shown in Figure 4-14(a).

■ *Solution:*

Studying the circuit in Figure 4-14(a), you can see that the upper AND gate has an inverted A input and an inverted B input, giving $\overline{A} \cdot \overline{B}$ at its output. The lower AND gate simply ANDs the A and B input, giving $A \cdot B$ at its output. The two AND gate outputs are then ORed, so the final equation for this circuit will be

$$Y = \overline{A} \cdot \overline{B} + A \cdot B \ \text{ or } \ Y = \overline{A}\,\overline{B} + AB$$

The truth table for this circuit is shown in Figure 4-14(b). Comparing this truth table to an XNOR gate's truth table, you can see that the two are equivalent since a 1 output is only present when an even number of 1s is applied to the input; therefore

$$\overline{A}\,\overline{B} + AB = \overline{A \oplus B}$$

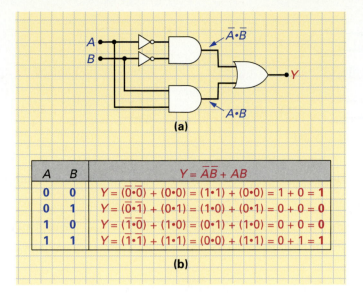

A	B	$Y = \overline{A}\overline{B} + AB$
0	0	$Y = (\overline{0} \cdot \overline{0}) + (0 \cdot 0) = (1 \cdot 1) + (0 \cdot 0) = 1 + 0 = 1$
0	1	$Y = (\overline{0} \cdot \overline{1}) + (0 \cdot 1) = (1 \cdot 0) + (0 \cdot 1) = 0 + 0 = 0$
1	0	$Y = (\overline{1} \cdot \overline{0}) + (1 \cdot 0) = (0 \cdot 1) + (1 \cdot 0) = 0 + 0 = 0$
1	1	$Y = (\overline{1} \cdot \overline{1}) + (1 \cdot 1) = (0 \cdot 0) + (1 \cdot 1) = 0 + 1 = 1$

(b)

FIGURE 4-14 (NOT *A* AND NOT *B*) OR (*A* AND *B*).

SELF-TEST EVALUATION POINT FOR SECTION 4-1

Use the following questions to test your understanding of Section 4-1.

1. Which logic gate would be needed to perform each of the following Boolean expressions?
   a. $A\,B$ **c.** $N + M$ **e.** $\overline{S + T}$
   b. $A \oplus B$ **d.** $\overline{X \cdot Y}$ **f.** $\overline{GH}$

2. State DeMorgan's first theorem using Boolean expressions.

3. Apply DeMorgan's theorem to the following:
   a. $\overline{A} + \overline{B}$ **b.** $Y = \overline{(A + B)} \cdot C$

4. Sketch the logic gates or circuits for the expressions:
   a. $A\,B\,C$ **c.** $\overline{A + B}$ **e.** $A \cdot (L \oplus D)$
   b. $\overline{L \oplus D}$ **d.** $(\overline{A} \cdot B) + \overline{C}$ **f.** $\overline{(A + B) \cdot C}$

4-2   BOOLEAN ALGEBRA LAWS AND RULES

In this section we will discuss some of the rules and laws that apply to Boolean algebra. Many of these rules and laws are the same as those of ordinary algebra and, as you will see in this section, are quite obvious.

4-2-1   *The Commutative Law*

Commutative Law

Combining elements in such a manner that the result is independent of the order in which the elements are taken.

The word *commutative* is defined as "combining elements in such a manner that the result is independent of the order in which the elements are taken." This means that the order in which the inputs to a logic gate are ORed or ANDed, for example, is not important since the result will be the same, as shown in Figure 4-15.

In Figure 4-15(a), you can see that ORing *A* and *B* will achieve the same result as reversing the order of the inputs and ORing *B* and *A*. As stated in the previous section, the OR

FIGURE 4-15   The Commutative Law. (a) Logical Addition. (b) Logical Multiplication.

function is described in Boolean algebra as logical addition, so the *commutative law of addition* can be algebraically written as

> Commutative Law of Addition: $A + B = B + A$

Figure 4-15(b) shows how this law relates to an AND gate, which in Boolean algebra is described as logical multiplication. The *commutative law of multiplication* can be algebraically written as

> Commutative Law of Multiplication: $A \cdot B = B \cdot A$

Associative Law

Combining elements such that when the order of the elements is preserved, the result is independent of the grouping.

4-2-2   *The Associative Law*

The word *associative* is defined as "combining elements such that when the order of the elements is preserved, the result is independent of the grouping." To explain this in simple terms, Figure 4-16 shows that how you group the inputs in an ORing process or ANDing process will have no effect on the output.

In Figure 4-16(a), you can see that ORing B and C and then ORing the result with A will achieve the same result as ORing A and B and then ORing the result with C. This *associative law of addition* can be algebraically written as

> Associative Law of Addition: $A + (B + C) = (A + B) + C$

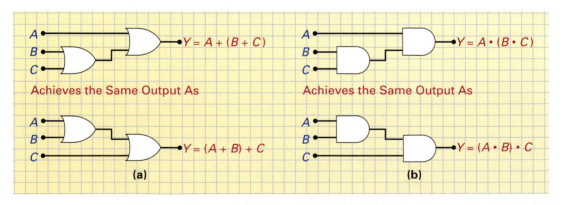

FIGURE 4-16   The Associative Law. (a) Logical Addition. (b) Logical Multiplication.

Figure 4-16(b) shows how this law relates to an AND gate, which in Boolean algebra is described as logical multiplication. The *associative law of multiplication* can be algebraically written as

Associative Law of Multiplication: $A \cdot (B \cdot C) = (A \cdot B) \cdot C$

4-2-3   *The Distributive Law*

By definition, the word *distributive* means "producing the same element when operating on a whole as when operating on each part, and collecting the results." It can be algebraically stated as

Distributive Law: $A \cdot (B + C) = (A \cdot B) + (A \cdot C)$

Figure 4-17 illustrates this law by showing that ORing two or more inputs and then ANDing the result, as shown in Figure 4-17(a), achieves the same output as ANDing the single variable (A) with each of the other inputs (B and C) and then ORing the results, as shown in Figure 4-17(b). To help reinforce this algebraic law, let us use an example involving actual values for A, B, and C.

■ **EXAMPLE:**

Prove the distributive law by inserting the values $A = 2$, $B = 3$ and $C = 4$.

■ *Solution:*

$$A \cdot (B + C) = (A \cdot B) + (A \cdot C)$$
$$2 \times (3 + 4) = (2 \times 3) + (2 \times 4)$$
$$2 \times 7 = 6 + 8$$
$$14 = 14$$

In some instances, we may wish to reverse the process performed by the distributive law to extract the common factor. For example, consider the following equation:

$$Y = \overline{A} B + AB$$

Since "$\cdot B$" (AND B) seems to be a common factor in this equation, by *factoring*, we could obtain

$$Y = \overline{A}B + AB \qquad \text{Original Expression}$$
$$Y = (\overline{A} + A) \cdot B \qquad \text{Factoring AND B}$$

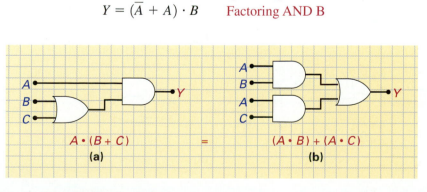

$$A \cdot (B + C) \qquad = \qquad (A \cdot B) + (A \cdot C)$$

(a)   (b)

FIGURE 4-17   The Distributive Law.

4-2-4 *Boolean Algebra Rules*

Now that we have covered the three laws relating to Boolean algebra, let us concentrate on these Boolean algebra rules, beginning with those relating to OR operations.

OR Gate Rules

The first rule can be seen in Figure 4-18(a). This shows what happens when one input to an OR gate is always 0 and the other input (*A*) is a variable. If $A = 0$, the output equals 0, and if $A = 1$, the output equals 1. Therefore, a variable input ORed with 0 will always equal the variable input. Stated algebraically:

$$A + 0 = A$$

Another OR gate Boolean rule is shown in Figure 4-18(b), and states that

$$A + A = A$$

As you can see in Figure 4-18(b), when a variable is ORed with itself, the output will always equal the logic level of the variable input.

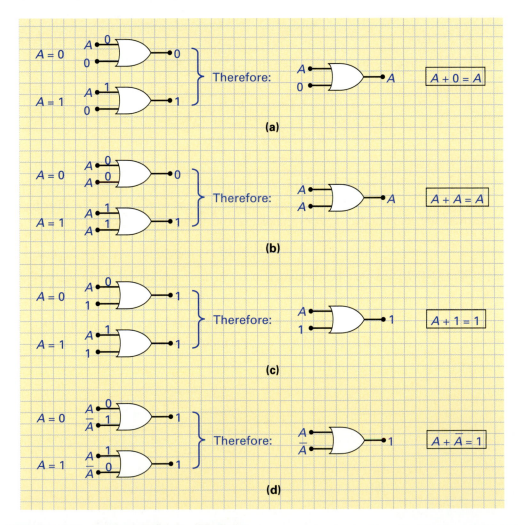

FIGURE 4-18 **Boolean Rules for OR Gates.**

CHAPTER 4 / LOGIC CIRCUIT SIMPLIFICATION

Figure 4-18(c) shows the next OR gate Boolean rule. This rule states that any 1 input to an OR gate will result in a 1 output, regardless of the other input. Stated algebraically:

$$A + 1 = 1$$

The final Boolean rule for OR gates is shown in Figure 4-18(d). This rule states that when any variable (A) is ORed with its complement ($\overline{A}$) the result will be a 1.

$$A + \overline{A} = 1$$

AND Gate Rules

As with the OR gate, there are four Boolean rules relating to AND gates. The first is illustrated in Figure 4-19(a) and states that if a variable input A is ANDed with a 0, the output will be 0 regardless of the other input.

$$A \cdot 0 = 0$$

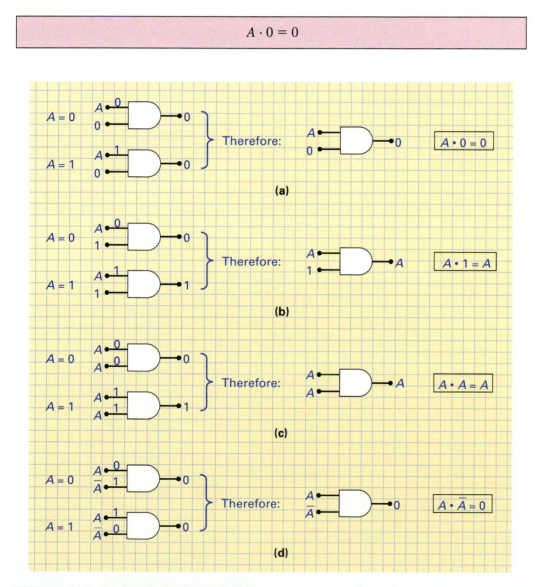

FIGURE 4-19   Boolean Rules for AND Gates.

The second Boolean rule for AND gates is shown in Figure 4-19(b). In this illustration you can see that if a variable is ANDed with a 1, the output will equal the variable.

$$A \cdot 1 = A$$

Another AND gate Boolean rule is shown in Figure 4-19(c). In this instance, any variable that is ANDed with itself will always give an output that is equal to the variable.

$$A \cdot A = A$$

The last of the AND gate Boolean rules is shown in Figure 4-19(d). In this illustration you can see that if a variable (A) is ANDed with its complement ($\overline{A}$), the output will always equal 0.

$$A \cdot \overline{A} = 0$$

Double-Inversion Rule

The *double-inversion rule* is illustrated in Figure 4-20(a) and states that if a variable is inverted twice, then the variable will be back to its original state. To state this algebraically, we use the double bar as follows:

$$\overline{\overline{A}} = A$$

In Figure 4-20(b), you can see that if the NOR gate were replaced with an OR gate, the INVERTER would not be needed since the double inversion returns the logic level to its original state.

DeMorgan's Theorems

DeMorgan's first and second theorems were discussed earlier in the previous section and are repeated here since they also apply as Boolean rules.

DeMorgan's First Theorem: $\overline{A} \cdot \overline{B} = \overline{A + B}$

DeMorgan's Second Theorem: $\overline{A} + \overline{B} = \overline{A \cdot B}$

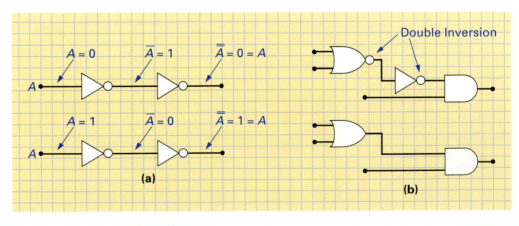

(a)

(b)

FIGURE 4-20 Double-Inversion Rule.

CHAPTER 4 / LOGIC CIRCUIT SIMPLIFICATION

■ **EXAMPLE:**

Apply DeMorgan's theorem and the double-inversion rule to find equivalent logic circuits of the following:

$$\text{a. } \overline{\overline{A} + B} \qquad \text{b. } \overline{A \cdot \overline{B}}$$

■ *Solution:*

With both equations we simply "break the line and change the sign" and then cancel any double inversions to find the equivalent logic circuits, as follows:

$$\text{a. } \overline{\overline{A} + B} = \overline{\overline{A}} \cdot \overline{B} = A \cdot \overline{B}$$

In the first example we found that a NOR gate with a bubbled A input is equivalent to an AND gate with a bubbled B input.

$$\text{b. } \overline{A \cdot \overline{B}} = \overline{A} + \overline{\overline{B}} = \overline{A} + B$$

In the second example we found that a NAND gate with a bubbled B input is equivalent to an OR gate with a bubbled A input.

The Duality Theorem

The *duality theorem* is very useful since it allows us to change Boolean equations into their *dual* and to produce new Boolean equations. The two steps to follow for making the change are very simple.

    a. First, change each OR symbol to an AND symbol and each AND symbol to an OR symbol.

    b. Second, change each 0 to a 1 and each 1 to a 0.

To see if this works, let us apply the duality theorem to a simple example.

■ **EXAMPLE:**

Apply the duality theorem to $A + 1 = 1$ and then state whether the resulting rule is true.

■ *Solution:*

Changing the OR symbol to an AND symbol, and changing the 1s to 0s, we arrive at the following:

Original Rule: $A + 1 = 1$
Dual Rule: $A \cdot 0 = 0$

OR symbol is changed to AND symbol. ⎯⎯⎿↑↑⎽↑⎯⎯ 1s are changed to 0s.

The dual Boolean rule states that any variable ANDed with a 0 will always yield a 0 output, which is true.

■ **EXAMPLE:**

Determine the dual Boolean equation for the following:

$$A \cdot (B + C) = (A \cdot B) + (A \cdot C)$$

Once you have determined the new Boolean equation, sketch circuits and give the truth tables for both the left and right parts of the equation.

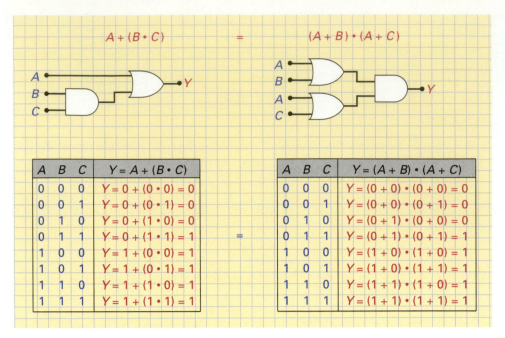

FIGURE 4-21 *A* OR (*B* AND *C*) Equals (*A* OR *B*) AND (*A* OR *C*).

■ *Solution:*

Applying the duality theorem to $A \cdot (B + C) = (A \cdot B) + (A \cdot C)$ means that we will have to change all the AND symbols to OR symbols, and vice versa.

Original Equation: $A \cdot (B + C) = (A \cdot B) + (A \cdot C)$

Dual Equation: $A + (B \cdot C) = (A + B) \cdot (A + C)$

Figure 4-21 shows the dual equation and its circuits and truth tables. As you can see by comparing the truth tables, the left and right parts of the equation are equivalent.

As a summary, here is a list of the Boolean laws and rules described in this section, with their dual equations.

	ORIGINAL EQUATION	DUAL EQUATION
1. Commutative Law	$A + B = B + A$	$A \cdot B = B \cdot A$
2. Associative Law	$A \cdot (B \cdot C) = (A \cdot B) \cdot C$	$A + (B + C) = (A + B) + C$
3. Distributive Law	$A + (B \cdot C) = (A + B) \cdot (A + C)$	$A \cdot (B + C) = (A \cdot B) + (A \cdot C)$
4. OR-AND Rules	$A + 0 = A$	$A \cdot 1 = A$
	$A + A = A$	$A \cdot A = A$
	$A + 1 = 1$	$A \cdot 0 = 0$
	$A + \overline{A} = 1$	$A \cdot \overline{A} = 0$
5. DeMorgan's Laws	$\overline{A} \cdot \overline{B} = \overline{A + B}$	$\overline{A} + \overline{B} = \overline{A \cdot B}$

Use the following questions to test your understanding of Section 4-2.

1. Give the answers to the following Boolean algebra rules.
   a. $A + 1 = ?$ **b.** $A + \overline{A} = ?$ **c.** $A \cdot 0 = ?$ **d.** $A \cdot A = ?$
2. What is the dual equation of $A + (B \cdot C) = (A + B) \cdot (A + C)$?
3. Apply the associative law to the equation $A \cdot (B \cdot C)$.
4. Apply the distributive law to the equation $(A \cdot B) + (A \cdot C) + (A \cdot D)$.

FROM TRUTH TABLE TO GATE CIRCUIT

Now that you understand the Boolean rules and laws for logic gates, let us put this knowledge to some practical use. If you need a logic gate circuit to perform a certain function, the best way to begin is with a truth table that details what input combinations should drive the output HIGH and what input combinations should drive the output LOW. As an example, let us assume that we need a logic circuit that will follow the truth table shown in Figure 4-22(a). This means that the output (Y) should only be 1 when

$$A = 0, B = 0, C = 1$$
$$A = 0, B = 1, C = 1$$
$$A = 1, B = 0, C = 1$$
$$A = 1, B = 1, C = 1$$

In the truth table in Figure 4-22(a), you can see the **fundamental products** listed for each of these HIGH outputs. A fundamental product is a Boolean expression that describes what the inputs will need to be in order to generate a HIGH output. For example, the first fundamental product ($Y = \overline{A} \cdot \overline{B} \cdot C$) states that A must be 0, B must be 0, and C must be 1 for the output Y to be 1 ($Y = \overline{A} \cdot \overline{B} \cdot C = \overline{0} \cdot \overline{0} \cdot 1 = 1 \cdot 1 \cdot 1 = 1$). The second fundamental product ($Y = \overline{A} \cdot B \cdot C$) states that A must be 0, B must be 1, and C must be 1 for the output Y to be 1 ($Y = \overline{A} \cdot B \cdot C = \overline{0} \cdot 1 \cdot 1 = 1 \cdot 1 \cdot 1 = 1$). The third fundamental product ($Y = A \cdot \overline{B} \cdot C$) states that A must be 1, B must be 0, and C must be 1 for Y to be 1 ($Y = A \cdot \overline{B} \cdot C = 1 \cdot \overline{0} \cdot 1 = 1 \cdot 1 \cdot 1 = 1$). The fourth and final fundamental product ($Y = A \cdot B \cdot C$) states that A must be 1, B must be 1, and C must be 1 for Y to be 1 ($Y = A \cdot B \cdot C = 1 \cdot 1 \cdot 1 = 1$). The fundamental products for each of the HIGH outputs are listed in the truth table in Figure 4-22(a). By ORing all of these fundamental products, a Boolean equation can be derived, as follows:

$$Y = \overline{A}\,\overline{B}C + \overline{A}BC + A\overline{B}C + ABC$$

From this Boolean equation we can create a logic network that is the circuit equivalent of the truth table. To complete this step, we will need to study in detail the different parts of the Boolean equation. First, the Boolean equation states that the outputs of four three-input AND gates are connected to the input of a four-input OR gate. The first AND gate is connected to inputs $\overline{A}\,\overline{B}$ and C, the second to inputs $\overline{A}\,B$ and C, the third to inputs $A\,\overline{B}$ and C, and finally the fourth to inputs $A\,B$ and C. The resulting logic circuit equivalent for the truth table in Figure 4-22(a) is shown in Figure 4-22(b).

The Boolean equation for the logic circuit in Figure 4-22 is in the **sum-of-products (SOP) form.** To explain what this term means, we must recall once again that in Boolean

Fundamental Products

The truth table input combinations that are essential since they produce a HIGH output.

Sum-of-Products (SOP) Form

A Boolean expression that describes the ORing of two or more AND functions.

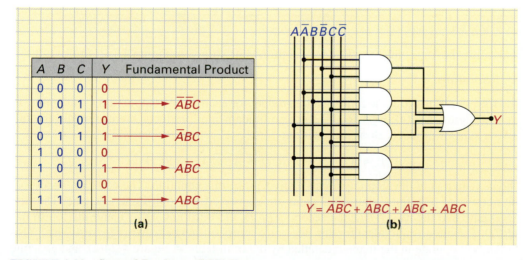

A	B	C	Y	Fundamental Product
0	0	0	0	
0	0	1	1	$\overline{A}\,\overline{B}C$
0	1	0	0	
0	1	1	1	$\overline{A}BC$
1	0	0	0	
1	0	1	1	$A\overline{B}C$
1	1	0	0	
1	1	1	1	ABC

(a)

$$Y = \overline{A}\,\overline{B}C + \overline{A}BC + A\overline{B}C + ABC$$

(b)

FIGURE 4-22   **Sum-of-Products (SOP) Form.**

algebra, an AND gate's output is the *logical product* of its inputs. On the other hand, the OR gate's output is the *logical sum* of its inputs. A sum-of-products expression, therefore, describes the ORing together of (sum of) two or more AND functions (products). Here are some examples of sum-of-product expressions.

$$Y = AB + \overline{A}B$$
$$Y = A\overline{B} + C\overline{D}E$$
$$Y = \overline{A}BCD + AB\overline{C}D + A\overline{B}C\overline{D}$$

Product-of-Sums (POS) Form

A Boolean expression that describes the ANDing of two or more OR functions.

The other basic form for Boolean expressions is the **product-of-sums (POS) form.** This form describes the ANDing of (product of) two or more OR functions (sums). Here are some examples of product-of-sums expressions.

$$Y = (A + B + \overline{C}) \cdot (C + D)$$
$$Y = (\overline{A} + \overline{B}) \cdot (A + B)$$
$$Y = (A + B + C + \overline{D}) \cdot (\overline{A} + B + \overline{C} + D) \cdot (A + \overline{B} + \overline{C} + \overline{D})$$

■ **EXAMPLE:**

Is the circuit shown in Figure 4-23 an example of sum-of-products or product-of-sums, and what would this circuit's Boolean expression be?

FIGURE 4-23   SOP or POS?

■ *Solution:*

The circuit in Figure 4-23 is a product-of-sums equivalent circuit, and its Boolean expression is

$$Y = (A + B) \cdot (C + D)$$

■ **EXAMPLE:**

Determine the logic circuit needed for the truth table shown in Figure 4-24(a).

■ *Solution:*

The first step is to write the fundamental product for each HIGH output in the truth table, as seen in Figure 4-24(a). By ORing (+) these fundamental products, we can obtain the following sum-of-products Boolean equation:

$$Y = \overline{A}\,\overline{B}\,CD + \overline{A}BC\overline{D} + A\overline{B}\,\overline{C}D + AB\overline{C}D$$

The next step is to develop the circuit equivalent of this equation. The equation states that the outputs of four four-input AND gates are connected to a four-input OR gate. The first AND gate had inputs $\overline{A}\,\overline{B}\,CD$, the second $\overline{A}BC\overline{D}$, the third has $A\overline{B}\,CD$, and the fourth has $AB\overline{C}D$. The circuit equivalent for this Boolean expression can be seen in Figure 4-24(b). This circuit will operate in the manner detailed by the truth table in Figure 4-24(a), and generate a 1 output only when

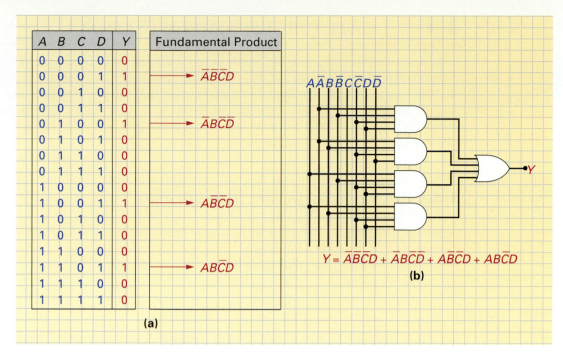

FIGURE 4-24 Sum-of-Products Example.

$$A = 0, B = 0, C = 0, D = 1$$
$$A = 0, B = 1, C = 0, D = 0$$
$$A = 1, B = 0, C = 0, D = 1$$
$$A = 1, B = 1, C = 0, D = 1$$

In all other instances, the output of this circuit will be 0.

■ EXAMPLE:

Apply the distributive law to the following Boolean expression in order to convert it into a sum-of-products form:

$$Y = A\overline{B} + C(D\overline{E} + F\overline{G})$$

■ Solution:

$$Y = A\overline{B} + C(D\overline{E} + F\overline{G}) \quad \text{Original Expression}$$
$$Y = A\overline{B} + CD\overline{E} + CF\overline{G} \quad \text{Distributive Law Applied}$$

SELF-TEST EVALUATION POINT FOR SECTION 4-3

Use the following questions to test your understanding of Section 4-3.

1. A two-input AND gate will produce at its output the logical _____ of A and B, while an OR gate will produce at its output the logical _____ of A and B.

2. What are the fundamental products of each of the following input words?
   a. A, B, C, D = 1011   b. A, B, C, D = 0110

3. The Boolean expression $Y = AB + \overline{A}B$ is an example of a/an _____ (SOP/POS) equation.

4. Describe the steps involved to develop a logic circuit from a truth table.

In the last section we saw how we could convert a desired set of conditions in a truth table to a sum-of-products equation and then to an equivalent logic circuit. In this section we will see how we can simplify the Boolean equation using Boolean laws and rules and therefore simplify the final logic circuit. As to why we would want to simplify a logic circuit, the answer is: A simplified circuit performs the same function, but is smaller, easier to construct, consumes less power, and costs less.

4-4-1   *Boolean Algebra Simplification*

Generally, the simplicity of one circuit compared to another circuit is judged by counting the number of logic gate inputs. For example, the logic circuit in Figure 4-25(a) has a total of six inputs (two on each of the AND gates and two on the OR gate). The Boolean equation for this circuit would be

$$Y = \overline{A}B + AB$$

Since "$\cdot\ B$" (AND B) seems to be a common factor in this equation, by factoring we could obtain the following:

$$Y = \overline{A}B + AB \qquad \text{Original Expression}$$
$$Y = (\overline{A} + A) \cdot B \qquad \text{Factoring AND B}$$

This equivalent logic circuit is shown in Figure 4-25(b), and if you count the number of logic gate inputs in this circuit (two for the OR and two for the AND, a total of four), you can see that the circuit has been simplified from a six-input circuit to a four-input circuit.

   This, however, is not the end of our simplification process, since another rule can be applied to further simplify the circuit. Remembering that any variable ORed with its complement results in a 1, we can replace $\overline{A} + A$ with a 1, as follows:

$$Y = (\overline{A} + A) \cdot B$$
$$Y = (1) \cdot B$$
$$Y = 1 \cdot B$$

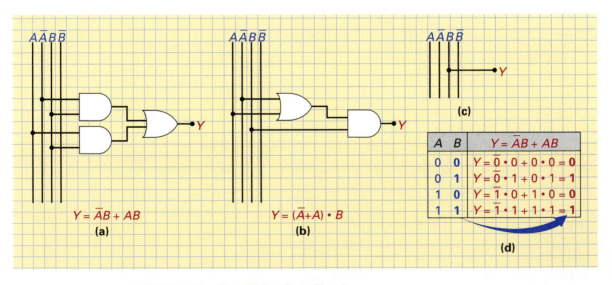

$$Y = \overline{A}B + AB$$
(a)

$$Y = (\overline{A}+A) \cdot B$$
(b)

(c)

A	B	$Y = \overline{A}B + AB$
0	0	$Y = \overline{0} \cdot 0 + 0 \cdot 0 = 0$
0	1	$Y = \overline{0} \cdot 1 + 0 \cdot 1 = 1$
1	0	$Y = \overline{1} \cdot 0 + 1 \cdot 0 = 0$
1	1	$Y = \overline{1} \cdot 1 + 1 \cdot 1 = 1$

(d)

FIGURE 4-25 **Simplifying Gate Circuits.**

The resulting equation $Y = 1 \cdot B$ can be further simplified since any variable ANDed with a 1 will always cause the output to equal the variable. Therefore, $Y = 1 \cdot B$ can be replaced with B, as follows:

$$Y = 1 \cdot B$$
$$Y = B$$

This means that the entire circuit in Figure 4-25(a) can be replaced with a single wire connected from the B input to the output, as seen in Figure 4-25(c). This fact is confirmed in the truth table in Figure 4-25(d), which shows that output Y exactly follows the input B.

EXAMPLE:

Simplify the Boolean equation shown in Figure 4-26(a).

Solution:

The Boolean equation for this circuit is

$$Y = A\overline{B}\,\overline{C} + \overline{A}B\overline{C} + AB\overline{C} + \overline{A}\,\overline{B}\,\overline{C}$$

The first step in the simplification process is to try and spot common factors in each of the ANDed terms and then rearrange these terms as shown in Figure 4-26(b). The next step is to factor out the common expressions, which are $A\overline{B}$ and $\overline{A}\,\overline{C}$, resulting in

$$Y = A\overline{B}\,\overline{C} + \overline{A}B\overline{C} + AB\overline{C} + \overline{A}\,\overline{B}\,\overline{C} \qquad \text{Original Equation}$$
$$Y = A\overline{B}\,\overline{C} + A\overline{B}C + \overline{A}\,\overline{B}\,\overline{C} + \overline{A}B\overline{C} \qquad \text{Rearranged Equation}$$
$$Y = A\overline{B}(\overline{C} + C) + \overline{A}\,\overline{C}(B + \overline{B}) \qquad \text{Factored Equation}$$

FIGURE 4-26 Boolean Equation Simplification.

Studying the last equation, you can see that we can apply the OR rule $(\overline{A} + A) = 1$ and simplify the terms within the parentheses to obtain the following:

$$Y = A\overline{B}(\overline{C} + C) + \overline{A}\,\overline{C}(B + \overline{B})$$
$$Y = A\overline{B} \cdot (1) + \overline{A}\,\overline{C} \cdot (1)$$

Now we can apply the $A \cdot 1 = A$ rule to further simplify the terms, since $A\overline{B} \cdot 1 = A\overline{B}$ and $\overline{A}\,\overline{C} \cdot 1 = \overline{A}\,\overline{C}$; therefore:

$$Y = A\overline{B} + \overline{A}\,\overline{C}$$

This equivalent logic circuit is illustrated in Figure 4-26(c) and has only six logic gate inputs compared to the original circuit shown in Figure 4-26(a), which has sixteen logic gate inputs. The truth table in Figure 4-26(d) compares the original equation with the simplified equation and shows how the same result is obtained.

4-4-2   *Karnaugh Map Simplification*

Karnaugh Map

Also called K-map, it is a truth table that has been rearranged to show a geometrical pattern of functional relationships for gating configurations. Using this map, essential gating requirements can be more easily recognized and reduced to their simplest form.

In most instances, engineers and technicians will simplify logic circuits using a **Karnaugh map,** or **K-map.** The Karnaugh map, named after its inventor, is quite simply a rearranged truth table, as shown in Figure 4-27. With this map, the essential gating requirements can be more easily recognized and reduced to their simplest form.

The total number of boxes or cells in a K-map depends on the number of input variables. For example, in Figure 4-27, only two inputs (A and B) and their complements ($\overline{A}$ and $\overline{B}$) are present, and therefore the K-map contains (like a two-variable truth table) only four combinations (00, 01, 10, and 11).

Each cell of this two-variable K-map represents one of the four input combinations. In practice, the input labels are placed outside of the cells, as shown in Figure 4-28(a), and apply to either a column or row of cells. For example, the row label $\overline{A}$ applies to the two upper

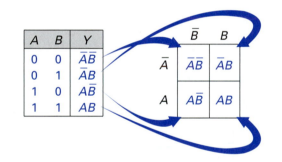

FIGURE 4-27   The Karnaugh Map—A Rearranged Truth Table.

FIGURE 4-28   Karnaugh Maps. (a) Two-Variable. (b) Three-Variable. (c) Four-Variable.

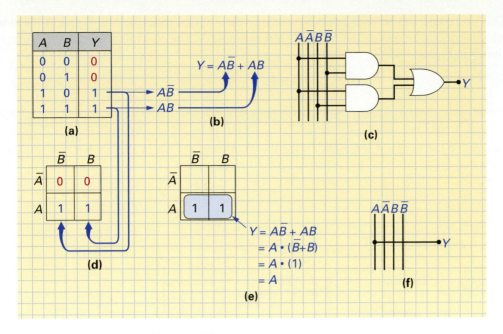

FIGURE 4-29 Karnaugh Map Simplification.

cells, while the row label A applies to the two lower cells. Running along the top of the K-map, the label $\overline{B}$ applies to the two left cells, while the label B applies to the two right cells. As an example, the upper right cell represents the input combination $\overline{A}B$. Figures 4-28(b) and (c) show the formats for a three-variable ($2^3 = 8$ cells) and four-variable ($2^4 = 16$ cells) K-map.

Now that we have an understanding of the K-map, let us see how it can be used to simplify a logic circuit. As an example, imagine that we need to create an equivalent logic circuit for the truth table given in Figure 4-29(a). The first step is to develop a sum-of-products Boolean expression. This is achieved by writing the fundamental product for each HIGH output in the truth table and then ORing all of the fundamental products, as shown in Figure 4-29(b). The equivalent logic circuit for this equation is shown in Figure 4-29(c). The next step is to plot this Boolean expression on a two-variable K-map, as seen in Figure 4-29(d). When plotting a sum-of-products expression on a K-map, remember that each cell corresponds to each of the input combinations in the truth table. A HIGH output in the truth table should appear as a 1 in its equivalent cell in the K-map, and a LOW output in the truth table should appear as a 0 in its equivalent cell. A 1, therefore, will appear in the lower left cell (corresponding to $A\overline{B}$) and in the lower right cell (corresponding to AB). The other input combinations ($\overline{A}\,\overline{B}$ and $\overline{A}B$) both yield a 0 output, and therefore a 0 should be placed in these two upper cells.

Reducing Boolean equations is largely achieved by applying the rule of complements, which states that $A + \overline{A} = 1$. Now that the SOP equation has been plotted on the K-map shown in Figure 4-29(d), the next step is to group terms and then factor out the common variables. If you study the K-map in Figure 4-29(d), you will see that adjacent cells differ by only one input variable. This means that if you move either horizontally or vertically from one cell to an adjacent cell, only one variable will change. By grouping adjacent cells containing a 1, as seen in Figure 4-29(e), cells can be compared and simplified (using the rule of complements) to create one-product terms. In this example, cells $A\overline{B}$ and AB contain B and $\overline{B}$, so these opposites or complements cancel, leaving A, as follows:

$$Y = A\overline{B} + AB \qquad \textcolor{red}{\text{Grouped Pair}}$$
$$Y = A \cdot (\overline{B} + B) \qquad \textcolor{red}{\text{Factoring } A \text{ AND}}$$
$$Y = A \cdot 1$$
$$Y = A$$

This procedure can be confirmed by studying the original truth table in Figure 4-29(a) in which you can see that output Y exactly follows input A. The equivalent circuit, therefore, is shown in Figure 4-29(f).

■ EXAMPLE:

Determine the simplest logic circuit for the truth table shown in Figure 4-30(a). Illustrate each step in the process.

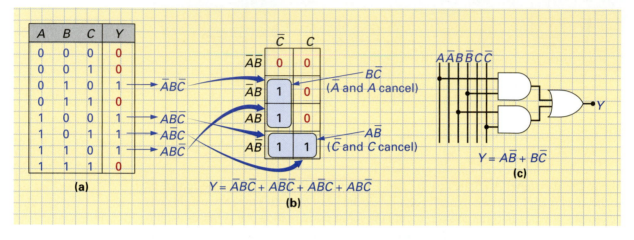

FIGURE 4-30 Three-Variable K-Map Simplification.

■ *Solution:*

Since this example has three input variables, the first step is to draw a three-variable K-map, as shown in Figure 4-30(b). The next step is to look for the HIGH outputs in the truth table in Figure 4-30(a) and plot these 1s in their equivalent cells in the K-map, as shown in Figure 4-30(b). After inserting 0s in the remaining cells, group the 1s into pairs, as shown in Figure 4-30(b), and then study the row and column variable labels associated with that grouped pair to see which variable will drop out due to the rule of complements. In the upper group, $\overline{A}$ and A will cancel, leaving $B\overline{C}$, and in the lower group $\overline{C}$ and C will cancel, leaving $A\overline{B}$. These reduced products will form an equivalent Boolean equation and logic circuit, as seen in Figure 4-30(c). In this example, the original sixteen-input equation in Figure 4-30(b) has been simplified to the equivalent six-input logic gate circuit shown in Figure 4-30(c).

The 1s in a K-map can be grouped in pairs (groups of two), quads (groups of four), octets (groups of eight), and all higher powers of two. Figure 4-31 shows some examples of grouping and how the K-map has been used to reduce large Boolean equations. Notice that the larger groups will yield smaller terms and, therefore, gate circuits with fewer inputs. For this reason, you should begin by looking for the largest possible groups, and then step down in group size if none are found (for example, begin looking for octets, then quads, and finally pairs). Also notice that you can capture 1s on either side of a map by wrapping a group around behind the map.

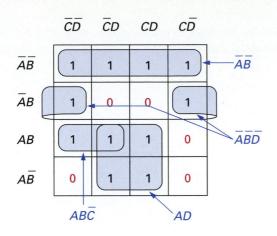

Before : $Y = \overline{A}\overline{B}\overline{C}\overline{D} + \overline{A}\overline{B}\overline{C}D + \overline{A}\overline{B}CD + \overline{A}\overline{B}C\overline{D} +$
$\overline{A}BC\overline{D} + \overline{A}B\overline{C}\overline{D} + AB\overline{C}\overline{D} + ABC\overline{D} +$
$AB\overline{C}D + A\overline{B}CD + A\overline{B}C\overline{D}$

After : $Y = AB\overline{C} + AD + \overline{A}B\overline{D} + \overline{A}\overline{B}$

(a)

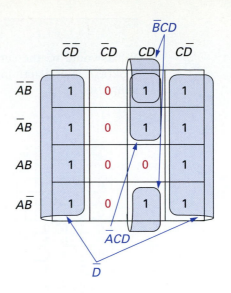

Before : $Y = \overline{A}\overline{B}\overline{C}\overline{D} + \overline{A}\overline{B}CD + \overline{A}\overline{B}C\overline{D} +$
$\overline{A}B\overline{C}\overline{D} + \overline{A}BCD + \overline{A}BC\overline{D} +$
$AB\overline{C}\overline{D} + AB\overline{C}D + ABC\overline{D} +$
$A\overline{B}\overline{C}\overline{D} + A\overline{B}C\overline{D}$

After : $Y = \overline{A}CD + \overline{B}CD + \overline{D}$

(b)

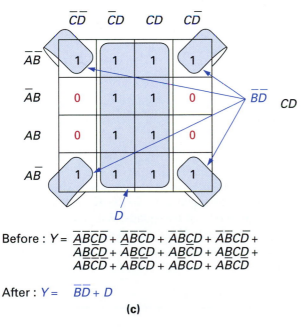

CD

Before : $Y = \overline{A}\overline{B}\overline{C}\overline{D} + \overline{A}\overline{B}\overline{C}D + \overline{A}\overline{B}CD + \overline{A}\overline{B}C\overline{D} +$
$\overline{A}B\overline{C}D + \overline{A}BCD + A\overline{B}\overline{C}D + A\overline{B}CD +$
$ABCD + AB\overline{C}\overline{D} + ABC\overline{D} + AB\overline{C}D$

After : $Y = \overline{B}\overline{D} + D$

(c)

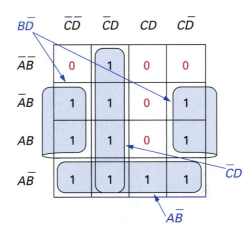

Before : $Y = \overline{A}\overline{B}C\overline{D} + \overline{A}B\overline{C}\overline{D} + \overline{A}B\overline{C}D + \overline{A}B\overline{C}\overline{D} +$
$A\overline{B}\overline{C}\overline{D} + A\overline{B}\overline{C}D + A\overline{B}\overline{C}D + AB\overline{C}\overline{D} +$
$AB\overline{C}D + ABC\overline{D} + ABCD$

After : $Y = A\overline{B} + B\overline{D} + \overline{C}D$

(d)

FIGURE 4-31   K-Map Grouping Examples.

What would be the sum-of-products Boolean expression for the truth table in Figure 4-32(a)? After determining the Boolean expression, plot it on a K-map to see if it can be simplified.

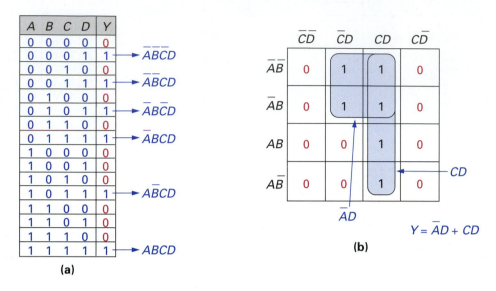

FIGURE 4-32   Example of Four-Variable K-Map.

■   *Solution:*

The first step in developing the sum-of-products Boolean expression is to write down the fundamental products for each HIGH output in the truth table, as seen in Figure 4-32(a). From this, we can derive an SOP equation, which will be

$$Y = \overline{A}\,\overline{B}\,\overline{C}D + \overline{A}\,\overline{B}CD + \overline{A}BC\overline{D} + \overline{A}BCD + A\overline{B}CD + ABCD$$

The next step is to draw a four-variable K-map, as seen in Figure 4-32(b), and then plot the HIGH outputs of the truth table in their equivalent cells in the K-map. Looking at Figure 4-32(b), you can see that the ls can be grouped into two quads. With the square-shaped quad, row labels B and $\overline{B}$ will cancel and column labels C and $\overline{C}$ will cancel, leaving $\overline{A}D$. With the rectangular-shaped quad, all of its row labels will cancel, leaving CD. The simplified Boolean expression, therefore, will be

$$Y = \overline{A}D + CD$$

Referring to the truth table in Figure 4-33(a), you can see that input words 0000 through 1001 have either a 0 or 1 at the output Y. Input words 1010 through 1111 on the other hand, have an "X" written in the output column to indicate that the output Y for these input combinations is unimportant (output can be either a 0 or 1). These Xs in the output are appropriately called "don't-care conditions" and can be treated as either a 0 or 1. In K-maps, these Xs can be grouped with 1s to create larger groups, and therefore a simpler logic circuit, as we will see in the next example.

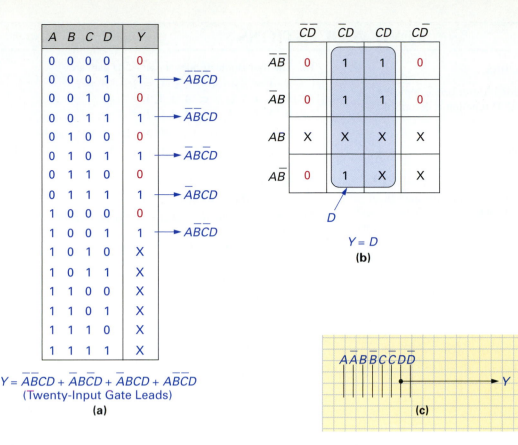

A	B	C	D	Y	
0	0	0	0	0	
0	0	0	1	1	→ $\overline{A}\,\overline{B}\,\overline{C}D$
0	0	1	0	0	
0	0	1	1	1	→ $\overline{A}\,\overline{B}CD$
0	1	0	0	0	
0	1	0	1	1	→ $\overline{A}B\overline{C}D$
0	1	1	0	0	
0	1	1	1	1	→ $\overline{A}BCD$
1	0	0	0	0	
1	0	0	1	1	→ $A\overline{B}\,\overline{C}D$
1	0	1	0	X	
1	0	1	1	X	
1	1	0	0	X	
1	1	0	1	X	
1	1	1	0	X	
1	1	1	1	X	

$Y = \overline{A}\,\overline{B}\,\overline{C}D + \overline{A}\,\overline{B}CD + \overline{A}B\overline{C}D + \overline{A}BCD$
(Twenty-Input Gate Leads)
(a)

$Y = D$
(b)

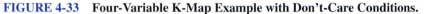
(c)

FIGURE 4-33 Four-Variable K-Map Example with Don't-Care Conditions.

■ EXAMPLE:

Sketch an equivalent logic circuit for the truth table given in Figure 4-33(a) using the least amount of inputs.

■ *Solution:*

Figure 4-33(b) shows a four-variable K-map with the 0s, 1s, and Xs (don't-care conditions) inserted into their appropriate cells. Since a larger group will always yield a smaller term, visualize the Xs as 1s and then try for as big a group as possible. In this example, the Xs have allowed us to group an octet, resulting in the single product D. As a result, the twenty-input Boolean equation in Figure 4-33(a) can be replaced with a single line connected to D, as shown in Figure 4-33(c).

SELF-TEST EVALUATION POINT FOR SECTION 4-4

Use the following questions to test your understanding of Section 4-4.

1. A two-variable K-map will have _____ cells.
2. Simplify the equation $Y = \overline{A}B + AB$.
3. When grouping 1s in a K-map, the _____ (smaller/larger) the groups, the smaller the terms.
4. Can the don't-care conditions in a K-map be grouped with 0s or with 1s, or doesn't it matter?

Multiple-Choice Questions

1. In Boolean, the OR function is described as logical _____ and the AND function is described as logical _____.
   a. addition, division
   b. multiplication, addition
   c. addition, multiplication
   d. multiplication, subtraction

2. The OR gate is said to produce at its output the logical _____ of its inputs, while the AND gate is said to produce at its output the logical _____ of its inputs.
   a. sum, product
   c. sum, subtrahend
   b. product, quotient
   d. product, sum

3. If inputs A and B are ANDed together and then the result is ORed with C, the Boolean expression equation describing this operation would be
   a. $Y = AB + BC$
   c. $Y = AB + C$
   b. $Y = (A + B)C$
   d. $Y = AB + AC$

4. The Boolean expression $Y = (A \cdot B) + (A \cdot C)$ describes a logic circuit that has
   a. six inputs
   b. two OR and one AND gates
   c. two AND and one OR gates
   d. four inputs

5. The Boolean expression $Y = \overline{(A + B)}$ describes a two-input _____.
   a. NAND gate
   c. AND gate
   b. NOR gate
   d. OR gate

6. DeMorgan's first theorem states that
   a. $\overline{A} \cdot \overline{B} = \overline{A + B}$
   c. $\overline{A} + \overline{B} = \overline{A \cdot B}$
   b. $\overline{A + B} = \overline{A} + \overline{B}$
   d. $\overline{\overline{A} \cdot \overline{B}} = A \cdot B$

7. DeMorgan's second theorem states that
   a. $\overline{A} \cdot \overline{B} = \overline{A + B}$
   c. $\overline{A} + \overline{B} = \overline{A \cdot B}$
   b. $\overline{A + B} = \overline{A} + \overline{B}$
   d. $\overline{A \cdot B} = A \cdot B$

8. In Boolean algebra, which symbol is used to describe the exclusive OR action?
   a. "+"
   c. "×"
   b. "."
   d. "⊕"

9. Which of the following algebraically states the commutative law?
   a. $A + B = B + A$
   b. $A \cdot (B \cdot C) = (A \cdot B) \cdot C$
   c. $A + (B \cdot C) = (A + B) \cdot (A + C)$
   d. None of the above

10. $A + (B \cdot C) = (A + B) \cdot (A + C)$ is the algebraic definition of the _____.
    a. associative law
    c. commutative law
    b. distributive law
    d. complements law

11. The four Boolean rules for an OR gate are $A + 0 = ?, A + A = ?, A + 1 = ?, A + \overline{A} = ?$.
    a. $A, 0, 1, 1$
    c. $A, A, 1, 1$
    b. $A, A, 0, 0$
    d. $A, A, 0, 1$

12. The four Boolean rules for an AND gate are $A \cdot 1 = ?, A \cdot A = ?, A \cdot 0 = ?, A \cdot \overline{A} = ?$.
    a. $A, 0, 1, 1$
    c. $A, A, 1, 1$
    b. $A, A, 0, 0$
    d. $A, A, 0, 1$

13. If you applied DeMorgan's theorem and the double-inversion rule to the equation $Y = \overline{\overline{A} + B\overline{C}}$, what would be the result?
    a. $A\overline{B} + C$
    c. $\overline{A}B + C$
    b. $A + B\overline{C}$
    d. $AB + \overline{C}$

14. What would be the dual equation of $A + (B \cdot C) = (A + B) \cdot (A + C)$?
    a. $A \cdot (B + C) = (A \cdot B) + (A \cdot C)$
    b. $\overline{A + B} = \overline{AB} + \overline{C}$
    c. $A \cdot (B \cdot C) = (A \cdot B) + (A \cdot C)$
    d. $A + (B \cdot C) = (A \cdot B) \cdot (A + C)$

15. $Y = AB + BC + AC$ is an example of a _____.
    a. product-of-sums expression
    b. sum-of-products expression
    c. quotient-of-sums expression
    d. sum-of-quotients expression

16. What would be the fundamental product of the input word $ABCD = 1101$?
    a. $ABCD$
    c. $AB\overline{C}D$
    b. $\overline{A}\,\overline{B}C\overline{D}$
    d. $ABC\overline{D}$

17. How many cells would a three-variable Karnaugh map have?
    a. Four
    c. Eight
    b. Six
    d. Sixteen

18. What would be the Boolean expression for a four-variable logic circuit that NANDs inputs A and B and also NANDs inputs C and D and then NANDs the results?
    a. $\overline{\overline{AB} + \overline{CD}}$
    c. $\overline{\overline{(AB)} \cdot \overline{(CD)}}$
    b. $\overline{(AB)} \cdot \overline{(CD)}$
    d. $\overline{(A + B) \cdot (C + D)}$

19. If you applied DeMorgan's theorem and the double-inversion rule to the answer in question 18, you will obtain which of the following equations?
    a. $AB + CD$
    b. $A + B + C + D$
    c. $(A + B) \cdot (C + D)$
    d. $(A + B) + (CD)$

20. In reference to questions 18 and 19, you can see that the DeMorgan equivalent of a NAND-NAND circuit is a/an _____.
    a. OR-AND
    c. AND-OR
    b. AND-NOR
    d. NOR-AND

Practice Problems

21. Determine the Boolean expressions for the logic circuits shown in Figure 4-34.

22. Which of the expressions in question 21 would be considered POS, and which SOP?

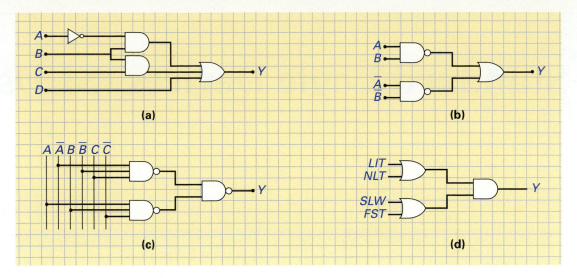

FIGURE 4-34   **From Logic Circuit to Boolean Equation.**

23. Sketch the equivalent logic circuit for the following Boolean equations:

   a. $Y = (AB) + (\overline{AB}) + (\overline{A + B}) + (A \oplus B)$

   b. $Y = (\overline{A} + B) \oplus ABC$

24. Determine the Boolean expression for the Y_2 output in Table 4-1.

25. Use a K-map to simplify the Y_2 expression from question 24.

26. How much was the Y_2 logic circuit simplified in question 25?

27. Referring to Table 4-1, (a) determine the Boolean expression for the Y_1 output. (b) Simplify the expression using a K-map. (c) Describe how much the logic circuit was simplified.

28. Determine the Boolean expression for the Y_3 output in Table 4-1, simplify the expression using a K-map, and then describe how much the logic circuit was simplified.

29. Sketch the simplified logic circuits for the Y_1, Y_2, and Y_3 outputs listed in Table 4-1.

30. If the don't-care condition in the Y_3 output in Table 4-1 were 0s, what would the simplified expression be? Is the expression more or less simplified with don't-care conditions?

Web Site Questions

Go to the web site http://www.prenhall.com/cook, select the textbook *Practical Digital Electronics,* select this chapter, and then follow the instructions when answering the multiple choice practice problems.

TABLE 4-1   **The Y_1, Y_2, and Y_3 Truth Table**

A	B	C	D	Y_1	Y_2	Y_3
0	0	0	0	1	1	1
0	0	0	1	1	1	0
0	0	1	0	1	0	1
0	0	1	1	1	0	0
0	1	0	0	0	0	1
0	1	0	1	0	0	1
0	1	1	0	0	1	1
0	1	1	1	0	1	1
1	0	0	0	0	1	0
1	0	0	1	1	1	0
1	0	1	0	0	0	0
1	0	1	1	1	0	0
1	1	0	0	1	0	1
1	1	0	1	1	0	x
1	1	1	0	1	0	x
1	1	1	1	1	0	x

Standard Logic Devices (SLDs)

A Noyce Invention

On June 3, 1990, Robert Noyce, who was best known as the inventor of the silicon integrated circuit, or IC, died of a heart attack at the age of 62.

Noyce was the son of a Congregational minister and became fascinated by computers and electronics while studying to earn his bachelor's degree at Grinnell College in Grinnell, Iowa. Luckily, he was the student of Grant Gale, who in 1948 was given one of the first transistors by one of its inventors, John Bardeen. Using this device in his lectures. Gale taught one of the first courses in solid-state physics to a class of 18 physics majors, one of whom was Noyce.

Wanting to pursue further interests in electronics, Noyce went on to obtain his doctorate in physics from the Massachusetts Institute of Technology in 1953. Later in that same year, Noyce took his first job as a research engineer at Philco Corp. in Philadelphia. He left in 1956 to join Shockley Semiconductor Laboratory in Mountain View, a company that was founded by another one of the inventors of the transistor, William Shockley.

A year later Noyce and seven other colleagues, whom Shockley called "the traitorous eight," resigned in what was to become a common pattern in Silicon Valley to join Fairchild Camera and Instruments Corp.'s semiconductor division. While at Fairchild, Noyce invented a process for interconnecting transistors on a single silicon chip, which was officially named an IC and was nicknamed "chip." This technological breakthrough, which brought the miniaturization of electronic circuits, was to be used in almost every electronic product, but probably the first to make use of the process was the digital electronic computer.

The honor for this discovery is actually shared by Jack Kilby, a staff scientist at Texas Instruments, who independently developed the same process and also holds patents for the invention.

In 1968 Noyce and More, another Fairchild scientist, founded Intel, a company which grew to become the nation's leading semiconductor company and a pioneer in the development of memory and microprocessor circuits.

In 1979 Noyce was awarded the National Medal of Science by President Jimmy Carter, and in 1987 he received the National Medal of Technology from President Ronald Reagan. He held more than a dozen patents and in 1983 was inducted into the National Inventors Hall of Fame.

While Noyce's name never became widely known, his invention, the IC, is used in every electronic product today, and he was an instrumental figure in creating the $50 billion semiconductor industry, which is at the heart of the $500 billion electronics industry. Sadly, he did not possess a feeling of success or real accomplishment, and this was made clear when he once described his career as "the result of a succession of dissatisfactions."

Introduction

In Chapter 3, we saw how diodes and transistors could be used to construct digital-logic gate circuits, which are the basic decision-making elements in all digital circuits. In reality, many more components are included in a typical logic gate circuit to obtain better input and output characteristics. In this chapter, we will see how the bipolar transistor and the MOS transistor are used to construct the two most frequently used digital IC logic families.

BIPOLAR LOGIC FAMILY	MOS LOGIC FAMILY

TTL (Transistor-Transistor Logic)
  Standard TTL
  Low-Power TTL
  High-Speed TTL
  Schottky TTL
  Low-Power Schottky TTL
  Advanced Low-Power Schottky TTL
 	
ECL (Emitter-Coupled Logic)
 	
IIL (Integrated-Injection Logic)
	PMOS (P-Channel MOSFET)
	
	NMOS (N-Channel MOSFET)
	
	CMOS (Complementary MOSFET)
	  High-Speed CMOS
	  High-Speed CMOS TTL Compatible
	  Advanced CMOS Logic
	  Advanced CMOS TTL Compatible

A logic family is a group of digital circuits with nearly identical characteristics. Each of these two groups or families of digital ICs has its own characteristics and, therefore, advantages. For instance, as you will discover in this chapter, logic gates constructed using MOS transistors use less space due to their simpler construction, have a very high noise immunity, and consume less power than equivalent bipolar transistor logic gates; however, the high input impedance and input capacitance of the E-MOSFET transistor due to its insulated gate means that time constants are larger and, therefore, the transistor ON/OFF switching speeds are slower than equivalent bipolar gates. There are, therefore, trade-offs between the two logic families, and those are

a. Bipolar circuits are faster than MOS circuits, but their circuits are larger and consume more power.

b. MOS circuits are smaller and consume less power than bipolar circuits, but they are generally slower.

In this chapter, we will be examining the operation and characteristics of these bipolar and MOS circuits within digital ICs. This understanding is important if you are going to be able to determine the use of digital ICs within electronic circuits, interpret the characteristics of manufacturer data sheets, test digital ICs, and troubleshoot digital electronic circuits.

5-1 THE BIPOLAR FAMILY OF DIGITAL INTEGRATED CIRCUITS

Dual-In-Line Package (DIP)

An electronic package having two rows of connecting pins.

Transistor-Transistor Logic (TTL)

A line of digital ICs that have logic gates containing several interconnected transistors.

In Chapter 3, diodes and bipolar transistors were used to construct each of the basic logic gate types. These simplified logic gate circuits were used to show how we could use the bipolar transistor to perform each logic gate function. In reality, many more components are included in a typical logic gate circuit to obtain better input and output characteristics. For example, Figure 5-1 shows a typical digital IC containing four logic gates. Referring to the inset in this figure, you can see that a logic gate is actually constructed using a number of bipolar transistors, diodes, and resistors. All of these components are formed and interconnected on one side of a silicon chip. This single piece of silicon actually contains four logic gate transistor circuits, with the inputs and output of each logic gate connected to the external pins of the IC package. For example, looking at the top view of the IC in Figure 5-1, you can see that one logic gate has its inputs on pins 1 and 2 and its output on pin 3. Since all four logic gate circuits will need a $+V_{CC}$ supply voltage (typically $+5$ V) and a ground (0 V), two pins are assigned for this purpose (pin 14 = $+V_{CC}$, pin 7 = ground). In the example in Figure 5-1, the IC contains four two-input NAND gates. This IC package type is called a **dual-in-line package (DIP)** because it contains two rows of connecting pins.

In 1964 Texas Instruments introduced the first complete range of logic gate integrated circuits. They called this line of products **transistor-transistor logic (TTL) ICs** because the circuit used to construct the logic gates contained several interconnected transistors, as shown in the example in the inset in Figure 5-1. These ICs became the building blocks for all digital cir-

FIGURE 5-1 Basic Logic Gate IC.

TABLE 5-1 **A Few of the 7400 Standard TTL Series of Digital ICs**

DEVICE NUMBER	LOGIC GATE CONFIGURATION
7400	Quad (four) two-input NAND gates
7402	Quad two-input NOR gates
7404	Hex (six) NOT or INVERTER gates
7408	Quad two-input AND gates
7432	Quad two-input OR gates
7486	Quad two-input XOR gates
74135	Quad two-input XOR or XNOR gates

cuits. They were in fact, in the true sense of the word, building blocks—no knowledge of electronic circuit design was necessary because the electronic circuits had already been designed and fabricated on a chip. All that had to be done was connect all of the individual ICs in a combination that achieved the specific or the desired circuit operation. All of these TTL ICs were compatible, which meant they all responded and generated the same logic 1 and 0 voltage levels, and they all used the same value of supply voltage; therefore, the output of one TTL IC could be directly connected to the input of one or more other TTL ICs in any combination.

By modifying the design of the bipolar transistor circuit within the logic gate, TTL IC manufacturers can change the logic function performed. The circuit can also be modified to increase the number of inputs to the gate. The most common type of TTL circuits are the 7400 series, which were originally developed by Texas Instruments but are now available from almost every digital IC semiconductor manufacturer. As an example, Table 5-1 lists all of the 7400 series of TTL ICs discussed in Chapter 3.

The TTL IC example shown in Figure 5-1, therefore, would have a device part number of "7400" since it contains four (quad) two-input NAND gates.

5-1-1 *Standard TTL Logic Gate Circuits*

It is important that you understand the inner workings of a logic gate circuit, since you will be able to diagnose if a logic gate is working or not working only if you know how a logic gate is supposed to operate. In this section, we will examine the bipolar transistor circuits within two typical **standard TTL** logic gates.

Bipolar Transistor NOT Gate Circuit

Figure 5-2(a) shows the pin assignment for the six (hex) NOT gates within a 7404 TTL IC. All six NOT gates are supplied power by connecting the $+V_{CC}$ terminal (pin 14) to +5 V and the ground terminal (pin 7) to 0 V. Connecting power and ground to the 7404 will activate all six of the NOT gate circuits, so they are ready to function as inverters.

The inset in Figure 5-2(a) shows the circuit for a bipolar TTL NOT or logic INVERTER gate. As expected, the INVERTER has only one input, which is applied to the **coupling transistor** Q_1, and one output, which is developed by the **totem pole circuit** made up of Q_3 and Q_4. This totem pole circuit derives its name from the fact that Q_3 sits on top of Q_4, like the elements of a Native American totem pole. Transistor Q_2 acts as a **phase splitter** since its collector and emitter outputs will be out-of-phase with one another (base-to-emitter = 180° phase shift, base-to-collector = 0° phase shift), and it is these two opposite outputs that will drive the bases of Q_3 and Q_4.

Let us now see how this circuit will respond to a LOW input. A 0 V (binary 0) input to the NOT gate circuit will apply 0 V to the emitter of Q_1. This will forward-bias the PN base-emitter junction of Q_1 (base is connected to +5 V, emitter has an input of 0 V), causing electron flow from the 0 V input through Q_1's base emitter, through R_1 to $+V_{CC}$. Since the base of Q_1 is only 0.7 V above the emitter voltage, the collector diode of Q_1 will be reverse-biased,

Standard TTL

The original transistor-transistor logic circuit type.

Coupling Transistor

A transistor connected to couple or connect one part of a circuit to another.

Totem Pole Circuit

A circuit arrangement in which two devices are stacked so that the operation of one affects the operation of the other.

Phase Splitter

A circuit designed to generate two out-of-phase outputs.

FIGURE 5-2 **Standard TTL Logic Gate Circuits. (a) NOT (Inverter) Gate. (b) NAND Gate.**

so the base of Q_2 will receive no input voltage and therefore turn OFF. With Q_2 cut OFF, all of the +5 V supply voltage will be applied via R_2 to the base of Q_3, sending Q_3 into saturation. With Q_3 saturated, its collector-to-emitter junction is equivalent to a closed switch between collector and emitter. As a result, the +5 V supply voltage ($+V_{CC}$) will be connected to the output via the low output resistance path of R_3, Q_3's collector-to-emitter, and D_2. A LOW input, therefore, will result in a HIGH output. In this condition, transistor Q_4 is cut OFF and equivalent to an open switch between collector and emitter because Q_2 is OFF and therefore the emitter of Q_2 and the base of Q_4 are at 0 V.

Let us now see how this circuit will respond to a logic 1 input. A HIGH input (+5 V) to the emitter of Q_1 will reverse-bias the PN base-emitter junction of Q_1 (base is connected to +5 V, emitter has an input of +5 V). The +5 V at the base of Q_1 will forward-bias Q_1's base-collector junction, applying +5 V to the base of Q_2, sending it into saturation. In the last condition, when Q_2 was cut OFF, all of the $+V_{CC}$ supply voltage was applied to the base of Q_3. In this condition, Q_2 is heavily ON and therefore equivalent to a closed switch between collector and emitter, so the $+V_{CC}$ supply voltage will be proportionally developed across R_2, Q_2's collector-to-emitter, and R_4. Transistor Q_4 will turn ON due to the positive voltage drop across R_4 and switch its emitter voltage of 0 V to the collector and, therefore, the output. A HIGH input will therefore result in a LOW output. In this condition, Q_3 is kept OFF by a sufficiently low voltage level on Q_2's collector and by diode D_2, which adds an additional diode to the emitter diode of Q_3, ensuring Q_3 stays OFF when Q_4 is ON.

The only other component that has not been discussed is the input diode D_1, which is included to prevent any negative input voltage spikes (negative transient) from damaging Q_1. D_1 will conduct these negative input voltages to ground.

Bipolar Transistor NAND Gate Circuit

Figure 5-2(b) shows the pin-out for the standard TTL 7400 IC and the bipolar transistor circuitry needed for each of the four NAND gates within this IC. The circuit is basically the same as the INVERTER or NOT gate circuit previously discussed, except for the **multiple-emitter input transistor,** Q_1. This transistor is used in all two-input logic gates, and if three inputs were needed, you would see a three-emitter input transistor. Q_1 can be thought of as having two transistors with separate emitters, but with the bases and collectors connected, as shown adjacent to the NAND circuit in Figure 5-2(b). Any LOW input to either A or B will cause the respective emitter diode to conduct current from the LOW input, through Q_1's base emitter and R_1, to $+V_{CC}$. The LOW voltage at the base of Q_1 will reverse-bias Q_1's collector diode, preventing any voltage from reaching Q_2's base, and so Q_2 will cut OFF. The remainder of the circuit will operate in exactly the same way as the NOT gate circuit. With Q_2 OFF, Q_3 will turn ON and connect $+V_{CC}$ through to the output. Any LOW input, therefore, will generate a HIGH output, which is in keeping with the NAND gate's truth table. Only when both inputs are HIGH will both of Q_1's emitter diodes be OFF and the HIGH at the base of Q_1 be applied via the forward-biased collector diode of Q_1 to the base of Q_2. With Q_2 ON, Q_4 will receive a positive base voltage and therefore switch ground, or a LOW voltage, through to the output.

Multiple-Emitter Input Transistor

A bipolar transistor having more than one emitter control terminal.

TTL Logic Gate Characteristics

Almost every digital IC semiconductor manufacturer has its own version of the standard TTL logic gate family or series of ICs. For instance, Fairchild has its 9300 series of TTL ICs, Signetics has the 8000 series, and so on. No matter what manufacturer is used, all of the TTL circuits are compatible in that they all have the same input and output characteristics. To examine the compatibility of these bipolar transistor logic gate circuits, we will have to connect two logic gates to see how they will operate when connected to form a circuit. Figure 5-3 shows a **TTL driver,** or logic source, connected to a **TTL load.** (In this example a NAND gate output is connected to the input of a NOT gate.)

Figure 5-4 lists all of the standard TTL logic gate characteristics. Let us now step through the items mentioned in this list of characteristics and explain their meaning.

a. For TTL circuits, the valid output logic levels from a gate and valid input logic levels to a gate are shown in the blocks in Figure 5-4. Looking at the block on the left you can see that a gate must generate an output voltage that is between the minimum and maximum points in order for that output to be recognized as a valid logical 1 or 0 (valid LOW output = 0 V to +0.4 V, valid HIGH output = +2.4 V to +5 V). Referring to the block on the right

TTL Driver

A transistor-transistor logic source device.

TTL Load

A transistor-transistor logic device that will encumber or burden the source.

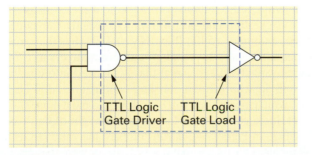

FIGURE 5-3   Connecting Logic Gates.

Output Logic Levels:

Input Logic Levels:

Power Dissipation per Gate : 1 milliwatt to 22 milliwatts (typically 10 milliwatts per gate)

Propagation Delay per Gate : 3 nanoseconds to 33 nanoseconds (typically 10 ns per gate)

Supply Voltage ($+V_{CC}$) :

Temperature Range :

7400 (Commercial)	5400 (Military)
4.75 V to 5.25 V	4.5 V to 5.5 V
0°C to 70°C	−55°C to 125°C

Noise Immunity : High

Fanout : 10 Gates

FIGURE 5-4 **Standard TTL Logic Gate Characteristics.**

Power Dissipation

The power dissipated by a device or circuit.

Propagation Delay Time

The time it takes for an output to change after a new input has been applied.

Noise Immunity

The ability of a device to not be affected by noise.

Noise Margin

The range in which noise will not affect the operation of a circuit.

of Figure 5-4, you can see that as long as the input voltages to a gate are between the minimum and maximum points, the gate will recognize the input as a valid logical 1 or 0 (valid LOW input = +0 V to +0.8 V, valid HIGH input = +2.0 V to +5 V).

b. The second item on our list of TTL characteristics in Figure 5-4 is **power dissipation.** A standard TTL circuit will typically dissipate 10 mW per gate.

c. The third item on our list of TTL characteristics in Figure 5-4 is the **propagation delay time** of a gate, which is the time it takes for the output of a gate to change after the inputs have changed. This delay time is typically about 10 nanoseconds (ns) for a standard TTL gate. For example, it would normally take 10 ns for the output of an AND gate to go HIGH after all of its inputs go HIGH.

d. As you can see from the next two items mentioned in Figure 5-4, there are two standard TTL series of ICs available. The 7400 TTL series is used for all commercial applications, and it will operate reliably when the supply voltage is between 4.75 V and 5.25 V and when the ambient temperature is between 0°C and 70°C. The 5400 series is a more expensive line used for military applications because of its increased supply voltage tolerance (4.5 V to 5.5 V) and temperature tolerance (−55°C to 125°C). A 5400 series TTL IC will have the same pin configuration and perform almost exactly the same logic function as a 7400 series device. For example, a 5404 (hex INVERTER) has the same gate input and output characteristics, pin numbers, $+V_{CC}$, and ground pin numbers as a 7404.

e. Noise is an unwanted voltage signal that is induced into conductors due to electromagnetic radiation from adjacent current-carrying conductors. All TTL logic circuits have a high **noise immunity,** which means they will be immune to most noise fluctuations at the gate's inputs. A standard TTL logic gate actually has a **noise margin** of 0.4 V between a TTL driver and TTL load, as is shown in Figure 5-4, if you compare the valid HIGH and LOW outputs to the HIGH and LOW inputs. This means that if a TTL driver were to output its worst-case LOW output (0.4 V) and induced noise increased this signal further away from 0.4 V to 0.8 V, the

CHAPTER 5 / STANDARD LOGIC DEVICES (SLDs)

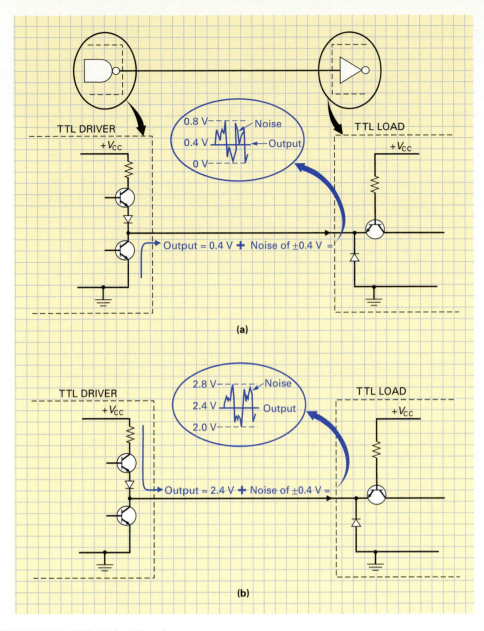

FIGURE 5-5   TTL Noise Margin.

TTL load would still accept this input as a valid LOW input since it is between 0 V and 0.8 V. This is shown in the simplified diagram in Figure 5-5(a). Similarly, if a TTL driver were to output its worst-case HIGH output (2.4 V) and induced noise decreased this signal further away from 2.4 V to 2.0 V, the TTL load would still accept this input as a valid logic 1 since it is between 2.0 V and 5 V. This is shown in the simplified diagram in Figure 5-5(b).

    f. The last specification that has to be discussed in Figure 5-4 is the listing called **fanout.** To explain this term, Figure 5-6 shows how a TTL driver has to act as both a current **sink** and a current **source.** Looking at the direction of current in this illustration, you can see that this characteristic is described using conventional current flow (current travels from $+V \rightarrow -V$) instead of electron current flow (current travels from $-V \rightarrow +V$). Whether the current flow is from ground to $+V_{CC}$ (electron flow) or from $+V_{CC}$ to ground (conventional flow), the value of current is always the same if the same circuit exists between these two points. The only reason we have changed to conventional at this point is that most semiconductor manufacturers' specifications for the terms "sinking" and "sourcing" are detailed using conventional current flow.

Fanout

The number of parallel loads that can be driven simultaneously by one logic source, keeping logic levels within spec.

Sink

A power-consuming device or circuit.

Source

A circuit or device that supplies power.

FIGURE 5-6   **Sinking and Sourcing Current.**

In Figure 5-6(a), a single logic gate load has been connected to a TTL logic gate driver, which is producing a LOW output (Q_4 in driver's totem pole is ON). In this instance, Q_4 will "sink" the single gate load current of 1.6 mA from Q_1 to ground. Manufacturer's data sheets state that Q_4 can sink a maximum of 16 mA and still produce a LOW output voltage of 0.4 V or less (the voltage drop across Q_4 will always be 0.4 V or less when it is sinking 16 mA). Since the maximum sinking current of 16 mA is ten times larger than the single gate load current of 1.6 mA, we can connect ten TTL loads to each TTL driver or source, as seen in Figure 5-6(b), and still stay "within spec" (within the parameters defined in the data sheet).

To examine both sides of the coin, Figure 5-6(c) shows a single TTL load connected to a TTL driver, which is producing a HIGH output (Q_3 in driver's totem pole is ON). In this instance, Q_3 will "source" the single gate load current of 40 µA. Manufacturers' data sheets state that Q_3 can source a maximum of 400 µA and still maintain a HIGH output voltage of 2.4 V or more. Since the maximum source current of 400 µA is ten times larger than the single gate load current of 40 µA, we can connect ten TTL loads to each TTL driver, as seen in Figure 5-6(d), and still stay within spec.

As we have seen in Figure 5-6, the maximum number of TTL loads that can be reliably driven by a single TTL driver is called the fanout. Referring to the standard TTL logic gate characteristics in Figure 5-4, you will now understand why a standard TTL gate has a fanout of 10.

g. Using Figure 5-6, we can describe another TTL circuit characteristic. When a TTL load receives a LOW input, as seen in Figure 5-6(a), a large current exists in the emitter of Q_1. On the other hand, when a TTL load receives a HIGH input, as seen in Figure 5-6(c), the emitter current in Q_1 is almost zero. This action explains why a TTL logic gate acts like it is receiving a HIGH input when its input lead is not connected. For example, if the input lead to a NOT gate was not connected to a circuit (an open input) and power was applied to the gate, the output of the NOT gate would be LOW. An unconnected input is called a **floating input.** In this condition, no emitter current will exist in Q_1 since the input is disconnected; so we can say *a floating TTL input is equivalent to a HIGH input.*

5-1-2   *Low-Power and High-Speed TTL Logic Gates*

A low-power TTL circuit is basically the same as the standard TTL circuit previously discussed except that all of the internal resistance values in the logic gate circuit are ten times larger. This means that the power dissipation of a low-power TTL circuit is one-tenth that of a standard TTL circuit (low-power TTL logic gates will typically have a power dissipation of about 1 mW per gate).

Increasing the internal resistances, however, will decrease the circuit's internal currents, and since the bipolar transistor is a current-operated device, the response or switching time of these logic gate types will decrease. (Low-power TTL logic gates will typically have a propagation delay time of 33 ns per gate.) With low-power TTL circuits, therefore, high speed is sacrificed for low power consumption. Low-power logic gate ICs can be identified because the device number will have the letter "L" following the series number 74 or 54; for example, 74L00, 74L01, 74L02, and so on.

By decreasing the resistances within a standard logic gate, manufacturers can lower the logic gate circuit's internal time constants and therefore decrease the gate's propagation delay time (typically 6 ns). The lower resistances, however, increase the gate's power dissipation (typically 22 mW), so the trade-off is therefore once again speed versus power. High-speed logic gate ICs can be identified because the device number will have the letter "H" following the series number 74 or 54; for example, 74H00, 74H01, 74H02, and so on.

5-1-3   *Schottky TTL Logic Gates*

With standard TTL and low-power TTL logic gates, transistors go into saturation when they are switched ON and are flooded with extra carriers. When a transistor is switched OFF, it

Floating Input
An input that is not connected.

Low-Power TTL
A line of transistor-transistor logic ICs that have a low power-dissipation figure.

High-Speed TTL
A line of transistor-transistor logic ICs that have a high speed or low propagation-delay-speed figure.

Schottky TTL
A line of transistor-transistor logic ICs that use Schottky transistors.

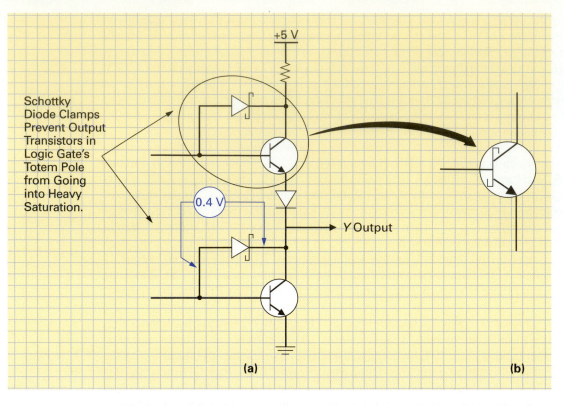

Schottky Diode Clamps Prevent Output Transistors in Logic Gate's Totem Pole from Going into Heavy Saturation.

+5 V

0.4 V

Y Output

(a)

(b)

FIGURE 5-7 **Schottky Clamped Bipolar Transistors. (a) Connecting a Schottky Diode across a Transistor Collector-Base Junction. (b) Schottky Transistor Symbol.**

Saturation Delay Time

The additional time it takes a saturated transistor to turn off.

Schottky Transistor

A transistor having a Schottky diode connected between its collector and base.

takes a short time for these extra carriers to leave the transistor, and therefore it takes a short time for the transistor to cut OFF. This **saturation delay time** accounts for the slow propagation delay times, or switching times, of standard and low-power TTL logic gates.

Schottky TTL logic gates overcome this problem by fabricating a Schottky diode between the collector and base of each bipolar transistor, as seen in Figure 5-7(a). This transistor is called a **Schottky transistor** and uses the symbol shown in Figure 5-7(b). Let us now see how this Schottky diode will affect the operation of the transistor and decrease saturation delay time. When the transistor begins to turn ON, its collector voltage will fall. When the collector drops below 0.4 V of the base voltage, the Schottky diode will conduct and shunt current away from the collector junction. This action effectively clamps the collector to 0.4 V of the base voltage, keeping the collector diode of the transistor reverse-biased. Keeping the collector diode reverse-biased prevents the transistor from slipping into heavy saturation and therefore decreases the transistor's saturation delay time.

Schottky TTL devices are fast, with propagation delay times of typically 3 ns. This fast switching time means that the input signals can operate at extremely high frequencies of typically 100 MHz. Schottky TTL ICs are labeled with an "S" after the first two digits of the series number; for example, 74S04, 74S08, and so on.

By using Schottky transistors and increasing the internal resistances in a logic gate circuit, manufacturers compromised the speed of Schottky to obtain a lower dissipation rating. Schottky TTL devices will typically have a power dissipation of 19 mW per gate, whereas *low-power Schottky TTL devices* typically have a power dissipation of 2 mW per gate and propagation delay times of 9.5 ns. These low-power Schottky TTL devices include the letters "LS" in the device number; for example, 74LS32, 74LS123, and so on.

Special Schottky TTL devices are available that use a new integration process called oxide isolation, in which the transistor's collector diode is isolated by a thin oxide layer instead of a reverse-bias junction. These *advanced Schottky (AS) TTL devices* (labeled 74AS00) have a typical power dissipation of 8.5 mW per gate and a propagation delay time of 1.5 ns. By increasing internal resistances once again, manufacturers came up with an *advanced low-power*

DEVICE: SN7400, SN5400 (Quad Two-Input NAND)

recommended operating conditions (the levels of voltage, current, and temperature at which the IC should be run under normal conditions)

		SN5400			SN7400			UNIT
		MIN	NOM	MAX	MIN	NOM	MAX	
V_{CC}	Supply voltage (min. to max. supply voltage range)	4.5	5	5.5	4.75	5	5.25	V
V_{IH}	High-level input voltage (the min. input voltage for a HIGH)	2			2			V
V_{IL}	Low-level input voltage (the max. input voltage for a LOW)			0.8			0.8	V
I_{OH}	High-level output current (the max. output current for a HIGH)			−0.4			−0.4	mA
I_{OL}	Low-level output current (the max. output current for a LOW)			16			16	mA
T_A	Operating free-air temperature (min. to max. operating temp.)	−55		125	0		70	°C

DC Specifications (Electrical Characteristics)

HIGH Output Voltage

LOW Output Voltage

PARAMETER	TEST CONDITIONS †		SN5400			SN7400			UNIT
			MIN	TYP‡	MAX	MIN	TYP‡	MAX	
V_{IK}	V_{CC} = MIN,	I_I = −12 mA			−1.5			−1.5	V
V_{OH}	V_{CC} = MIN, V_{IL} = 0.8 V,	I_{OH} = −0.4 mA	2.4	3.4		2.4	3.4		V
V_{OL}	V_{CC} = MIN, V_{IH} = 2 V,	I_{OL} = 16 mA		0.2	0.4		0.2	0.4	V
I_I	V_{CC} = MAX,	V_I = 5.5 V			1			1	mA
I_{IH}	V_{CC} = MAX,	V_I = 2.4 V (HIGH Input Current)			40			40	µA
I_{IL}	V_{CC} = MAX,	V_I = 0.4 V (LOW Input Current)			−1.6			−1.6	mA
I_{OS} §	V_{CC} = MAX (Current when output is shorted to ground)		−20		−55	−18		−55	mA
I_{CCH}	V_{CC} = MAX,	V_I = 0 V		4	8		4	8	mA
I_{CCL}	V_{CC} = MAX,	V_I = 4.5 V		12	22		12	22	mA

Not a recommended condition

Current supply to IC when all outputs are HIGH or LOW

Highest rate at which a signal can change when applied to an input

Minimum time a pulse must be present at the input to cause a reliable change at the output

Minimum amount of time data must be available at an input

Minimum amount of time a data input must be maintained

Maximum Clock Frequency (f_{max})	15 MHz
Input Pulse Width (t_w)	20 ns
Data Setup Time (t_{su})	15 ns
Data Hold Time (t_h)	5 ns

FIGURE 5-8   A Standard TTL Data Sheet.

Schottky TTL line of products (labeled 74ALS00) which have power dissipation and propagation time figures of 1 mW and 4 ns. Some manufacturers, such as Fairchild, use the letter "F" for fast in their AS series of ICs; for example, 74F10, 74F14, and so on.

As you have seen, an IC contains quite a complex circuit, and this circuit has a detailed list of circuit characteristics. These characteristics or specifications are listed in a manufacturer's data sheet or spec sheet. As an example, Figure 5-8 shows a typical data sheet for a 7400 (quad two-input NAND) TTL IC.

Most data sheets can be broken down into five basic blocks:

a. pin-out and logic symbols

b. maximum ratings

c. recommended ratings

d. DC specifications or electrical characteristics

e. AC specifications or switching characteristics

Study all of the annotations included in this data sheet, because they explain the meaning of the terms used and the characteristics.

Open-Collector Output

A logic circuit in which the final output, which is a transistor's collector, is open or floating.

Pull-up Resistor

The name given to a resistor that is connected between a power line and a signal line. Its function is to pull up or make HIGH a signal line when it is not switched LOW.

5-1-4   *Open-Collector TTL Gates*

Some TTL devices have an **open-collector output** instead of the totem pole output circuit previously discussed. Figure 5-9(a) shows a standard TTL INVERTER circuit with an open-collector output, and Figure 5-9(b) shows the logic symbol used to represent an open-collector output gate.

In order to get a HIGH or LOW output from an open-collector gate, an external **pull-up resistor** (R_P) must be connected between $+V_{CC}$ and the collector of Q_3 (which is the output) as seen in Figure 5-9(c). In this circuit, two open-collector inverters are driving the same output, *Y*. If two standard TTL logic gates with totem pole outputs were wired together in this way and the *A* INVERTER produced a HIGH output and the *B* INVERTER a LOW output,

(a)   **(b)**

(c)   **(d)**

FIGURE 5-9   Open Collector Gates. (a) Bipolar Transistor Open Collector INVERTER Circuit. (b) Open Collector Symbol. (c) Wired INVERTER Open Collector Gates. (d) Open Collector NOR Driver Gate.

Symbol	Parameter	54/74	54/74 H	54/74 S	54/74 LS
I_{OH} (max)	Power Supply	12 mA	26 mA	19.8 mA	2.4 mA
I_{OL} (max)	Current	33 mA	58 mA	54 mA	6.6 mA
t_{PLH} (max)	Propagation Delay	55 ns	18 ns	7.5 ns	22 ns
t_{PHL} (max)		15 ns	15 ns	7.0 ns	18 ns

7401 - QUADRUPLE TWO-INPUT
POSITIVE-NAND GATES
WITH OPEN-COLLECTOR OUTPUTS

Pin 14 = $+V_{CC}$
Pin 7 = GND

A Standard TTL open-collector gate can sink a maximum of 33 mA when supplying a LOW output, whereas a high-speed TTL open-collector gate can sink a maximum of 58 mA.

FIGURE 5-10 An Open-Collector Output Data Sheet.

there would be a direct short between the two, causing both gates to burn out. With open-collector gates, the outputs can be wired together or connected to a common output line without causing any HIGH-LOW output conflicts between gates. With the circuit in Figure 5-9(c), only when both inputs are LOW will the output transistors of both inverters be floating and the output be pulled up by R_P to +5 V ($A = 0$, $B = 0$, $Y = 1$). If either A input or B input is HIGH, one of the output transistors will turn ON and pull down or ground the output Y ($A = 1$ or $B = 1$, $Y = 0$). You may recognize this logic operation as that of a NOR gate.

Open-collector gates can sink around 40 mA, which is a vast improvement over standard TTL gates that only sink 16 mA (74ASXX gates can sink 20 mA).

In Figure 5-9(d), you can see how the high sinking current ability of an open-collector NOR gate can be used to drive the high current demand of a light emitting diode (LED) load. The current limiting resistor (R_S) is being used to limit the current through the LED to about 25 mA, which produces a good level of brightness from an LED. In many applications, open-collector gates are used to drive external loads that require increased current, such as LED displays, lamps, relays, and so on. As an example, Figure 5-10 shows typical open-collector current characteristics.

5-1-5   *Three-State (Tri-State) Output TTL Gates*

The **three-state output TTL gate** has, as its name implies, three output states or conditions. These three output states are LOW, HIGH, and FLOATING, or a "high-impedance output" state. Before we discuss why we need this third output state, let us see how this circuit operates.

Figure 5-11(a) illustrates the internal bipolar transistor circuit for a three-state NOT gate. The circuit is basically the same as the previously discussed INVERTER gate, except for the additional control input applied to transistor Q_4. When the enable input is LOW, Q_4 is OFF and has no effect on the normal inverting operation of the circuit. However, when the enable input is HIGH, Q_4 is turned ON, and the ground on its emitter will be switched through to the second emitter of Q_1. The ON Q_1 will switch ground from its emitter through to the base of Q_2, turning Q_2 OFF and therefore Q_5 OFF. Transistor Q_4 will also ground the base of Q_3 via D_1, and so Q_3 will turn OFF. The result will be that both totem pole transistors are cut OFF, as seen in the equivalent circuit shown in the inset in Figure 5-11(a). This high-impedance output state, therefore, is created by turning OFF both totem pole output transistors so that the output is isolated from ground and the supply voltage. With the output of the gate completely

Three-State Output TTL Gate

A line of transistor-transistor logic gates that can have three possible outputs: LOW, HIGH, or OPEN.

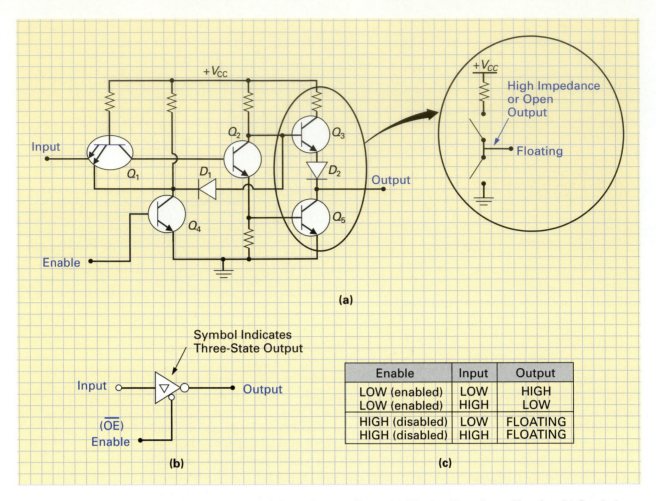

FIGURE 5-11   **Three-State Output Gates. (a) Bipolar Transistor Circuits. (b) Symbol. (c) Truth Table.**

Enable	Input	Output
LOW (enabled)	LOW	HIGH
LOW (enabled)	HIGH	LOW
HIGH (disabled)	LOW	FLOATING
HIGH (disabled)	HIGH	FLOATING

Active-LOW Control Line

A control line that goes LOW when activated.

disconnected from the circuit, this lead is said to be a disconnected, or a FLOATING high-impedance line.

Figure 5-11(b) shows the schematic symbol for this three-state output NOT gate. The bubble attached to the enable input is used to indicate an **active-LOW control line,** which means that if you want to "activate the gate," you must make this control line "LOW." To disable the gate, therefore, we would not make this line active, so the enable line would be HIGH. Studying the truth table or function table in Figure 5-11(c), you can see this gate's operation summarized. When the enable control line is LOW, the circuit is enabled for normal INVERTER operation. When the enable control line is HIGH, however, the output is always FLOATING, no matter what input logic level is present. The label for active-LOW control lines will usually have a bar over the label's letters, as shown with the output enable ($\overline{OE}$) in Figure 5-11(b). This bar over the label is used to distinguish an active-LOW control line label from an active-HIGH control line label.

Like the previously discussed open-collector gates, these devices are used in applications where two or more gate outputs are connected to a single output line. These common signal lines are called "signal buses," and they allow logic devices to individually take control of a common set of signal lines without causing signal interferences with other devices on the bus. To avoid HIGH/ LOW output line conflicts between gates, only one gate is enabled at one time to drive the output line, with all other gates being disabled and therefore their outputs FLOATING. These applications will be discussed in more detail in future chapters.

5-1-6   *Buffer/Driver TTL Gates*

A **buffer** is a device that isolates one circuit from another. In order to achieve this isolation, a buffer should have a high input impedance (low input current) and a low output impedance (high output current). A standard TTL gate provides a certain amount of isolation or buffering, since the output current is ten times that of the input current.

	Input	Output	
Standard TTL	$I_{IL} = -1.6$ mA	$I_{OL} = -16$ mA	(output is ten times larger)
	$I_{IH} = 40$ μA	$I_{OH} = 400$ μA	(output is ten times larger)

A buffer or driver logic gate is a basic gate that has been slightly modified to increase the output current. For example, a 7428 is a "quad two-input NOR gate buffer" whose output current is thirty times greater than the input currents, enabling it to drive heavier loads.

	Input	Output	
Buffer/Driver TTL	$I_{IL} = -1.6$ mA	$I_{OL} = -48$ mA	(output is thirty times larger)
	$I_{IH} = 40$ μA	$I_{OH} = 1.2$ mA	(output is thirty times larger)

5-1-7   *Schmitt-Trigger TTL Gates*

The **Schmitt-trigger circuit,** named after its inventor, is a two-state device that is used for pulse shaping. Its schematic is shown in Figure 5-12.
Several Schmitt-trigger TTL ICs are available, including 7413 (dual four-input NAND Schmitt-trigger), 7414 (hex Schmitt-trigger inverters), 74132 (quad two-input NAND Schmitt-trigger), and so on. These logic circuits will have a Schmitt-trigger circuit connected to the output of each logic gate circuit so that the gate will perform its normal function and then sharpen the rise and fall times of pulses that have been corrupted by noise and attenuation.

5-1-8   *Emitter-Coupled Logic (ECL) Gate Circuits*

Schottky TTL circuits improved the switching speed of digital logic circuits by preventing the bipolar transistor from saturating. **Emitter-coupled logic,** or **ECL,** is another non-saturating bipolar transistor logic circuit arrangement; however, ECL is the fastest of all logic circuits. It is used in large digital computer systems that require fast propagation delay times (typically 0.8 ns per gate) and are able to sacrifice power dissipation (typically 40 mW per gate).
Figure 5-13(a) shows a typical ECL gate circuit. The two inputs to this circuit, A and B, are applied to the transistors Q_1 and Q_2, which are connected in parallel. If additional inputs are needed, more input transistors can be connected in parallel across Q_1 and Q_2, as shown in the dashed lines in Figure 5-13(a). The two collector outputs of Q_2 and Q_3 drive the two low output impedance emitter-followers Q_5 and Q_6. These emitter-follower outputs provide an OR output from Q_6 and its complement, or opposite, which is a NOR output from Q_5. This dual output is one of the advantages of ECL gates.

FIGURE 5-12

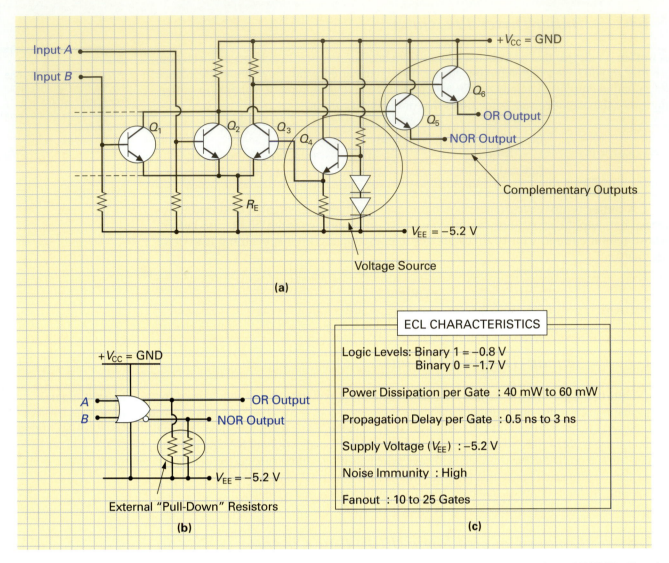

FIGURE 5-13 Emitter-Coupled Logic (ECL) Gates. (a) Bipolar Transistor OR/NOR Circuit. (b) Logic Symbol. (c) Characteristics.

Pull-Down Resistor

A name given to a resistor that is connected between a signal line and ground. Its function is to pull down or make LOW a signal line when it is not switched HIGH.

External **pull-down resistors** are required, as seen in Figure 5-13(b), which also shows the logic symbol for this OR/NOR ECL gate. The open outputs of ECL gates mean that two or more gate outputs can drive the same line without causing any HIGH/LOW voltage conflicts. The V_{EE} pin of an ECL IC is typically connected to −5.2 V, and the V_{CC} pin is connected to ground. Although these dc supply values seem to not properly bias the circuit, remember that the V_{CC} power line is still positive (since it is at ground) relative to the V_{EE} power line (which is at −5.2 V). The transistor Q_4 has a constant base bias voltage and an emitter resistor and will produce a temperature-stabilized constant voltage, or reference voltage, of approximately −1.3 V to the base of Q_3. The bipolar transistors in the circuit simply conduct more or conduct less depending on whether they represent a LOW or HIGH, ensuring that none of the transistors saturate.

Preventing transistors from going into saturation accounts for the fast speed of ECL logic gate circuits, and the fact that transistors, for the most part, are conducting and therefore drawing a current accounts for the ECL's increase in power dissipation. The typical characteristics of ECL gates are listed in Figure 5-13(c). Notice that the binary 0 and binary 1 voltage levels are slightly different from TTL circuits; however, the binary 1 voltage level (−0.8 V) is still positive relative to binary 0 (−1.7 V).

Let us now examine this ECL circuit's operation. When either one or both of the inputs go to a HIGH (−0.8 V), that transistor will conduct more (since its base is now positive with respect to the emitter, which is at -1.3 V). In so doing, the input transistor will make transistor Q_3 turn more OFF (since the base-emitter junction of the input transmitter will make Q_3's emitter positive with respect to its large negative base voltage). The conducting input transistor will switch its negative emitter voltage through to the collector and therefore to the base of Q_5. Q_5 will turn more OFF, and its output will be pulled LOW. For a NOR gate output, any binary 1 input will produce a binary 0 output.

On the other hand, with Q_3 more OFF, its collector will be more positive (closer to 0 V) and Q_6 will turn more ON and connect the more positive V_{CC} to the OR output. For an OR gate output, any binary 1 input will produce a binary 1 output. When all inputs are LOW (-1.7 V), all of the input transistors will be more OFF, producing a HIGH output to Q_5 (all inputs are 0, output from NOR is 1). With Q_1 and Q_2 more OFF, Q_3 will conduct more, producing a LOW output to Q_6 (all inputs are 0, output of OR is 0).

The high speed, high power consumption, and high-cost logic of these circuits account for why they are only used when absolutely necessary. The most widely used ECL family of ICs is Motorola's ECL (MECL) 10K and 100K series, in which devices are numbered MC10XXX or MC100XXX.

5-1-9 *Integrated-Injection Logic (I^2L) Gate Circuits*

Integrated-injection logic (IIL) or, as they are sometimes abbreviated, I^2L (pronounced "I squared L"), is a series of digital bipolar transistor circuits that have good speed, low power consumption, and very good circuit **packing density.** In fact, one I^2L gate is one-tenth the size of a standard TTL gate, making it ideal for compact or high-density circuit applications.

A typical I^2L logic NOT gate is shown in Figure 5-14(a). Transistor Q_1 acts as a current source and the multiple-collector transistor Q_2 acts as the INVERTER with two outputs. Studying the circuit, you can see that the base of Q_1 is connected to the emitter of Q_2, and the collector of Q_1 is connected to the base of Q_2. These common connections enable the entire I^2L gate to be constructed as one transistor with two emitters and two collectors and use the same space on a silicon chip as one standard TTL multiple-emitter transistor.

The circuit in Figure 5-14(a) operates in the following way. The emitter of Q_1 is connected to an external supply voltage ($+V_S$), which can be anywhere from 1 V to 15 V, depending on the injector current required. The constant base-emitter bias of Q_1 means that a constant current is available at the collector of Q_1 (called the **current injector transistor**), and this current will either be applied to the base of Q_2 or out of the input, depending on the input voltage logic level. To be specific, a LOW input will pull the injector current out of the input and away from the base of Q_2. This will turn Q_2 OFF, and its outputs will be open, which is equivalent to a HIGH output. On the other hand, a HIGH input will cause the injector current of Q_1 to drive the base of Q_2. This will turn Q_2 ON and switch the LOW on its emitter through to the collector outputs.

Figure 5-14(b) illustrates an I^2L NAND gate. As you can see, the only difference between this circuit and the INVERTER gate circuit in Figure 5-14(a) is the two inputs A and B. Transistors Q_3 and Q_4 are shown in this circuit; however, they are the multiple-collector transistors of the gates, driving inputs A and B of this NAND gate circuit. If either one of these transistors is ON (a low input to the NAND), the injector current of Q_1 will be pulled away from Q_2's base, turning OFF Q_2 and causing a HIGH output. Only when both inputs are HIGH (Q_3 and Q_4 are OFF) will the injector current from Q_1 turn ON Q_2 and produce a LOW output.

Figure 5-14(c) lists the characteristics of I^2L logic gate circuits. These circuits are used almost exclusively in digital watches, cameras, and almost all compact, battery-powered circuit applications. As mentioned earlier, its good speed, low power consumption, and small size make it ideal for any portable system. Also, the simplicity of I^2L circuits means that it is easy to combine analog or linear circuits (such as amplifiers and oscillators) with digital circuits to create complete electronic systems on one chip. The circuit complexity of standard TTL digital circuits means that it is normally too difficult and not cost-effective (due to fabrication difficulties and therefore high cost) to place digital and linear circuits on the same IC chip.

Integrated-Injection Logic (IIL)

A bipolar digital logic family that uses a current injector transistor.

Packing Density

A term used to describe the number of devices that can be packed into a given area.

Current Injector Transistor

A transistor connected to generate a constant current.

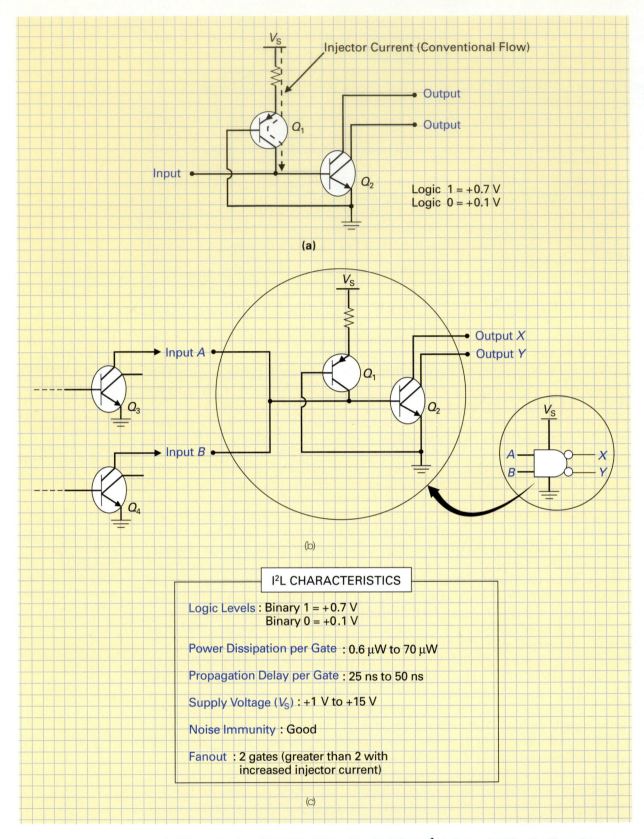

FIGURE 5-14 Integrated-Injection Logic (IIL or I²L) Gates. (a) Bipolar Transistor INVERTER (NOT) Gate Circuit. (b) Bipolar Transistor NAND Gate Circuit. (c) Characteristics.

Use the following questions to test your understanding of Section 5-1.

1. TTL is an abbreviation for _____.
2. What is the fanout of a standard TTL gate?
3. A floating TTL input is equivalent to a _____ (HIGH/LOW) input.
4. The three output states of a tri-state logic gate are _____, _____, and _____.
5. In what application would ECL circuits be used?
6. What is the advantage of I²L logic circuits?

5-2   THE MOS FAMILY OF DIGITAL INTEGRATED CIRCUITS

In this section we will examine the three basic types of MOS ICs, which are called PMOS (pronounced "pea-moss"), NMOS (pronounced "en-moss"), and CMOS (pronounced "sea-moss").

The E-MOSFET is ideally suited for digital or two-state circuit applications. One reason is that it naturally operates as a "normally-OFF voltage controlled switch," since it can be turned ON when the gate voltage is positive and turned OFF when the gate voltage falls below a threshold level. This threshold level is a highly desirable characteristic because it prevents noise from false triggering, or accidentally turning ON, the device. The other E-MOSFET advantage is its extremely high input impedance, which means that the device's circuit current, and therefore power dissipation, is low. This enables us to densely pack or integrate many thousands of E-MOSFETs onto one small piece of silicon, forming a high-component-density IC. These low-power and high-density advantages make the E-MOSFET ideal in battery-powered, small-sized (portable) applications such as calculators, wristwatches, notebook computers, hand-held video games, digital cellular phones, and so on.

In the following sections we will see how the E-MOSFET can be used to construct digital logic gate circuits and then compare these circuits and their characteristics to the previously discussed bipolar logic gate circuit characteristics.

5-2-1   *PMOS (P-Channel MOS) Logic Circuits*

The name **PMOS logic** is used to describe this logic circuitry because the logic gates are constructed using P-channel E-MOSFETs. Figure 5-15(a) shows how three P-channel E-MOSFETs can be used to construct a NOR logic gate. Figure 5-15(b) shows the logic levels used in PMOS circuits, and Figure 5-15(c) reviews the operation of a P-channel E-MOSFET.

The PMOS NOR gate circuit in Figure 5-15(a) will operate as follows. Transistor Q_3 can be thought of as a "current limiting resistor" between source and drain because its gate is constantly connected to $-V_{GG}$, and therefore Q_3 is always ON. Any HIGH input will turn OFF its respective E-MOSFET, resulting in a negative or LOW (-8 V) output due to the pull-down action of Q_3. Only when both inputs are LOW, will both Q_1 and Q_2 be ON, allowing $+V_{DD}$ (0 V, which is a HIGH) to be connected through Q_1 and Q_2 to the output.

PMOS Logic
A digital logic family of ICs that use P-channel MOSFETs.

5-2-2   *NMOS (N-Channel MOS) Logic Circuits*

The name **NMOS logic** is used to describe this logic circuitry because the logic gates are constructed using N-channel E-MOSFETs. Figure 5-16(a) shows how three N-channel E-MOSFETs can be used to construct a NAND gate. Figure 5-16(b) shows the logic levels used in NMOS circuits, and Figure 5-16(c) reviews the operation of an N-channel E-MOSFET.

NMOS Logic
A digital logic family of ICs that use N-channel MOSFETs.

FIGURE 5-15 A P-Channel MOS (PMOS) NOR Gate.

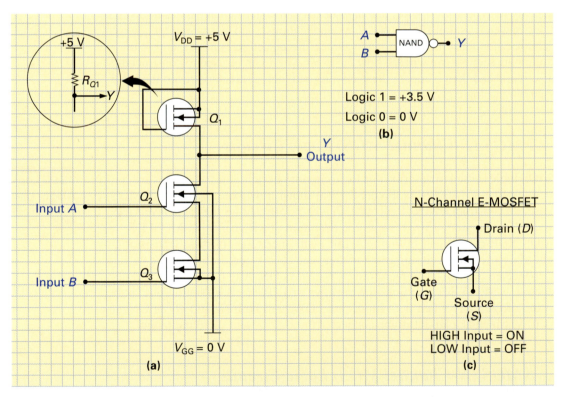

FIGURE 5-16 An N-Channel MOS (NMOS) NAND Gate.

114

The NMOS NAND gate circuit in Figure 5-16(a) will operate as follows. Transistor Q_1 functions as a "current limiting resistor" between source and drain because its gate is constantly connected to $+V_{DD}$, and therefore Q_1 is always ON. A LOW (0 V) on either or both of the inputs will turn OFF one or both of the transistors Q_2 and Q_3, causing the output to be pulled up to a HIGH ($+3.5$ V) by Q_1. Only when both inputs are HIGH, will Q_1 and Q_2 turn ON and switch the ground on Q_3's source through Q_3 and Q_2 to the output.

5-2-3   *CMOS (Complementary MOS) Logic Circuits*

The name **CMOS logic** is used to describe this logic circuitry because the logic gates are constructed using both a P-channel E-MOSFET (Q_1) and its complement, or opposite, an N-channel E-MOSFET (Q_2).

A CMOS NOT Gate Circuit

Figure 5-17(a) shows how a CMOS NOT or INVERTER gate can be constructed. Notice the N-channel E-MOSFET (Q_2) has its source connected to 0 V, and therefore its gate needs to be positive (relative to the 0 V source) for it to turn ON. On the other hand, the P-channel E-MOSFET has its source connected to $+10$ V, and therefore its gate needs to be negative (relative to the $+10$ V source) for it to turn ON. The table in Figure 5-17(a) shows how this circuit will function. When V_{IN} or A is LOW (0 V), Q_1 will turn ON (N-P gate-source is forward-biased; $G = 0$ V, $S = +10$ V) and Q_2 will turn OFF (P-N gate-source is reverse-biased; $G = 0$ V, $S = 0$ V). A LOW input will therefore turn ON Q_1 creating a channel between source and drain, connecting the $+10$ V supply voltage to the output, or Y. On the other hand, when V_{IN} or A is HIGH ($+10$ V), Q_1 will turn OFF (N-P gate-source is reverse-biased; $G = +10$ V, $S = +10$ V) and Q_2 will turn ON (P-N gate-source is forward-biased; $G = +10$ V, $S = 0$ V). A HIGH input will therefore turn ON Q_2, creating a channel between source and drain connecting the 0 V at Q_2's source to the output, or Y.

A CMOS NAND Gate Circuit

Figure 5-17(b) shows how two complementary MOSFET pairs (two N-channel and two P-channel) could be connected to form a two-input CMOS NAND gate. If either input A or B is LOW, one or both of the P-channel MOSFETs (Q_1 and Q_2) will turn ON and switch $+V_{DD}$ or a HIGH through to the output. Only when both inputs A and B are HIGH will Q_1 and Q_2 be OFF, and Q_3 and Q_4 will be ON, switching ground or a LOW through to the output.

CMOS Logic Gate Characteristics

CMOS logic is probably the best all-round logic circuitry because of its low power consumption, very high noise immunity, large power supply voltage range, and high fanout. These characteristics are given in detail in Figure 5-18. The low power consumption achieved by CMOS circuits is due to the "complementary pairs" in each CMOS gate circuit. As was shown in Figure 5-17, during gate operation there is never a continuous path for current between the supply voltage and ground. When the P-channel device is OFF, $+V_{DD}$ is disconnected from the circuit, and when the N-channel device is OFF, ground is disconnected from the circuit. It is only when the output voltage switches from a HIGH to LOW, or LOW to HIGH, that the P-channel and N-channel E-MOSFET are momentarily ON at the same time. It is during this time of switching over from one state to the next that a small current will flow due to the complete path from $+V_{DD}$ to ground. This is why the power dissipation of CMOS logic gates increases as the operating frequency of the circuit increases.

CMOS Logic

A digital logic family that makes use of both P-channel and N-channel MOSFETs.

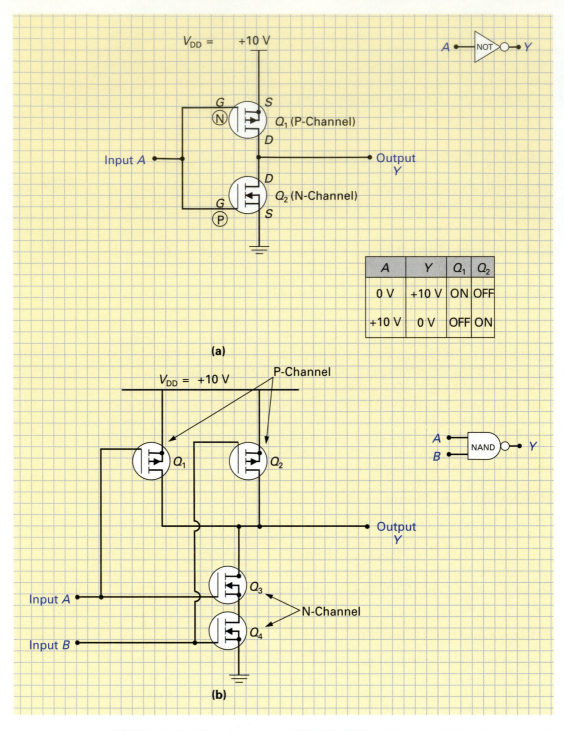

FIGURE 5-17 Complementary MOS (CMOS) Logic Gates. (a) An Inverter. (b) A NAND Gate.

The CMOS Series of ICs

There is a full line of CMOS ICs available that can provide almost all the same functions as their TTL counterparts. These different types are given different series numbers to distinguish them from others. Some of the more frequently used CMOS ICs are described in the following paragraphs.

Output Logic Levels:

HIGH Output Voltage – V_{OH}
$\left\{ \begin{array}{l} V_{OH(MAX)} \\ V_{OH(MIN)} \end{array} \right.$

LOW Output Voltage – V_{OL}
$\left\{ \begin{array}{l} V_{OL(MAX)} \\ V_{OL(MIN)} \end{array} \right.$

5 V — Valid Logic 1 or HIGH Output
4.9 V — Invalid Output
0.1 V — Valid Logic 0 or LOW Output
0 V

Input Logic Levels:

5 V — Valid Logic 1 or HIGH Input
3.5 V — Invalid Input
1.5 V — Valid Logic 0 or LOW Input
0 V

HIGH Input Voltage – V_{IH}
$\left. \begin{array}{l} V_{IH(MAX)} \\ V_{IH(MIN)} \end{array} \right\}$

LOW Input Voltage – V_{IL}
$\left. \begin{array}{l} V_{IL(MAX)} \\ V_{IL(MIN)} \end{array} \right\}$

Power Dissipation per Gate: 2.7 nW (static) to 170 μW (at 100 kHz)
Propagation Delay per Gate: 10 ns
Supply Voltage: +3 V to +6 V
Noise Immunity: Very High
Fanout: 10

FIGURE 5-18   **CMOS Logic Gate Characteristics (High-Speed CMOS—74HCXX).**

a. *4000 Series*   This first line of CMOS digital ICs was originally made by RCA and uses the numbers 40XX, with the number in the "XX" position indicating the circuit type. For example, the 4001 is a quad two-input NOR gate, while the 4011 is a quad two-input NAND gate. This series was eventually improved and called the 4000B series, and the original was labeled the 4000A series. Both use a V_{DD} supply of between $+3$ V and $+15$ V, with a logic 0 (LOW) = ⅓ of V_{DD} and logic 1 (HIGH) = ⅔ of V_{DD}. Propagation delay times for this series are typically 40 ns to 175 ns.

b. *40H00 Series*   This "high-speed" CMOS series improved the propagation delay to approximately 20 ns; however, it was still not able to match the speed of the bipolar TTL 74LS00 series.

c. *74C00 Series*   This series was designed to be "pin-compatible" with its TTL counterpart. This meant that all of the ICs used the same input and output pin numbers. Supply voltages can be anywhere from 3 V_{dc} to 15 V_{dc}, and propagation delays are long (90 ns, typically) and so applications are limited to signal frequency applications below 2 MHz. Low frequency and dc input signals result in almost no power dissipation; however, at high frequencies the gate power dissipation can exceed 10 mW to 15 mW.

d. *74HC00 and 74HCT00 Series*   This high-speed CMOS series of ICs matches the propagation delay times of the 74LS00 bipolar ICs and still maintains the CMOS advantage of low power consumption. It was introduced to overcome some of the disadvantages that plagued older CMOS series. The HC and HCT series are both pin-compatible with TTL. The HCT (high-speed CMOS TTL-compatible) series is a truly compatible CMOS line of ICs since it uses the same input and output voltage levels for a HIGH and LOW. This advantage makes it easier to swap a TTL IC for a CMOS IC and not cause any interfacing problems. The 74HC series logic gates can operate from a supply voltage of 2 V_{dc} to 6 V_{dc} (typically 5 V_{dc} so that it is identical to TTL levels). They have a typical gate propagation delay of about 20 ns and a power dissipation of about 1 mW. At a supply voltage of 5 V_{dc}, a 74HCT gate will typically have a propagation delay of 40 ns and a power dissipation of 1 mW.

e. *74AC00 and 74ACT00 Series*   This "advanced CMOS logic" series and "advanced CMOS, TTL-compatible" series has even better characteristics than the 74HC00 and 74HCT00 series. When operated from a supply voltage of 5 V_{dc}, a 74AC gate has a 3 ns propagation delay, and a 74ACT gate has a 5 ns propagation delay, enabling both families to operate beyond 100 MHz yet dissipate only 0.5 mW of power per gate. The 74AC series is not completely compatible with TTL; however, the 74ACT series is fully compatible.

5-2-4 MOSFET Handling Precautions

Certain precautions must be taken when handling any MOSFET devices. The very thin insulating layer between the gate and the substrate of a MOSFET can easily be punctured if an excessive voltage is applied. Your body can build up extremely large electrostatic charges due to friction. If this charge were to come in contact with the pins of a MOSFET device, an electrostatic-discharge (ESD) would occur, resulting in a possible arc across the MOSFET's thin insulating layer, causing permanent damage. Most MOSFETs presently manufactured have zeners internally connected between the gate and source to bypass high voltage static or in-circuit potentials to protect the MOSFET; however, it is important that

a. All MOS devices are shipped and stored in a "conductive foam" when not in use so that all of the IC pins are kept at the same potential, and therefore electrostatic voltages cannot build up between terminals.

b. When MOS devices are removed from the conductive foam, be sure not to touch the pins, since your body may have built up an electrostatic charge.

c. When MOS devices are removed from the conductive foam, always place them on a grounded surface such as a metal tray.

d. When continually working with MOS devices, use a wrist-grounding strap, which is a length of cable with a 1 MΩ resistor in series to prevent electrical shock if you were to come in contact with a voltage source.

e. All test equipment, soldering irons, and work benches should be properly grounded.

f. All power in equipment should be off before MOS devices are removed or inserted into printed circuit boards.

g. Any unused MOSFET terminals must be connected, since an unused input left open can build up an electrostatic charge and float to high voltage levels.

h. Any boards containing MOS devices should be shipped or stored with the connection side of the board in conductive foam.

SELF-TEST EVALUATION POINT FOR SECTION 5-2

Use the following questions to test your understanding of Section 5-2.

1. True or false: The E-MOSFET naturally operates as a voltage-controlled switch.
2. List three advantages that MOS logic gates have over bipolar logic gates.
3. Which type of E-MOSFETs are used in CMOS logic circuits?
4. What is the key advantage of CMOS logic?

5-3 DIGITAL IC PACKAGE TYPES AND COMPLEXITY CLASSIFICATION

There are four basic methods of packaging integrated circuits, and these are illustrated in Figure 5-19.

5-3-1 Early Digital IC Package Types

Transistor Outline
A package type used to house a transistor or circuit that resembles a small can.

Figure 5-19(a) shows the first type of package used for ICs, called the **transistor outline** or **TO can** package. This transistor package was modified from the basic three-lead transistor package to have the extra leads needed and is still used to house linear ICs due to its good

118 CHAPTER 5 / STANDARD LOGIC DEVICES (SLDs)

TO Can
(Transistor Outline)

(a)

Flat Pack

(b)

DIP (Dual-in-Line Package)

(c)

SOIC (Small Outline IC)
Package

PLCC (Plastic-Leaded
Chip Carrier)

LCCC (Leadless Ceramic
Chip Carrier)

(d) Surface-Mount Technology (SMT) Packages

FIGURE 5-19   IC Package Types.

heat dissipation ability but is rarely used for digital ICs. Figure 5-19(b) illustrates another early IC package called the flat pack. The leads of the package were designed to be soldered directly to the tracks of the printed circuit board. This package allowed circuits to be placed close together for small-size applications, such as avionics and military systems. The package is generally made of ceramic to withstand the high temperatures due to the densely packaged circuits.

5-3-2   *Present-Day Digital IC Package Types*

Figure 5-19(c) shows the dual-in-line package, or DIP, which has been a standard for many years. Called a DIP because of its two (dual) lines of parallel connecting pins, its leads feed through holes punched in the printed circuit board. Connection pads circle the holes both on the top and the bottom of the printed circuit board (PCB) so that when the DIP package is soldered in place it makes a connection.

Figure 5-19(d) illustrates the three types of **surface-mount technology (SMT)** packages. The key advantage of SMT over DIP is that a DIP package needs both a hole and a connecting pad around the hole. With SMT, no holes are needed since the package is mounted directly onto the printed circuit board, and the space can be saved when a surface-mount resistor is used instead of a through-hole axial lead resistor. Without the need for holes, pads can be placed closer together, resulting in a considerable space savings. Figure 5-19(d) shows the three basic types of SMT packages. The "small outline IC" (SOIC) package uses L-shaped leads, the "plastic-leaded chip carrier" (PLCC) uses J-shaped leads, and the "leadless ceramic chip carrier" (LCCC) has metallic contacts molded into its ceramic body.

**Surface-Mount
Technology**

A package type that has connections that connect to the surface of a printed circuit board.

5-3-3 *Digital IC Circuit Complexity Classification*

Although ICs are generally classified as being either linear (analog) or digital, another method of classification groups ICs based on their internal circuit complexity. The four categories are called small-scale integration (SSI), medium-scale integration (MSI), large-scale integration (LSI) and very-large-scale integration (VLSI).

Small-Scale Integration (SSI) ICs

These circuits are the simplest form of IC, such as single-function digital logic gate circuits. They contain less than 180 interconnected components on a single chip.

Medium-Scale Integration (MSI) ICs

These ICs contain between 180 and 1,500 interconnected components, and their advantage over an SSI IC is that the number of ICs needed in a system is less, resulting in a reduced cost and assembly time.

Large-Scale Integration (LSI) ICs

These circuits contain several MSI circuits all integrated on a single chip to form larger functional circuits or systems such as memories, microprocessors, calculators, and basic test instruments. These ICs contain between 1,500 and 15,000 interconnected components.

Very-Large-Scale Integration (VLSI) ICs

VLSI circuits contain extremely complex circuits such as large microprocessors, memories, and single-chip computers. These ICs contain more than 15,000 interconnected components.

SELF-TEST EVALUATION POINT FOR SECTION 5-3

Use the following questions to test your understanding of Section 5-3.

1. Give the full names for the following abbreviations:
   a. TO Can   **b.** DIP   **c.** SMT

2. Which package type saves the most printed circuit board space?

3. An IC containing four digital bipolar logic gates with about 44 components would be classified as a/an _____ IC.
   a. SSI   **b.** MSI   **c.** LSI   **d.** VLSI

4. The low circuit current and therefore low heat dissipation of _____ circuits make it possible for small-size circuits, and this is why _____ technology dominates the VLSI circuit market.

5-4   COMPARING AND INTERFACING BIPOLAR AND MOS LOGIC FAMILIES

Table 5-2(a) summarizes the differences between the two main digital IC logic families.

5-4-1 *The Bipolar Family*

The bipolar family of digital ICs has three basic types, called TTL (transistor-transistor logic), ECL (emitter-coupled logic), and IIL, or I^2L (integrated-injection logic). The characteristics of these three basic bipolar transistor logic gate types are summarized in

TABLE 5-2   **Comparing Logic Types. (a) Logic Families (b) TTL and CMOS Series**

	BIPOLAR FAMILY		
	LS TTL	**ECL**	**IIL**
Cost	Low	High	Medium
Fanout	20	10 to 25	2
Power Dissipation/Gate	2 mW	40 to 60 mW	0.06 to 70 μW
Propagation Delay/Gate	8 ns	0.5 to 3 ns	25 to 50 ns
Typical Input Signal Frequency	15 to 120 MHz	200 to 1000 MHz	1 to 10 MHz
External Noise Immunity	Good	Good	Fair–Good
Typical Supply Voltage	+5 V	−5.2 V	1 V to 15 V
Temperature Range	−55 to 125°C 0 to 70°C	−55 to 125°C	0 to 70°C
Internally Generated Noise	Medium–High	Low–Medium	Low

	MOS FAMILY		
	PMOS	**NMOS**	**HCMOS**
Cost	High	High	Medium
Fanout	20	20	10
Power Dissipation/Gate	0.2 to 10 mW	0.2 to 10 mW	2.7 nW to 170 μW
Propagation Delay/Gate	300 ns	50 ns	10 ns
Typical Input Signal Frequency	2 MHz	5 to 10 MHz	5 to 100 MHz
External Noise Immunity	Good	Good	Very Good
Typical Supply Voltage	−12 V	+5 V	+3 V to + 6 V
Temperature Range	−55 to 125°C 0 to 70°C	−55 to 125°C 0 to 70°C	−55 to 125°C −40 to 85°C
Internally Generated Noise	Medium	Medium	Low–Medium

(a)

Series	74	74L	74S	74LS	74AS	74ALS	74F	74C	4000B	74HC	74HCT	7AC	74ACT
Supply Voltage Range (V_{CC})	4.75 VDC to 5.25 VDC	4.75 VDC to 5.25 VDC	4.75 VDC to 5.25 VDC	4.75 VDC to 5.25 VDC	4.5 VDC to 5.5 VDC	4.5 VDC to 5.5 VDC	4.5 VDC to 5.5 VDC	3.0 VDC to 15.0 VDC	3.0 VDC to 18.0 VDC	2.0 VDC to 6.0 VDC	4.5 VDC to 5.5 VDC	3.0 VDC to 5.5 VDC	4.5 VDC to 5.5 VDC
Min. Logic-1 Input Voltage (V_{IN})	2.0 VDC	2.0 VDC	2.0 VDC	2.0 VDC	2.0 VDC	2.0 VDC	2.0 VDC	2.0 VDC	2.3 V_{CC}	3.15 VDC	2.0 VDC	3.15 VDC	2.0 VDC
Max. Logic-0 Input Voltage (V_{IL})	0.8 VDC	0.8 VDC	0.8 VDC	0.8 VDC	0.8 VDC	0.8 VDC	0.8 VDC	1.5 VDC	1.3 V_{CC}	0.9 VDC	0.8 VDC	1.35 VDC	0.8 VDC
Min. Logic-1 Output Voltage (V_{ON})	2.4 VDC	2.4 VDC	2.7 VDC	2.7 VDC	2.7 VDC	2.7 VDC	2.7 VDC	4.5 VDC	≈ V_{CC}	4.4 VDC	3.84 VDC	4.2 VDC	3.8 VDC
Max. Logic-0 Output Voltage (V_{OL})	0.4 VDC	0.4 VDC	0.5 VDC	0.4 VDC	0.4 VDC	0.4 VDC	0.5 VDC	0.5 VDC	≈0 VDC	0.1 VDC	0.33 VDC	0.5 VDC	0.5 VDC
Max. Logic-0 Input Current (I_G)	−1.6 mA	−0.18 mA	−2.0 mA	−0.36 mA	−0.5 mA	−0.1 mA	−0.6 mA	−0.5 nA	±1.0 μA	±1.0 μA	±1.0 μA	±1.0 μA	±1.0 μA
Max. Logic-1 Input Current (I_{IK})	40.0 μA	10.0 μA	50.0 μA	20.0 μA	20.0 μA	20.0 μA	20.0 μA	5.0 nA	±1.0 μA	±1.0 μA	±1.0 μA	±1.0 μA	±1.0 μA
Max. Logic-0 Output Current (I_{OS})	16.0 mA	3.6 mA	20.0 mA	4.0 mA	20.0 mA	8.0 mA	20.0 mA	0.4 mA	3.0 mA	20.0 μA	20.0 μA	24.0 mA	24.0 mA
Max. Logic-1 Output Current (I_{OH})	−400.0 μA	−200.0 μA	−1.0 mA	−400.0 μA	−2.0 mA	−400.0 μA	−1.0 mA	−0.36 mA	−3.0 mA	−20.0 μA	−20.0 μA	−24.0 mA	−24.0 mA
Propagation Delay (High to Low) (t_{PH})	8.0 ns	30.0 ns	5.0 ns	8.0 ns	1.5 ns	7 ns	3.7 ns	90.0 ns	50.0 ns	20.0 ns	40.0 ns	3.0 ns	5.0 ns
Propagation Delay (Low to High) (t_{PLH})	13.0 ns	60.0 ns	5.0 ns	8.0 ns	1.5 ns	5 ns	3.2 ns	90.0 ns	65.0 ns	20.0 ns	40.0 ns	3.0 ns	5.0 ns
Max. Operating Frequency (f_{max})	35 MHz	3 MHz	125 MHz	45 MHz	80 MHz	35 MHz	100 MHz	2 MHz	6 MHz	20 MHz	24 MHz	125 MHz	125 MHz
Power Dissipation Per Gate (mW)	10.0 mW	1.0 mW	20.0 mW	2.0 mW	4.0 mW	1.0 mW	4.0 mW	Note 1	Note 1	25.0 μW	80.0 μW	440 μW/ device	440 μW/ device

NOTE 1: Value is Frequency-Dependent

(b)

the upper table in Table 5-2(a). Generally, the power dissipation of TTL and ECL logic circuits limits their packing density and therefore they are only used for SSI and MSI circuit applications. On the other hand, I²L digital circuits are easy to combine with linear circuits (such as op-amps and oscillators) to create LSI and VLSI systems on a single chip.

5-4-2   *The MOS Family*

The lower table in Table 5-2(a) shows the three basic E-MOSFET digital IC types, which are more frequently called the MOS family of digital ICs. The three basic types of MOS ICs are called PMOS, NMOS, and CMOS. Comparing the characteristics, you can see that MOS logic gates use less space due to their simpler construction, have a very high noise immunity, and consume less power than equivalent bipolar logic gates. However, the high input impedance and capacitance due to the E-MOSFET's insulated gate means that time constants are larger and therefore transistor ON/OFF switching speeds are slower than equivalent bipolar gates. Both PMOS and NMOS ICs can be used in LSI and VLSI applications such as large memory circuits, single-chip microcomputers, and test instruments. On the other hand, CMOS ICs are generally used in SSI and MSI circuit applications.

The trade-offs, therefore, between the two logic families are

a. Bipolar circuits are faster than MOS circuits, but their circuits are larger and consume more power.

b. MOS circuits are smaller and consume less power than bipolar circuits, but they are generally slower.

5-4-3   *Interfacing Logic Families*

Interface

A device or circuit that connects an output to an input.

Compatible

In relation to ICs, it describes a circuit family that has input/output logic levels and operating characteristics that are compatible with other logic IC families.

Table 5-2(b) lists the key differences between the TTL series and CMOS series of digital ICs. In some circuit applications, there is a need to connect or **interface** one logic gate family type to another logic gate family type. For example, in some circuit applications we may need to interface a TTL logic gate to a CMOS logic gate or connect a CMOS gate to a TTL gate. In these instances, it is important that both devices are **compatible,** meaning that the HIGH and LOW logic levels at the output of the source logic gate will be of the correct amplitude and polarity so that they will be interpreted correctly at the input of the load logic gate. If we were to use the 74HCT00 CMOS series, there would be no interfacing problems at all, since this series generates and recognizes the TTL HIGH and LOW voltage levels (it is TTL-compatible). Other CMOS series, however, are not voltage- and current-compatible and therefore an interfacing device will have to be included between the two different types of logic gates to compensate for the differences.

Interfacing Standard TTL and 4000B Series CMOS

The voltage and current differences for the key TTL and CMOS series of logic gates are listed in Figure 5-20(a). Figure 5-20(b) shows the incompatibility between the HIGH and LOW voltage levels of standard TTL series and 4000B CMOS series logic gates. A LOW logic level output voltage from the TTL logic gate will be recognized at the CMOS input as a LOW, since even a worst-case LOW output of 0.8 V is still less than the worst-case LOW level CMOS input of 1.67 V. The problem arises with a HIGH output from the TTL gate, which even if it is at 2.4 V, will still be less than the minimum acceptable CMOS input voltage of 3.33 V. To compensate for this problem, a 10 kΩ pull-up resistor is connected between the TTL and CMOS gate, as shown in Figure 5-20(c). This pull-up resistor will pull a HIGH output from the TTL gate up towards 5 V so that it will be recognized at the input as a HIGH by the CMOS gate.

74TTL	74LSTTL	74ALSTTL	Parameter	4000B CMOS	74HCMOS	74HCTMOS
2.0	2.0	2.0	V_{IH} (min) (V)	3.33	3.5	2.0
0.8	0.8	0.8	V_{IL} (max) (V)	1.67	1.0	0.8
2.4	2.7	2.7	V_{OH} (min) (V)	4.95	4.9	4.9
0.4	0.4	0.4	V_{OL} (max) (V)	0.05	0.1	0.1
40	20	20	I_{IH} (max) (μA)	1	1	1
−1600	−400	−100	I_{IL} (max) (μA)	−1	−1	−1
−0.4	−0.4	−0.4	I_{OH} (max) (mA)	−0.51	−4	−4
16	8	4		0.51	4	4

(a)

(b)

(c)

(d)

FIGURE 5-20 Interfacing a TTL Gate and a 4000B CMOS Gate that Are Using the Same Supply Voltage Value.

Figure 5-20(d) shows the interfacing device that needs to be included when connecting a CMOS gate to a TTL gate. Referring to Figure 5-20(b), you can see that a HIGH output voltage of 4.95 V will be easily recognized as a HIGH by the TTL gate, and a LOW of 0.05 V will also easily be recognized as a LOW by the TTL gate. The voltage levels in this instance are not the problem, but the current levels are because CMOS devices are low-current devices. For example, referring to Figure 5-20(d) you can see that the LOW output current of the 4069 CMOS gate (0.51 mA) is not sufficient to drive the TTL LOW input current requirement (1.6 mA). Similarly, the HIGH output current of the 4069 CMOS gate (0.51 mA) is not sufficient to drive the TTL HIGH input current requirement (40 µA). Including the 40508 CMOS buffer/driver will ensure current compatibility, as shown in Figure 5-20(d), since the output LOW and HIGH current values from this buffer are more than sufficient for two standard TTL loads.

Interfacing TTL and 4000B CMOS Using Level-Shifters

In the previous example, the 4000B CMOS gates and the standard TTL gates were using the same supply voltage (5 V). The 4000B CMOS series can use a supply voltage that is anywhere between +3 V and +15 V. To interface two different logic gate types with different supply voltages, we will have to include a **level-shifter** or **translator,** as shown in Figure 5-21. The 4050B level-shifter will convert 0 V to 15 V CMOS logic to 0 V to 5 V TTL logic, as shown in Figure 5-21(a), enabling us to interface CMOS to TTL. The 4504B level-shifter

FIGURE 5-21   Using Level-Shifters or Translators to Interface TTL and 4000B CMOS Gates with Different Supply Voltages.

FIGURE 5-22 Using Level-Shifters or Translators to Interface ECL and TTL.

will convert 0 V to 5 V TTL logic to 0 V to 15 V CMOS logic, as shown in Figure 5-21(b), enabling us to interface TTL to CMOS.

Interfacing ECL and TTL Using Level-Shifters

The 0 V to +5 V logic levels of TTL are very much different than the −5.2 V to 0 V logic levels of ECL. Figure 5-22(a) shows how a 10124 level-shifter can be used to interface a TTL logic gate to an ECL logic gate, and Figure 5-22(b) shows how a 10125 level-shifter can be used to interface an ECL logic gate to a TTL logic gate.

5-4-4 *Other Logic Gate Families*

Semiconductor manufacturers are continually searching for new semiconductor circuits that have faster propagation delay times (since this enables them to operate at higher frequencies), lower power consumption, better noise immunity, and increased packing density. Some of the new hopefuls are gallium arsenide (GaAs) circuits, BiCMOS circuits (which is a combination of bipolar and CMOS on the same chip, an example of which was featured in Figure 5-12), silicon-on-sapphire (SOS) circuits, and Josephson junction circuits. It is therefore important in this ever-changing semiconductor industry that you read electronics magazines and manufacturers' technical publications to stay abreast of the latest technological advances. Most of the semiconductor manufacturers have their own home page on the Internet with access to all data sheets, circuit applications, technical support, and new product information.

SELF-TEST EVALUATION POINT FOR SECTION 5-4

Use the following questions to test your understanding of Section 5-4.

1. What is the speed/power trade-off with regard to the selection of TTL logic gates?
2. Which logic gate type is probably the best all-round logic because of its low power consumption, very high noise immunity, wide power supply voltage range, and high fanout?
3. What interfacing problems would occur between a 74HCT00 gate and a 74ALS00 gate?
4. What is a level-shifter gate?
5. Which logic family combines both bipolar and CMOS on a single chip?
6. Which logic gate series would be best suited if a circuit is processing input signal frequencies in the 110 MHz range?

REVIEW QUESTIONS

Multiple-Choice Questions

1. Which stage within a standard TTL gate provides a low output impedance for both a HIGH and LOW output?
   a. Multiple-emitter
   b. Phase splitter
   c. Totem pole
   d. Injection emitter

2. A floating TTL input has exactly the same effect as a/an
   a. LOW input.
   b. HIGH input.
   c. invalid input.
   d. both (a) and (c).

3. A standard TTL logic gate can sink a maximum of _____ and source a maximum of

   _____.
   a. 400 µA, 1.6 mA
   b. 40 µA, 16 mA
   c. 16 mA, 400 µA
   d. 1.6 mA, 40 µA

4. Which TTL logic gates require an external pull-up resistor?
   a. ECL
   b. Open-collector
   c. Schottky TTL
   d. Both (a) and (b)

5. The output current of a standard TTL gate is _____ times greater than the input current, while the output current of a buffer/driver TTL gate is typically _____ times greater than the input.
   a. 10, 30
   b. 10, 15
   c. 30, 10
   d. 15, 30

6. _____ logic gates are generally incorporated in digital circuits to sharpen the rise and fall times of input pulses.
   a. Buffer/driver
   b. Schmitt trigger
   c. Open-collector
   d. Tri-state

7. The fanout of a standard TTL logic gate is

   _____.
   a. 10
   b. 20
   c. 2
   d. 40

8. The output current of a standard TTL logic gate is _____ times greater than the input current.
   a. 10
   b. 20
   c. 2
   d. 40

9. Which is the fastest of all the logic gate types?
   a. CMOS
   b. Open-collector
   c. Schottky TTL
   d. ECL

10. Which of the following CMOS series types is both pin- and voltage-compatible with TTL?
    a. 4000
    b. 74C00
    c. 74HCT00
    d. 74HC00

11. Which logic gate type contains both a P-channel and N-channel E-MOSFET?
    a. CMOS
    b. NMOS
    c. PMOS
    d. TREEMOS

12. A _____ _____ gate is used to convert a sine wave into a rectangular wave or to sharpen the rise and fall times of a rectangular wave.
    a. open-collector
    b. high-speed
    c. Schmitt trigger
    d. MOS logic

13. _____ circuits have a combination of bipolar and CMOS logic circuits on the same chip.
    a. GaAs
    b. SOS
    c. Josephson junction
    d. BiCMOS

14. _____ circuits can combine linear circuits (such as amplifiers and oscillators) with digital circuits to create complete electronic systems on one chip.
    a. CMOS
    b. IIL
    c. ECL
    d. NMOS

15. _____ circuits are faster, but their circuits are larger and consume more power, whereas _____ circuits are smaller and consume less power, but they are generally slower.
    a. Bipolar, MOS
    b. CMOS, ECL
    c. MOS, CMOS
    d. MOS, bipolar

Practice Problems

16. Referring to the TTL logic circuit shown in Figure 5-23, complete the truth table.

17. Sketch the logic symbol and describe the logic function being performed by the logic circuit in Figure 5-23.

18. What circuit arrangement is formed by Q_5 and Q_6 in Figure 5-23, and what is the circuit's purpose?

19. What does the symbol within the logic gate in Figure 5-24 indicate?

20. Referring to Figure 5-24(a), complete the truth table.

21. Referring to Figure 5-24(a), determine whether the 7401 can handle the LOW output sinking current for this circuit (assuming a 2 V LED drop).

FIGURE 5-23   A Standard TTL Logic Gate Circuit.

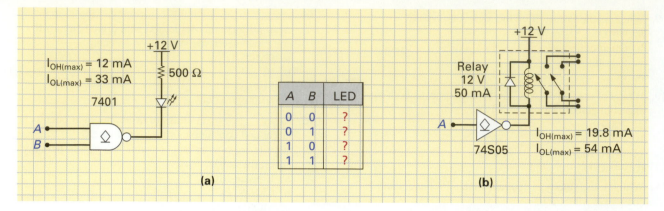

(a)

(b)

FIGURE 5-24   Application—Controlling an LED with a Logic Gate.

A	B	LED
0	0	?
0	1	?
1	0	?
1	1	?

22. Will the 74LS05 be able to handle the LOW output sinking current needed to energize the 12 V, 50 mA relay?

23. What open-collector output gate in Figure 5-10 would be ideal for a greater-than-50 mA load?

24. Identify the logic symbol used within the logic gate in Figure 5-25(a).

25. Referring to Figure 5-25(a), complete the truth table given.

26. Referring to Figure 5-25(b), complete the truth table given.

27. What is the difference between the 74LS125 shown in Figure 5-25(a) and the 74LS126 shown in Figure 5-25(b)?

28. Identify the logic symbol used within the logic gate in Figure 5-26(a).

29. Describe the operation of the complete circuit shown in Figure 5-26(a).

30. Comparing the inputs to the outputs in Figure 5-26(b), (c), (d), and (e), describe the circuit function being performed.

31. What type of E-MOSFET is being used in the logic circuit in Figure 5-27?

32. Complete the truth table for the logic gate shown in Figure 5-27.

33. Sketch the logic symbol and describe the logic function being performed by the logic circuit in Figure 5-27.

Web Site Questions

Go to the web site http://www.prenhall.com/cook, select the textbook *Pratical Digital Electronics,* select this chapter, and then follow the instructions when answering the multiple choice practice problems.

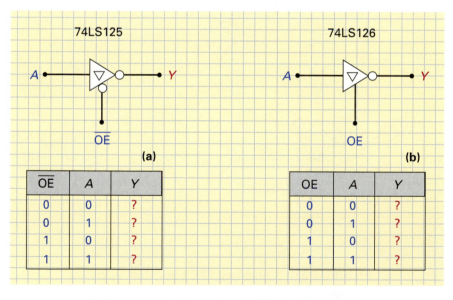

(a)

(b)

$\overline{OE}$	A	Y
0	0	?
0	1	?
1	0	?
1	1	?

OE	A	Y
0	0	?
0	1	?
1	0	?
1	1	?

FIGURE 5-25   Logic Gates With Control Inputs.

(a)

(b) Wave Shaper

(c) Pulse Restorer

(d) Noise Eliminator

(e) Threshold Detector

FIGURE 5-26   Applications—Logic Gate Circuits.

A	B	Y
0	0	?
0	1	?
1	0	?
1	1	?

FIGURE 5-27   A MOS Logic Gate.

Programmable Logic Devices (PLDs)

Don't Mention It!

Two years before the Japanese attacked Pearl Harbor, IBM president Thomas J. Watson decided to invest half a million dollars of the company's money in a project conceived by a young Harvard mathematician named Howard Aiken. Aiken's plan was to ignore the architecture of the sorters and calculators that were then available and concentrate on building a machine based on the nineteenth century concepts of Charles Babbage. Using Babbage's original description of his "analytical engine," Aiken set about to build a general-purpose programmable computer. In fact, Aiken stated at the time that "If Babbage had lived seventy-five years later, I would be out of a job."

The Second World War hampered Aiken's development of his machine because he was called to active duty in the navy soon after the attack on Pearl Harbor. In the navy, Aiken distinguished himself by disarming a new type of German torpedo single-handed. Meanwhile, at IBM, Watson did not appreciate the interference of the war and used his influence to get Aiken back on his computer project, arguing that it could greatly help the war effort.

In 1943, the Mark I computer was turned on and underwent its first successful test. It measured 51 feet long and 8 feet high, contained 750,000 devices, and 500 miles of wire. The 3,304 electromechanical relays clicked open and closed, and as one observer remarked, "The machine sounded like a room full of old ladies knitting away with steel needles." At this point, personality clashes began to erupt between the Mark I's inventor and the financial backer. Aiken wanted the Mark I's internal circuitry exposed so that scientists could examine it, while Watson wanted the machine encased in glass and stainless steel. Watson won on this point; however, Aiken retaliated by not even mentioning IBM or Watson's involvement in the project when the Mark I was introduced to the press in 1944.

The Mark I was leased to the navy and used to calculate the trajectory of cannon shells. It could add or subtract 23-digit numbers in a few tenths of a second, and multiply them in three seconds. This phenomenal speed—which was only slightly faster than Babbage had imagined 122 years earlier—enabled the navy, in a single day, to speed through calculations that previously took six months.

After the war Aiken stayed at Harvard to develop the second, third, and fourth generations of the Mark I, and had nothing further to do with Watson. Tom Watson was so angered by Aiken's failure to acknowledge him and IBM that he ordered all of his researchers to construct a better and faster system. Revenge launched IBM into the computer business and, when Watson died at the age of 82 in 1956, he had seen his wish fulfilled. IBM had overtaken all opposition, and the name had become so synonymous with computers that most people believed that IBM had invented them.

Introduction

Digital Technology has advanced, and continues to advance, almost meteorically each and every year. Proof of this is easily evident when we see the performance of personal computers (PCs) advance in leaps and bounds. To quote a recently posted statistic, microprocessor ICs are improving at a rate of 60% per year, while memory ICs are quadrupling their capacity every three years. These and other rapid changes in all the key building blocks of digital electronic systems mean that consumer products are generally obsolete in less than three years.

To keep up with this fast pace, electronic companies have to design and manufacture new products in a cycle of typically less than six months. To meet this accelerated schedule, engineers and technicians have looked for shortcuts that enable them to construct a digital prototype circuit and evaluate its performance in a much more timely manner.

In the previous chapter, you were introduced to most of the **standard logic devices (SLDs).** To construct a circuit using these ICs, you would first need to insert them into a protoboard and then connect them with a spaghetti-like maze of hookup wire. In this chapter, you will be introduced to **programmable logic devices (PLDs),** and as the industry has, you will see how much easier it is to construct your digital lab circuits using the highly versatile PLD.

Standard Logic Devices (SLDs)
Digital logic ICs that have a fixed design.

Programmable Logic Devices (PLDs)
Digital logic ICs that can have their function changed through programming.

To clearly demonstrate the PLD advantage, let us compare the differences that occur when we construct a circuit using standard logic devices and then the same circuit using a programmable logic device.

6-1-1   *Constructing a Circuit Using Standard Logic Devices*

As an example, Figure 6-1(b) shows an application in which two NOT gates, four AND gates, and four OR gates are connected to achieve the logic function described in the truth table in Figure 6-1(a). The two NOT gates and four AND gates form a 1-of-4 decoder. This decoder will make only one of its four outputs HIGH based on the binary value applied to inputs A and B, as shown in the table in Figure 6-1(b). When $AB = 00$, the AND gate 0's output is HIGH and, since a connection exists to the inputs of OR gates 0 and 1, the output will be $Q_3, Q_2, Q_1, Q_0 = 0011$. Referring to the second line of the truth table in Figure 6-1(a), when $AB = 01$, AND gate 1's output is HIGH and, since a connection exists to the inputs of OR gates 2 and 3, the output is $Q_3, Q_2, Q_1, Q_0 = 1100$. When $AB = 10$, AND gate 2's output is HIGH, giving a HIGH to the input of OR gate 0 and an output of 0001. Finally, when $AB = 11$, AND gate 3's output is HIGH, giving a HIGH to OR gate 2 and an output of 0100.

Figure 6-1(c) shows how this circuit could be constructed on a protoboard. Once completed, switches should be connected to the inputs and LEDs to the outputs so that the circuit can be tested to see if it performs the desired logic function specified in the function table.

This standard logic prototyping method has the following disadvantages:

- Hookup wire cutting and stripping is time-consuming.
- Wires can easily be inserted incorrectly, causing possible device damage and lengthy delays while the errors are isolated.
- A large and costly inventory of all standard logic ICs must be maintained.
- If the desired standard logic IC is not available, further delays will result.
- To modify or add to a working circuit, the wires and ICs will normally all have to be removed from the protoboard and the new design built again from scratch.

Like industry, you have had a good taste of this frustrating practice when you have had to construct the hands-on lab experiments and are probably quite ready to welcome an alternative.

6-1-2   *Constructing a Circuit Using Programmable Logic Devices*

By using an inexpensive **personal computer (PC),** a **computer-aided design (CAD)** software program, and a programmable logic device (PLD) IC, you can easily prototype a digital circuit.

Figure 6-2(a) and (b) shows the same truth table and application circuit from Figure 6-1, but in this instance a programmable logic device has been used to construct the circuit, as shown in Figure 6-2(c).

The single PLD contains a large inventory of logic gates, along with interconnecting devices, all within a single IC. Using the PC, you can write logic programs in a **hardware description language (HDL)** using a **text editor** or simply draw your logic circuit using a **schematic editor.** The HDL or schematic is then **compiled** by the CAD program to create a detailed logic circuit that will perform the actions you specified in your original HDL or schematic. The operation of this circuit can then be simulated to make sure that it is functioning as it should. If the **circuit simulation** is successful, the

---

Personal Computer

Abbreviated PC, it is a data processing digital system for personal use.

Computer-Aided Design (CAD)

Application software used in conjunction with a PC for design.

Hardware Description Language (HDL)

A programming language used to specify hardware designs.

Text Editor

An element of a software program used to design a circuit using a hardware description language.

Schematic Editor

An element of a software program used to graphically design a circuit.

Compiled

A term used to describe a process in which a program is converted into a format to make it compatible with the device it is being sent to.

Circuit Simulation

A process in which a theoretical model of a circuit design is checked to see how its outputs will respond to all possible inputs.

FIGURE 6-1 Constructing a Circuit Using Standard Logic Devices (SLDs).

design is then **downloaded** through the parallel port of the PC into the PLD. Switches are then connected to the inputs and LEDs are connected to the outputs to make a final test of the prototype.

In summary, the five-step process for creating a prototype using a PLD is

PLD Prototyping Process

Step 1: Create the new circuit using the software's schematic editor.

Step 2: Compile the circuit into a bitstream file that when loaded into the CPLD, will instruct it to act like the entered schematic.

Downloaded

A term used to describe a transfer of data from a computer to a destination.

(a)

Inputs		Outputs			
A	B	Q_3	Q_2	Q_1	Q_0
0	0	0	0	1	1
0	1	1	1	0	0
1	0	0	0	0	1
1	1	0	1	0	0

Function Table

A	B	High Out
0	0	AND 0
0	1	AND 1
1	0	AND 2
1	1	AND 3

(b)

(c)

FIGURE 6-2 **Constructing a Circuit Using Programmable Logic Devices (PLDs).**

Step 3: Verify the operation of your circuit using the software's functional and timing simulator.

Step 4: Download the circuit file from the PC to the PLD.

Step 5: Physically test the PLD by activating its inputs and monitoring its outputs.

This programmable logic prototyping method has the following advantages:

- With manual wiring reduced to a minimum, prototypes can be constructed, tested, and modified at a much faster rate.
- Wiring errors can be avoided.

- You can experiment with many digital IC types without having to stock them in your supply cabinet.
- Circuit designs can be saved as electronic files within the PC and used again when needed.
- Since the PLD can be used over and over, modifications can easily be made by altering the circuit in the PC and then downloading the new design into the PLD.
- Larger and more complex projects can be undertaken now that the tedious manual procedures are automated.

SELF-TEST EVALUATION POINT FOR SECTION 6-1

Use the following questions to test your understanding of Section 6-1.

1. What are the full names and brief definitions of the following abbreviations:
   - **a.** PLD
   - **b.** SLD
   - **c.** HDL
   - **d.** CAD
   - **e.** PC
2. What are five disadvantages of constructing a digital circuit prototype using standard logic devices?
3. What are five advantages of constructing a digital circuit prototype using programmable logic devices?
4. What are the basic steps you would follow to create and load a digital circuit into a PLD?

6-2 TYPES OF PROGRAMMABLE LOGIC DEVICES

The programmable logic devices of today evolved from the **programmable logic array (PLA),** which was first introduced in the early 1970s. To understand the advantages of today's powerful PLDs, let us first study the limitations of the first PLD devices.

6-2-1 *Early Programmable Logic Devices*

The basic programmable logic array, or PLA, contains a set of NOT gates, AND gates, and OR gates, all interconnected with programmable switches, as shown in Figure 6-3(a). Although it was an extremely flexible arrangement, having programmable AND gate inputs and programmable OR gate inputs was unnecessary and costly, and so industry soon followed up with the simpler **programmable array logic (PAL)** structure, shown in Figure 6-3(b). With the PAL, the AND gate inputs were programmable but the OR gate inputs were not.

The PLA and PAL gave the technician and engineer a versatile device that could be customized to fit any application. However, their small size meant that they still had to be placed on a protoboard and wired to each other, and so they still had many of the same disadvantages of standard logic devices.

Their other disadvantage was that they were **one-time programmable (OTP).** This meant that they included specialized circuitry within the IC that used a high-voltage input to burn out the unwanted fuses at the wire cross points. Once the fuses were blown, they remained that way, and so they could only be used once.

6-2-2 *Today's Programmable Logic Devices*

To combat the size limitation of the early PLDs, IC manufacturers came up with two extremely large PLD types, called **complex programmable logic devices (CPLDs)** and **field-programmable gate arrays (FPGA).** Figure 6-4 shows the internal structure of

Programmable Logic Array (PLA)

A PLD that has both a programmable AND array and a programmable OR array.

Programmable Array Logic (PAL)

A PLD that has a programmable AND array driving a hardwired OR array.

One-time Programmable (OTP)

A device that, after the initial programming, cannot be reprogrammed.

Complex Programmable Logic Devices (CPLDs)

A digital IC type that has several PLD blocks and a global interconnection matrix.

Field-Programmable Logic Arrays (FPGAs)

A digital IC type that has several programmable logic blocks that are interconnected by switches.

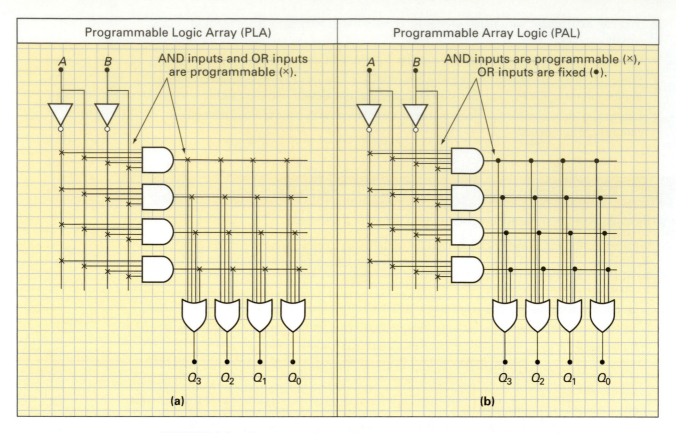

FIGURE 6-3 First Types of Programmable Logic Devices. (a) PLA. (b) PAL.

these two basic types, which can literally have a complete digital system programmed into a single IC.

The CPLD shown in Figure 6-4(a) contains several PLD blocks and a global interconnection matrix that determines the connection between the PLD blocks. To program a CPLD, you must first program each PLD block and then program the interconnections between the PLD blocks.

The FPGA shown in Figure 6-4(b) has a different internal structure. It has several programmable logic blocks that are interconnected by switches. To program an FPGA, you

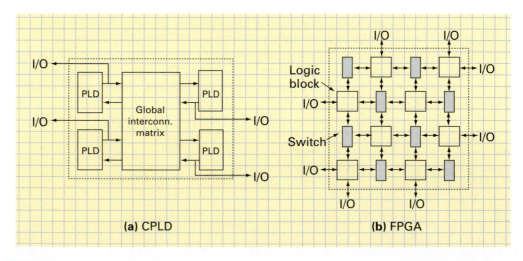

FIGURE 6-4 Programmable Logic Device Types. (a) Complex Programmable Logic Device. (b) Field Programmable Gate Array.

would first program the logic blocks to perform a logic function, such as AND, NOR, or XOR, and then program the switches to connect the logic blocks.

To overcome the one-time programmable disadvantage of the early PLDs, today's CPLDs and FPGAs contain reprogrammable switches at the cross points, which enable them to be used over and over.

SELF-TEST EVALUATION POINT FOR SECTION 6-2

Use the following questions to test your understanding of Section 6-2.

1. What are the full names and brief definitions of the following abbreviations:
   a. PLA   **d.** CPLD
   b. PAL   **e.** FPGA
   c. OTP

2. Which inputs are fuse-programmable in the PLA, and which are fuse-programmable in the PAL?

3. What are the difference(s) between the internal architecture of the CPLD and the FPGA?

4. What two key advantages do the present-day CPLDs and FPGAs have over the early PLAs and PALs?

REVIEW QUESTIONS

Multiple-Choice Questions

1. The abbreviation PLD stands for
   a. Program Logic Display
   b. Programmable Language Decoder
   c. Programmable Logic Device
   d. Programming Language Device

2. Which of the following would be considered disadvantages of the standard logic prototyping method?
   a. Wires can easily be inserted incorrectly, causing possible device damage and lengthy delays while the errors are isolated.
   b. A large and costly inventory of all standard logic ICs must be maintained.
   c. If the desired standard logic IC is not available, further delays will result.
   d. To modify or add to a working circuit, the wires and ICs will normally all have to be removed from the protoboard and the new design built again from scratch.
   e. All of the above.

3. The abbreviation CAD stands for
   a. Computer-Assisted Designer
   b. Code Activated Device
   c. Computer-Aided Design
   d. Computer-Activated Decoder

4. What three key devices are needed to construct a circuit using a programmable logic device?
   a. CAD, PLD, PC
   b. PLD, PC, SLDs
   c. PC, SLDs, CAD
   d. All of the above.

5. The abbreviation HDL stands for
   a. Hardware Description Language
   b. Hexadecimal Decoder Logic
   c. Hard Description Language
   d. Hardware Decoder Logic

6. Using the PC, you can write logic programs in a _____ using a text editor or simply draw your logic circuit using a _____.
   a. schematic editor, HAL
   b. PAL, HDL
   c. HDL, schematic editor
   d. HDL, PLA

7. Once the logic program or logic circuit has been created, it is then _____ by the CAD program to create a detailed logic circuit that will perform the actions you specified.
   a. compiled
   b. simulated
   c. downloaded
   d. tested

8. What is the correct order for prototyping a digital circuit using a PLD?
   a. Create, simulate, compile, download, test.
   b. Create, compile, simulate, download, test.
   c. Create, simulate, download, compile, test.
   d. Compile, create, simulate, download, test.

9. Which of the following would be considered advantages of the programmable logic prototyping method?
   a. Wiring errors can be avoided.
   b. You can experiment with many digital IC types without having to stock them in your supply cabinet.
   c. Circuit designs can be saved as electronic files within the PC and used again when needed.
   d. Since the PLD can be used over and over again, modifications can easily be made by altering the circuit in the PC and then downloading the new design into the PLD.
   e. All of the above.

10. The programmable logic devices of today evolved from the _____, which was first introduced in the early 1970s.
   a. PAL   c. PLA
   b. CPLD   d. FPGA

11. With the _____, the AND gate inputs were programmable but the OR gate inputs were not.
   a. PAL   c. PLA
   b. CPLD   d. FPGA

12. With _____ devices, once the fuses were blown they remained that way, and so they could only be used once.
   a. OTP   c. CAD
   b. HDL   d. PLD

13. The _____ contains several PLD blocks and a global interconnection matrix that determines the connection between the PLD blocks.
   a. PAL   c. PLA
   b. CPLD   d. FPGA

14. The _____ has several programmable logic blocks that are interconnected by switches.
   a. PAL   c. PLA
   b. CPLD   d. FPGA

15. To program a(n) _____ you would first program the logic blocks to perform a logic function, such as AND, NOR, or XOR, and then program the switches to connect the logic blocks.
   a. PAL   c. PLA
   b. CPLD   d. FPGA

16. To program a _____ you must first program each PLD block and then program the interconnections between the PLD blocks.
   a. PAL   c. PLA
   b. CPLD   d. FPGA

17. To overcome the OTP disadvantage of the early PLDs, today's CPLDs and FPGAs contain _____ at the cross points, which enable them to be used over and over again.
   a. programmable fuses
   b. reprogrammable switches
   c. reprogrammable circuit breakers
   d. Both a and c are true.

Practice Problems

To practice your device recognition and operation skills, refer to Figure 6-5 and answer the following questions.

18. How many inputs are being applied to the PLD, and how many outputs from the PLD are being applied to the 7-segment display?

19. Referring to the PLD's internal schematic diagram shown in part (b), would you say that this PLD is a PLA or a PAL?

20. What is the function of this PLD?

21. Describe the function table in part (a).

22. What do a through f refer to?

23. What would be the a through f outputs if the DCBA input is
   a. 0011   c. 0101
   b. LHLH   d. HLLH

24. What digit will be shown on the 7-segment display if the following DCBA inputs were applied:
   a. 0011   c. 0101
   b. LHLH   d. HLLH

25. Describe the similarity between the function table and the intact fuses in the PLD's schematic diagram.

26. What is the purpose of resistors R_1 through R_7?

Web Site Questions

Go to the web site http://www.prenhall.com/cook, select the textbook *Practical Digital Electronics,* select this chapter, and then follow the instructions when answering the multiple choice practice problems.

INPUTS				Digit Displayed	OUTPUTS						
D	C	B	A		a	b	c	d	e	f	g
L	L	L	L	0	ON	ON	ON	ON	ON	ON	OFF
L	L	L	H	1	OFF	ON	ON	OFF	OFF	OFF	OFF
L	L	H	L	2	ON	ON	OFF	ON	ON	OFF	ON
L	L	H	H	3	ON	ON	ON	ON	OFF	OFF	ON
L	H	L	L	4	OFF	ON	ON	OFF	OFF	ON	ON
L	H	L	H	5	ON	OFF	ON	ON	OFF	ON	ON
L	H	H	L	6	ON	OFF	ON	ON	ON	ON	ON
L	H	H	H	7	ON	ON	ON	OFF	OFF	OFF	OFF
H	L	L	L	8	ON	ON	ON	ON	ON	ON	ON
H	L	L	H	9	ON	ON	ON	OFF	OFF	ON	ON

ON = Logic 0
OFF = Logic 1

(a)

x = Fuse is intact, and therefore making connection.

(b)

TIL 312

7SEG, Common Anode Display

Segments activated to display digits 0 through 9

(c)

FIGURE 6-5   Application—Using a PLD to Drive a 7-Segment LED Display.

Testing and Troubleshooting

Space, the Final Frontier

On May 24, 1962 a deep silence and tension filled the Mercury Control Center at Cape Canaveral, Florida. The only voice to be heard was that of Gus Grissom as he repeatedly tried to make contact with fellow astronaut Scott Carpenter aboard the *Aurora 7,* which had just completed three orbits around the earth and was now attempting a reentry. The normal procedure at this stage of the mission is to fire the thrusters to correctly orient the craft to the proper angle for returning to earth. The next step is to slow down the capsule by applying the brakes or retrorockets, which causes the spacecraft to fall into the atmosphere and to earth. The final step is to pop open the parachutes at about 25,000 feet and lower the capsule into the sea. All of these steps are crucial, and if any one of the steps is not followed exactly, it can spell disaster. For example, if the spacecraft is pitched at the wrong angle, or if it is traveling at too great a speed, the craft and astronaut will burn up due to the friction of the earth's atmosphere.

This was the second American orbital flight; however, unlike the first, this mission had not gone at all according to plan. The problems began when Carpenter was asked to position *Aurora 7* for the return and the capsule's automatic stabilization system had failed. Carpenter responded immediately by switching to manual control, but due to the malfunction, the craft was misaligned by 25 degrees. To make matters worse, the retrorockets fired three seconds too late, failing to produce the expected braking power.

Nine minutes had now passed and Gus Grissom at the Cape urgently repeated the same message into his microphone, not knowing whether Carpenter was dead or alive: "*Aurora 7, Aurora 7,* do you read me? Do you read me?"

Hundreds of miles away, a pair of mainframe computers at Goddard Space Flight Center in Maryland knew exactly where he was. These machines had continually digested the direction and altitude data from a radar ground station in California that had tracked the capsule from its time of reentry. Eleven minutes after the last radar contact, the mainframe displayed an incredibly accurate estimate of the capsule's splashdown position. Aircraft were immediately dispatched and homed in on the capsule's signal beacon, where Carpenter was found riding in an inflatable life raft alongside his spacecraft. Luckily, the combined effects of the errors had brought Carpenter safely through the atmosphere, but some 250 miles away from the planned splashdown point and well out of range of Grissom's radio transmission and the awaiting naval recovery task force.

At this time in history, the computer was not completely accepted as a valued partner in the space business. Astronauts, aerospace engineers, and ground control personnel were wary of putting too much trust in a machine that at that time would sometimes fail to operate for no apparent reason.

Today, the computer has proved itself indispensable, lending itself to almost every phase of our exploration of the final frontier. Computers are used by aerospace engineers to design space vehicles and to simulate the designs to determine their flight characteristics. These digital electronic computer systems can also simulate missions into deep space;

an astronaut can take a flight simulation to almost any place in the galaxy, developing good flight skills on the way. Computers are also used to continually monitor the thousands of tests preceding a launch and will stop the countdown for any abnormality. Once the mission is underway, the on-board computer supplies the astronaut quickly and accurately with all information needed, enabling mankind to explore new worlds and new civilizations and to boldly go where no one has gone before!

Introduction

This chapter will introduce you to digital test equipment and digital troubleshooting techniques. An effective electronics technician or troubleshooter must have a thorough knowledge of electronics, test equipment, troubleshooting techniques, and equipment repair. Like analog circuits, digital circuits occasionally fail, and in most cases a technician is required to quickly locate the problem within the system and then make the repair. The procedure for fixing a failure can be broken down into three basic steps.

Step 1:   DIAGNOSE
       The first step is to determine whether a problem really exists. To carry out this step, a technician must collect as much information about the system, the circuit, and the components used, and then diagnose the problem.

Step 2:   ISOLATE
       The second step is to apply a logical and sequential reasoning process to isolate the problem. In this step, a technician will operate, observe, test, and apply troubleshooting techniques in order to isolate the malfunction.

Step 3:   REPAIR
       The third and final step is to make the actual repair and final test the circuit.

      Troubleshooting, by definition, is the process of locating and diagnosing breakdowns in equipment by means of systematic checking and analysis. To troubleshoot, you will need a thorough knowledge of troubleshooting techniques, a very good understanding of test equipment, documentation in the form of technical and service manuals, and experience. Troubleshooting experience can only be acquired with practice; therefore, in future chapters, complete troubleshooting sections will be included and applied to all new devices and circuits. As for practical experience, very few of your lab circuits will work perfectly the first time, so you will gain troubleshooting experience each and every time you experiment in lab. Although it seems very frustrating when a circuit is not operating correctly, remember that the more problems you have with a circuit, the better.

      In this chapter, you will first be introduced to digital test equipment, and then in the following section you will see how these test instruments can be used to isolate internal and external digital IC failures. In the final section we will discuss how to make repairs to an electronics circuit once the problem has been isolated.

7-1   DIGITAL TEST EQUIPMENT

A variety of test equipment is available to help you troubleshoot digital circuits and systems. Some are standard, such as the multimeter and oscilloscope, which can be used for either analog or digital circuits. Other test instruments, such as the logic clip, logic probe, logic pulser, and current tracer, have been designed specifically to test digital logic circuits. In this section we will be examining each of these test instruments, beginning with the multimeter.

7-1-1   *Testing with the Multimeter*

Multimeter

Test instrument able to perform multiple tasks in that it can be used to measure voltage, current, or resistance.

Both analog and digital **multimeters** can be used to troubleshoot digital circuits. In most cases, you will only use the multimeter to test voltage and resistance. The current settings are rarely used since a path needs to be opened to measure current, and as most circuits are soldered onto a printed circuit board, making current measurements is impractical. Some of the more common digital circuit tests using the multimeter include the following:

a. You can use the multimeter to test power supply voltages. On the AC VOLTS setting, you can check the 120 V ac input into the system's dc power supply, and using the DC VOLTS setting you can check all of the dc power supply voltages out of the power supply circuit, as seen in Figure 7-1(a). The multimeter can also be used to check that these dc supply voltages are present at each printed circuit board within the system, as in the case of the 5 V supply to the display board shown in Figure 7-1(b).

b. You can also use the multimeter on the DC VOLTS setting to test the binary 0 and binary 1 voltages throughout a digital circuit, as shown in Figure 7-1(b).

c. The multimeter is also frequently used on the OHMS setting to test components such as resistors, fuses, diodes, transistors, wires, cables, printed circuit board tracks, and other disconnected devices, as shown in Figure 7-1(c).

7-1-2   *Testing with the Oscilloscope*

Pulse Wave

A repeating wave that only alternates between two levels or values and remains at one of these values for a small amount of time relative to the other.

Since digital circuits manage two-state information, most of the signals within a digital system will be either square or rectangular. To review, the **pulse, or rectangular, wave** alternates between two peak values; however, unlike the square wave, the pulse wave does not remain at the two peak values for equal lengths of time. The *positive pulse waveform* seen in Figure 7-2(a), for instance, remains at its negative value for long periods of time and only momentarily pulses positive. On the other hand, the *negative pulse waveform* seen in Figure 7-2(b) remains at its positive value for long periods of time and momentarily pulses negative.

Pulse Repetition Frequency

The number of times a second a pulse is generated.

When referring to pulse waveforms, the term **pulse repetition frequency** (PRF) is used to describe the frequency or rate of the pulses, and the term **pulse repetition time** (PRT) is used to describe the period or time of one complete cycle, as seen in Figure 7-2(c). For example, if 1000 pulses are generated every second (PRF = 1000 pulses per second, pps or 1 kHz), each cycle will last 1/1000 of a second or 1 ms (PRT = 1/PRF = 1/1 kHz = 1 ms or 1000 μs).

Pulse Repetition Time

The time interval between the start of two consecutive pulses.

The duty cycle of a pulse waveform indicates the ratio of pulse width time to the complete cycle time (it compares ON time to OFF time) and is calculated with the formula

$$\text{Duty Cycle} = \frac{\text{Pulse Width } (P_w)}{\text{Period } (t)} \times 100\%$$

FIGURE 7-1 Testing with the Multimeter.

FIGURE 7-2 The Pulse Waveform. (a) Positive Pulse Waveform. (b) Negative Pulse Waveform. (c) Peak Voltage, Pulse Width, PRF, and PRT. (d) Rise Time and Fall Time. (e) Overshoot, Undershoot, and Ringing.

EXAMPLE:

Calculate the pulse width, PRT, PRF, and duty cycle for the waveform shown in Figure 7-2(c).

Solution:

In the example in Figure 7-2(c), a pulse width of 100 μs and a period of 1000 μs will produce a duty cycle of

$$\text{Duty Cycle} = \frac{\text{Pulse Width } (P_w)}{\text{Period } (t)} \times 100\%$$

$$= \frac{100 \text{ μs}}{1000 \text{ μs}} \times 100\% = 10\%$$

A duty cycle of 10% means that the positive pulse lasts for 10% of the complete cycle time, or PRT.

Since the pulse repetition time is 1000 μs, the pulse repetition frequency will equal the reciprocal of this value, or 1 kHz (PRF = 1/PRT = 1/1000 μs = 1 kHz).

Another area that needs to be reviewed with regard to digital signals is rise and fall time. The pulse waveforms seen in Figure 7-2(a), (b), and (c) all rise and fall instantly between their two states or peak values. In reality, there is a time lapse referred to as the *rise*

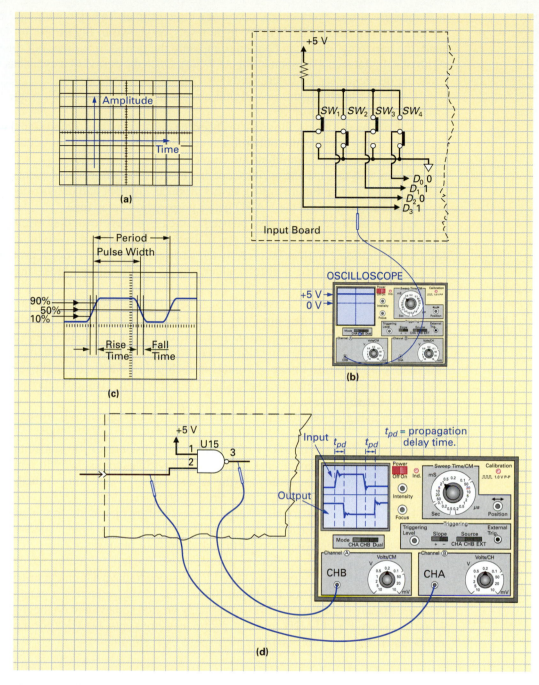

FIGURE 7-3   **Testing with the Oscilloscope.**

time and *fall time,* as seen in Figure 7-2(d). The rise time is defined as the time needed for the pulse to rise from 10% to 90% of the peak amplitude, while the fall time is the time needed for the pulse to fall from 90% to 10% of the peak amplitude.

    When carefully studying pulse waveforms on the oscilloscope, you will be able to measure pulse width, PRT, rise time, and fall time. You will also see a few other conditions known as *overshoot, undershoot,* and *ringing.* These unwanted conditions tend to always accompany high-frequency pulse waveforms due to imperfections in the circuit, as seen in Figure 7-2(e). As you can see, the positive rising edge of the waveform "overshoots" the peak, and then a series of gradually decreasing sine wave oscillations occur, called ringing. At the end of the pulse this ringing action reoccurs, as the negative falling edge "undershoots" the negative peak value.

    The graticules on an oscilloscope's display can be used to measure amplitude and time, as seen in Figure 7-3(a). This time/amplitude measuring ability of the oscilloscope enables us to measure almost every characteristic of a digital signal. To begin with, the vertical scale can be used to measure HIGH/LOW logic levels, as shown in Figure 7-3(b). On the other hand, the horizontal scale can be used to measure the period, rise time, fall time, and pulse width of a digital signal, as seen in Figure 7-3(c). Once the period of a cycle has been calculated, the frequency can be determined ($f = 1/t$).

Some of the more common digital circuit tests using the oscilloscope include the following:

a. You can use the oscilloscope to check the clock or master timing signal within a digital system to see that it is not only present at every point in the circuit but also that its frequency, wave shape, and amplitude are correct.

b. You will also use the oscilloscope frequently to monitor two or more signals simultaneously, as shown in Figure 7-3(d). In this instance, an input and output waveform are being compared to determine the NAND gate's propagation delay time.

c. The oscilloscope can also be used to measure waveform distortions, such as the ringing shown in Figure 7-3(d), to see if these undershoots or overshoots are causing any false operations, such as incorrectly triggering a logic gate. A noise problem in a digital circuit is often called a **glitch,** which stands for "gremlins loose in the computer housing."

From the previous discussion you can see that the oscilloscope is a very versatile test instrument, able to measure almost every characteristic of a digital signal. Its accuracy, however, does not match the multimeter for measuring logic level voltages, or the frequency counter for measuring signal frequency. The scope is a more visual instrument, able to display a picture image of signals at each point in a circuit and enabling its operator to see phase relationships, wave shape, pulse widths, rise and fall times, distortion, and other characteristics.

When using an oscilloscope to troubleshoot a digital circuit, try to remember the following:

a. Always use a *multitrace oscilloscope* when possible so you can compare input and output waveforms. Remember that a chopped or multiplexed display will not be accurate at high frequencies since one trace is trying to display two waveforms. A dual-beam oscilloscope is best for high-frequency digital circuit troubleshooting.

b. Trigger the horizontal sweep of the oscilloscope with the input signal or a frequency-related signal. Having a *triggered sweep* will ensure more accurate timing measurements and enable the oscilloscope to lock on to your input more easily.

c. Use an oscilloscope with a large *bandwidth.* The input signals are fed to vertical amplifiers within the oscilloscope, and these circuits need to have a high upper-frequency limit. Since all scopes can measure dc signals (0 Hz), this lower frequency is generally not listed. The upper-frequency limit, however, is normally printed somewhere on the front panel of the scope and should be 40 MHz for most digital measurements, 60 MHz for high-speed TTL, and between 100 and 1000 MHz for high-speed ECL circuits.

7-1-3   *Testing with the Logic Clip*

The **logic clip,** which is shown in Figure 7-4(a), was specially designed for troubleshooting digital circuits. It consists of a spring-loaded clip that clamps on to a standard IC package, where it makes contact with all the pins of the IC, as shown in Figure 7-4(b). These contacts are then available at the end of the clip, so a multimeter or oscilloscope probe can easily be connected.

Some logic clips simply have connections on the top of the clip, while others have light-emitting diodes (LEDs) and test points on the top of the clip, as seen in Figure 7-4(a). The LEDs give a quick indication of the binary level on the pins of the IC, with a logic 0 turning an LED OFF and a binary 1 turning an LED ON.

Glitch

An acronym for "gremlins loose in the computer housing." The term is used to describe a problem within an electronic system.

Logic Clip

A spring-loaded digital testing clip that can be clamped onto an IC package so that the signal present on any pin is available at the end of the clip. Some include LEDs to display logic levels present.

Testing Pins

Light-Emitting Diodes

14 PIN

Contacts make
connection with
pins of IC.

(a)

Probe

GND

16 PIN

DECODER
BOARD

(b)

FIGURE 7-4   Testing with the Logic Clip.

7-1-4   *Testing with the Logic Probe*

Another test instrument designed specifically for digital circuit troubleshooting is the **logic probe,** shown in Figure 7-5(a). In order for the probe to operate, its red and black power leads must be connected to a power source. This power source can normally be obtained from the circuit being tested, as shown in Figure 7-5(b). Once the power leads have been connected, the logic probe is ready to sense the voltage at any point in a digital circuit and give an indication as to whether that voltage is a valid binary 1 or 0. The HI and LO LEDs on the logic probe are used to indicate the logic level, as seen in Figure 7-5(b). Since the voltage thresholds for a binary 0 and 1 can be different, most logic probes have a selector switch to select different internal threshold detection circuits for TTL, ECL, or CMOS.

In addition to being able to perform **static tests** (constant or nonchanging HIGH and LOW signals) like those of the multimeter, the logic probe can also perform **dynamic tests** (changing signals) like those of the oscilloscope, such as detecting a single momentary pulse (single pulse) or a pulse train (sequence of pulses), such as a clock signal. Nearly all logic probes have an internal low-frequency oscillator that will flash the "PULSE" LED ON and OFF if a pulse train is detected at a frequency of up to 100 MHz. The logic probe can also detect if a single pulse has occurred, even if the width of the pulse is so small that the operator was unable to see the PULSE LED blink. In this instance, a memory circuit can be

Logic Probe

A digital test instrument
that can indicate whether a
HIGH, LOW, or PULSE is
present at a circuit point.

Static Test

A test performed on a
circuit that has been held
constant in one condition.

Dynamic Test

A test performed on a
circuit that is running at its
normal operating speed.

☀ ○ HI LO	Indicates a Logic 1.	
○ ☀ HI LO	Indicates a Logic 0.	
○ ○ HI LO	Indicates a FLOATING Logic Level.	
☀ PULSE PULSE ▮	Pulse Switch – Pulse Mode. PULSE LED – Flashing. Indicates Pulse Train.	
☀ ○ HI LO ☀ PULSE ▮ MEMORY	HI LED ON, PULSE LED ON, MEMORY MODE. Indicates Positive going single pulse has been detected.	

(c)

FIGURE 7-5 **Testing with the Logic Probe.**

enabled using the MEMORY switch, so that any pulse (even as narrow as 10 ns) will turn ON and keep ON the PULSE LED. This memory circuit within the logic probe senses the initial voltage level present at the test point (either HIGH or LOW) and then turns ON the PULSE LED if there is a logic transition or change. To reset this memory circuit, simply toggle (switch OFF and then ON) the memory switch. Figure 7-5(c) summarizes the operation and display interpretation of a typical logic probe.

Even though the logic probe is not as accurate as the multimeter, and cannot determine as much about a signal as the oscilloscope, it is ideal for digital circuit troubleshooting because of its low cost, small size, and versatility.

7-1-5   *Testing with the Logic Pulser*

Using the logic probe to test the static HIGH and LOW inputs and output of a gate does not fully test the operation of that logic gate. In most instances it is necessary to trigger the input of a circuit and then monitor its response to a dynamic test. A **logic pulser** is a signal generator designed to produce either a single pulse or a pulse train. Its appearance is similar to the logic probe, as shown in Figure 7-6(a), in that it requires a supply voltage to operate and has a probe which is used to apply pulses to the circuit under test, as shown in Figure 7-6(b).

The logic pulser is operated by simply touching the probe tip to any conducting point on a circuit and then pressing the SINGLE PULSE button for a single pulse or selecting the

Logic Pulser

A digital test instrument that can generate either single or repeating logic pulses.

FIGURE 7-6 Testing with the Logic Pulser.

PULSE TRAIN switch position for a constant sequence of pulses. The frequency of the pulse train, with this example model, can be either 1 pulse per second (1 pps) or 500 pulses per second (500 pps), based on the pps selector switch.

The logic probe is generally used in conjunction with the logic pulser to sense the pulse or pulses generated by the logic pulser, as shown in Figure 7-6(b). In this example, the logic pulser is being used to inject a single pulse into the input of an OR gate, and a logic probe is being used to detect or sense this pulse at the output of the OR gate. If the input to the OR gate was HIGH, the logic pulser would have automatically sensed the binary 1 voltage level and then pulsed the input to the OR gate LOW when the pulse button was pressed. In this example, however, the input logic level to the OR gate was LOW, so the logic pulser sensed the binary 0 voltage level and then pulsed the OR gate input HIGH. This ability of the logic pulser to sense the logic level present at any point and then pulse the line to its opposite state means that the operator does not have to determine these logic levels before a point is tested, making the logic pulser a fast and easy instrument to use. The logic pulser achieves this feature with an internal circuit that will override the logic level in the circuit under test and either source current or sink current when a pulse needs to be generated. This complete in-circuit testing ability means that components do not need to be removed and input and output paths do not need to be opened in order to carry out a test. Figure 7-6(c) summarizes the operation and display interpretation of a typical logic pulser.

Since TTL, ECL, and CMOS circuits all have different input and output voltages and currents, logic pulsers will need different internal circuits for each family of logic ICs. Typical pulse widths for logic pulsers are between 500 ns and 1 μs for TTL logic pulsers, and 10 μs for CMOS logic pulsers.

7-1-6   *Testing with the Current Tracer*

Current Tracer

A digital test instrument used to sense relative values of current in a conductor using an inductive pick-up tip.

The logic pulser can also be used in conjunction with another very useful test instrument called a **current tracer,** which is shown in Figure 7-7(a). Like the logic probe, the current tracer is also a sensing test instrument, but unlike the logic probe, the current tracer senses the relative values of current in a conductor. It achieves this by using an insulated inductive pick-up tip, which senses the magnetic field generated by the current in a conductor. By adjusting the sensitivity control on the current tracer and observing the lamp's intensity when

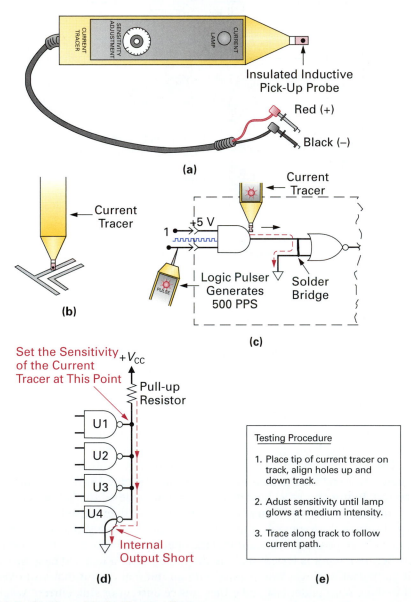

FIGURE 7-7   Testing with the Current Tracer.

the probe is placed on a pulsating logic signal line, a shorted path can be found by simply tracing the path of high current.

To troubleshoot a short in an IC's input or output circuit, a power supply, or a printed circuit board (PCB) track or cable, we would traditionally have to cut PCB tracks, snip pins, or open component leads. This would isolate the short, since the excessive current path or short would be broken when the path was opened. The current tracer allows the troubleshooter to isolate the shorted path without tampering with the circuit. Figure 7-7(b) illustrates how the current tracer should be correctly aligned on a conducting PCB track. Like the logic probe and logic pulser, the current tracer can use the same supply voltage as the circuit under test. Since the probe's tip is insulated, it can be placed directly on the track; however, be sure that the probe is always perpendicular to the board. The other important point to remember is to ensure that the small holes or dots on either side of the tip are aligned so they face up and down the track being traced at all times. This is so that the inductive coil in the insulated tip is oriented to pick up the maximum amount of magnetic flux.

Figure 7-7(c) shows an example of how a current tracer and logic pulser can be used to locate a short caused by a solder bridge. In this example, we will first connect the logic pulser so that it will generate a pulsating current, and therefore a changing magnetic field, which can be detected by the inductive tip of the current tracer. The next step is to touch the tip of the current tracer on the output of the AND gate and adjust the sensitivity level of the current tracer to light the current lamp (the sensitivity adjustment allows the current tracer to sense a wide range of current levels, from 1 mA to 1 A, typically). Once this reference level of brightness has been set, trace along the path between the output of the AND gate and the input of the OR gate. Since the current value should be the same at all points along this track, the lamp should glow at the same level of brightness. Once the current tracer moves past the solder bridge, the lamp will go out, since no current is present beyond this point. At this point you should, after a visual inspection, be able to locate the short.

The current tracer is also an ideal tool for troubleshooting a circuit in which many outputs are tied to a common point (such as open-collector gates), as seen in the example in Figure 7-7(d). If any of these gates were to develop an internal short, it would be difficult to isolate which of the parallel paths is causing the problem. In the past, you would have to remove each gate from the circuit (which is difficult with a soldered PCB) or clip the output IC pin of each gate until the short was removed and the faulty gate isolated. The current tracer can quickly isolate faults of this nature since it will highlight the current path provided by the short and therefore lead you directly to the faulty gate.

SELF-TEST EVALUATION POINT FOR SECTION 7-1

Use the following questions to test your understanding of Section 7-1.

1. What would be the most accurate instrument for measuring dc voltages?
2. Which test instrument would be best for measuring gate propagation delay time?
3. A logic _____ is used to generate a single or train of test pulses.
4. A _____ detects current pulses, while a _____ detects voltage levels and pulses.

7-2   DIGITAL CIRCUIT PROBLEMS

Since ICs account for almost 85% of the components within digital systems, it is highly likely that most digital circuit problems will be caused by a faulty IC. Digital circuit failures can be basically divided into two categories: failures within ICs, or failures outside of ICs. When troubleshooting digital circuits, therefore, you will have to first isolate the faulty IC and then determine whether the problem is internal or external to the IC. An internal IC problem cannot be fixed and therefore the IC will have to be replaced, whereas an external IC

FIGURE 7-8 Internal IC Failures.

problem can be caused by electronic devices, connecting devices, mechanical or magnetic devices, power supply voltages, system clock signals, and so on. To begin with, let us discuss some of the typical internal IC failures.

7-2-1   *Digital IC Problems*

Digital IC logic gate failures can basically be classified as either opens or shorts in either the inputs or output, as summarized in Figure 7-8. In all cases, these failures will result in the IC having to be discarded and replaced. The following examples will help you to isolate an internal IC failure.

Internal IC Opens

An internal open gate input or an open gate output are very common internal IC failures. These failures are generally caused by the very thin wires connecting the IC to the pins of the package coming loose or burning out due to excessive values of current or voltage.

Figure 7-9 illustrates how a logic probe and logic pulser can be used to isolate an internal open NAND gate input. In Figure 7-9(a) the open input on pin 10 has been jumper connected to $+V_{CC}$, and the other gate input (pin 9) is being driven by the logic pulser. As you know, any LOW input to a NAND gate will produce a HIGH output, and only when both inputs are HIGH will the output be LOW, as shown in the inset in Figure 7-9(a). The logic probe, which is monitoring the gate on pin 8, indicates that a negative-going pulse waveform is present at the output, and therefore the gate seems to be functioning normally. When the jumper and pulser inputs are reversed, however, as seen in

FIGURE 7-9   An Internal Open Gate Input.

Figure 7-9(b), the output remains permanently LOW. These conditions could only occur if the pin 10 input was permanently HIGH. As discussed previously, an open or floating input is equivalent to HIGH input, since both conditions have next-to-no input current. The cause, therefore, must be either an internal open at pin 10 (since this is equivalent to a HIGH input) or a $+V_{CC}$ short to the pin 10 input of this gate (which is less common than an open). In either case, however, the logic gate IC has an internal failure and will have to be replaced.

Figure 7-10 illustrates how a logic pulser and logic probe can be used to isolate an internal open gate output. In Figure 7-10(a), the logic gate's input at pin 10 has been jumper

FIGURE 7-10   An Internal Open Gate Output.

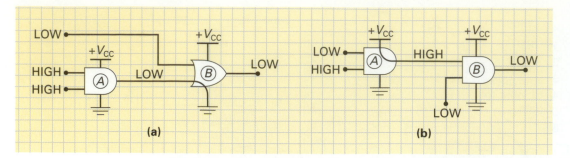

FIGURE 7-11   Internal IC Shorts.

connected to $+V_{CC}$, and the other input at pin 9 is being driven by the logic pulser. The logic probe, which is monitoring the output on pin 8, has none of its indicator LEDs ON, which indicates that the logic level is neither a valid logic 1 or a valid logic 0 (line is probably floating). In Figure 7-10(b), the inputs have been reversed; however, the logic probe is still indicating the same effect. This test highlights the failure, which is an open gate output. Once again, the logic gate IC has an internal failure and will have to be replaced.

Internal IC Shorts

Internal logic gate shorts to either $+V_{CC}$ or ground will cause the inputs or output to be stuck either HIGH or LOW. These internal shorts are sometimes difficult to isolate, so let us examine a couple of typical examples.

Figure 7-11(a) shows the effect an internal input lead short to ground will have on a digital logic gate circuit. The input to gate B is shorted to ground, so this input is being pulled LOW. This LOW between A output and B input makes it appear as though gate A is malfunctioning, since it has two HIGH inputs and therefore should be giving a LOW output. The problem is solved by disconnecting gate A from gate B. Since gate A's output will go HIGH when it is no longer being pulled down by gate B, the fault will be isolated to a short at the input of gate B.

In Figure 7-11(b), the output of gate A has shorted to $+V_{CC}$. This circuit condition gives the impression that gate A is malfunctioning; however, if the B gate input had shorted to $+V_{CC}$ we would get the same symptoms. Once again, we will have to isolate, by disconnecting the A output from the B input, to determine which gate is faulty.

Troubleshooting Internal IC Failures

Let us now try a few examples to test our understanding of digital test equipment and troubleshooting techniques.

■   EXAMPLE:

Determine whether a problem exists in the circuit in Figure 7-12(a) based on the logic probe readings indicated. If a problem does exist, indicate what you think could be the cause.

■   *Solution:*

Gates B and C seem to be producing the correct outputs based on their input logic levels. Gate A's output, on the other hand, should be LOW, and it is reading a HIGH. To determine whether the problem is the source gate A or the load gate B, the two gates have been isolated as shown in Figure 7-12(b). After isolating gate A from gate B, you can see that the pin 10 input to gate C seems to be stuck HIGH, probably due to an internal short to $+V_{CC}$.

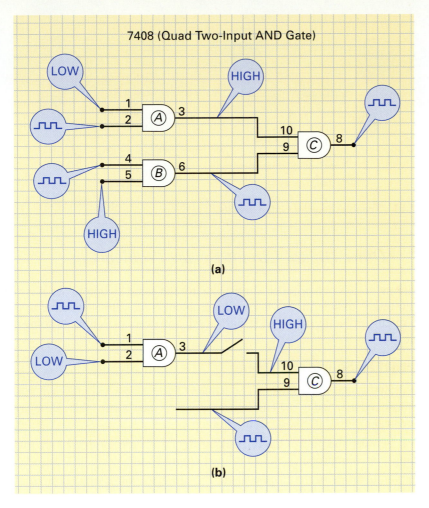

FIGURE 7-12 Troubleshooting Gate Circuits.

Even though gate *C* seems to be the only gate malfunctioning, the whole 7408 IC will have to be replaced. In situations like this, it is a waste of time isolating the problem to a specific gate when the problem has already been isolated down to a single component. Use your understanding of internal logic gate circuits to isolate the problem to a specific component and not to a device within that component.

■ **EXAMPLE:**

What could be the possible fault in the circuit shown in Figure 7-13?

■ *Solution:*

The output of NOT gate *A* should be HIGH; however, it is pulsating. This pulsating signal can only have come from pin 2 of the NAND gate. By disconnecting the pulsating input at pin 2 of the NAND gate and then single-pulsing this input using the logic pulser, you can use the logic probe to see if the pin 1 input of the NAND gate always follows what is on pin 2. Once an input short has been determined, visually inspect the NAND gate IC to be sure that there is no external short between pins 1 and 2.

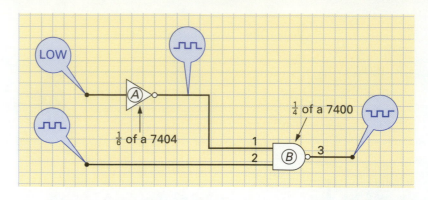

FIGURE 7-13 **Troubleshooting Gate Circuits.**

7-2-2   *Other Digital Circuit Device Problems*

Failures external to ICs can produce almost exactly the same symptoms as internal IC failures, so your task is to determine whether the problem is internal or external to the IC. The external IC circuit problems can be wide and varied; however, let us discuss some of the more typical failures.

External IC Shorts and Opens

Most digital signal line shorts can be isolated by injecting a signal onto the line using the logic pulser and then using the current tracer to find the shorted path. Here are some examples of typical external IC short circuit problems.

    a. A short between the pins of the IC package due to a solder bridge, sloppy wiring, wire clippings, or an improperly etched printed circuit board.

    b. A bending in or out of the IC pins as they are inserted into a printed circuit board.

    c. The shorting of an externally connected component, such as a shorted capacitor, resistor, diode, transistor, LED, switch, and so on.

    d. A shorted connector such as an IC socket, circuit board connector, or cable connector.

Most digital signal line opens are easily located since the digital signal does not arrive at its destination. The logic probe is ideal for tracing these open paths. Here are some examples of typical external IC open circuit problems.

    a. An open in a signal line due to an improperly soldered pin, a deep scratch in the etched printed circuit board, a bent or broken IC pin, a broken wire, and so on.

    b. An IC that is inserted into the printed circuit board backwards.

    c. The failure of externally connected components such as resistors, diodes, transistors, and so on.

    d. An open connector such as an IC socket, circuit board connector, or cable connector.

In summary, the logic pulser and current tracer are ideal for tracing shorts and the logic probe is ideal for tracing opens.

Power Supply Problems

Like analog circuits, digital circuits suffer from more than their fair share of power supply problems. When a digital circuit seems to have a power supply problem, remember to use the half-split method of troubleshooting first, as seen in Figure 7-14(a). First check the dc supply voltages from the dc power supply unit to determine whether the power problem is in the power supply (source) or in one of the digital circuits (load).

If the dc output voltages are not present at the output of the dc power supply, then the fault probably exists before this test point. In this instance, you should test the dc power supply circuit; however, a short in the digital circuits can pull down the dc supply voltage and

FIGURE 7-14 Power Supply Problems in Digital Circuits. (a) Using the Half-Spiit Method. (b) Tracing Shorts. (c) Tracing Opens.

make it appear as though the problem is in the dc power supply, as shown in Figure 7-14(b). Once again, the current tracer is ideal for tracing the faulty board and then the faulty device in that printed circuit board. If you cannot get power to the digital circuits because the short causes the dc power supply fuse to continually blow, use the logic pulser to inject a current into the supply line and then use the current tracer to follow the current path to the short.

Open power supply lines are easy to locate since power is prevented from reaching its destination. The logic probe is ideal for tracking down power line opens, as shown in Figure 7-14(c).

Clock Signal Problems

All digital systems use a master timing signal called a **clock.** This two-state timing signal is generated by a digital oscillator circuit. This clock or timing signal is distributed

Clock

A two-state timing signal that is distributed throughout a digital electronic system to ensure that all circuits operate in sync or at the correct time.

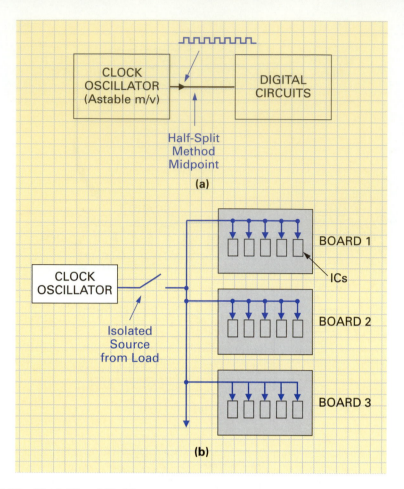

FIGURE 7-15   Clock Signal Problems.

throughout the digital electronic system to ensure that all circuits operate in sync or at their correct time. The presence or absence of this clock signal can be checked with a logic probe; however, in most cases it is best to check the quality of the signal with an oscilloscope. Even small variations in frequency, pulse width, amplitude, and other characteristics can cause many problems throughout the digital system.

To isolate a clock signal problem, first apply the half-split method to determine whether the fault is in the oscillator circuit (source) or in one of the digital circuits (load), as shown in Figure 7-15(a). If the clock signal from the oscillator is incorrect, or if no signal exists, the problem will generally be in the oscillator clock circuit; however, like the dc supply voltage, the clock signal is distributed throughout the digital system, and therefore you may have to disconnect the clock signal output from the oscillator to determine whether a short is pulling down the signal or whether the oscillator clock circuit is not generating the proper signal, as shown in Figure 7-15(b). If there is a short somewhere in one of the digital circuits, the current tracer can be used to find the shorted path, whereas an open in a clock signal line can easily be traced with a logic probe.

SELF-TEST EVALUATION POINT FOR SECTION 7-2

Use the following questions to test your understanding of Section 7-2.

1. How are most internal IC failures repaired?
2. An _____ is easier to isolate and will generally not affect other components while a _____ is harder to isolate and can often damage other components (short or open).
3. What is the half-split method?

4. What is a clock signal?

5. The _____ and _____ are ideal for tracing shorts, and the _____ is ideal for tracing opens.

7-3   CIRCUIT REPAIR

Isolating the faulty component will always take a lot longer than repairing the circuit once the problem has been found. Some repairs are simple, such as replacing a cable, removing a solder bridge from a PCB, reseating a connector, or adjusting a variable resistor. In most instances, however, the repair will involve replacing a component such as an IC, transistor, diode, resistor, or capacitor. This will mean that you will need to use the soldering iron to remove the component from the PCB. When soldering and desoldering components to and from a PCB, follow the techniques discussed in your dc/ac electronics course and also remember the following:

a. Always make a note of the component's orientation in the PCB since certain components such as ICs, diodes, transistors, and electrolytic capacitors have to be correctly oriented.

b. When desoldering components, use either a vacuum bulb or solder wick to remove the solder.

c. Be extremely careful not to overheat the PCB or the component being soldered or desoldered.

d. Avoid using too much or too little solder because this may cause additional problems.

e. Use a grounded soldering iron and a grounding strap when working with MOS devices to prevent any damage to the IC from static discharge.

Once the equipment has been repaired, always "final test" the equipment to see if it is now fully operational. Check that the system operates correctly in all aspects, especially in the area that was previously malfunctioning, and then reassemble the system.

SELF-TEST EVALUATION POINT FOR SECTION 7-3

Use the following questions to test your understanding of Section 7-3.

1. List the steps you would follow to remove an IC from a PCB.

2. What is a "final test"?

REVIEW QUESTIONS

Multiple-Choice Questions

1. Which test instrument would most accurately measure logic voltage levels?
 a. Logic probe
 b. Multimeter
 c. Oscilloscope
 d. Logic pulser

2. Which test instrument would be best suited for testing a clock signal's duty cycle?
 a. Logic probe
 b. Multimeter
 c. Oscilloscope
 d. Logic pulser

3. Which test instrument would be best suited for testing the HIGH and LOW logic levels at the inputs and outputs of a series of several logic gates?
 a. Logic probe
 b. Multimeter
 c. Oscilloscope
 d. Logic pulser

4. Which test instrument would be best suited for generating a test signal?
 a. Logic probe
 b. Multimeter
 c. Oscilloscope
 d. Logic pulser

5. The voltmeter can be used to make _____ tests, while the oscilloscope can be used to make _____ tests.
   a. dynamic, static
   b. static, dynamic

6. The logic pulser
   a. has an insulated tip.
   b. senses current pulses at different points in a circuit.
   c. is a signal generator.
   d. is a signal-sensing instrument.

7. Which digital test instrument can be used to sense the logic levels at different points in a circuit?
   a. Multimeter
   b. Logic probe
   c. Oscilloscope
   d. Current tracer

8. The current tracer is ideal for locating what type of circuit faults?
   a. Shorts
   b. Opens
   c. Invalid logic levels
   d. Rise/fall signal problems

9. A scratch in a PCB track will more than likely cause a/an _____.
   a. short
   b. open
   c. no problem
   d. any of the above

10. Which of the following is most likely to cause a short?
    a. Broken IC pin
    b. Low current
    c. Solder bridge
    d. IC pin not in an IC socket

Practice Problems

11. Identify the waveform shown in Figure 7-16.

12. What test instrument would you use to check the rise and fall time of the waveform shown in Figure 7-16?

13. Referring to Figure 7-16, calculate the following characteristics for the waveform:
    a. Pulse width
    b. Period
    c. Frequency
    d. Duty cycle

14. Calculate the PRF and PRT of the waveform shown in Figure 7-16.

15. Could a logic probe be used to identify whether the waveform in Figure 7-16 was a square or rectangular wave?

Troubleshooting Questions

16. Why is it important to be able to recognize different types of internal IC failures if the ICs themselves cannot be repaired?

17. In Figure 7-17(a), a logic probe has been used to check the circuit since it is not functioning as it should. In Figure 7-17(b), you can see what logic levels were obtained when the output of the NAND gate was isolated. Which IC should be replaced?

18. What do you suspect is wrong with the faulty IC in Figure 7-17?

19. In Figure 7-18, an oscilloscope has been used to test several points on the digital circuit. Which IC should be replaced, and what do you think is the problem?

20. Are the logic gates in Figure 7-19 operating as they should?

Web Site Questions

Go to the web site http://www.prenhall.com/cook, select the textbook *Practical Digital Electronics*, select this chapter, and then follow the instructions when answering the multiple choice practice problems.

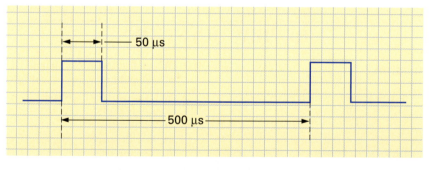

FIGURE 7-16   A Repeating or Pulse Waveform.

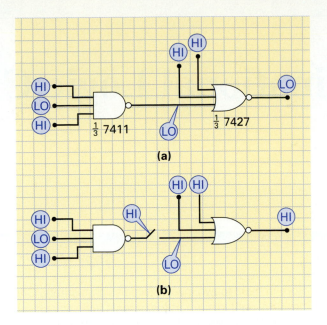

FIGURE 7-17   Troubleshooting Exercise 1.

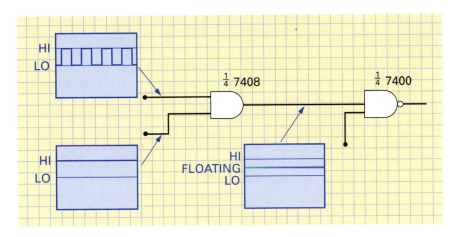

FIGURE 7-18   Troubleshooting Exercise 2.

FIGURE 7-19   Troubleshooting Exercise 3.

Combinational Logic Circuits

Copy Master

Born in 1906 to two invalid Swedish immigrants, Chester Carlson worked part-time day and night to support his parents and still achieved excellent grades at high school. Sadly, when Carlson was seventeen both of his parents died, so Carlson left New York and went to the California Institute of Technology where he graduated in 1930 with a degree in physics.

Carlson's first job was as a researcher for Bell Telephone Laboratories in New York. It was here that he saw the need for a machine that could copy documents—the method at that time was to have someone in a typing pool retype the original. Carlson realized that available photographic methods were too messy and time consuming. He left Bell Labs in 1935 and began developing a clean and quick copying machine in the small rented bedroom that he had converted into a laboratory.

On the morning of October 23, 1938, Carlson statically charged a metal sulfur-coated plate by rubbing it with his handkerchief. He then exposed the plate to a glass slide that had on it the date and place "10-23-38 ASTORIA." When dry black powder on paper was pressed against the metal plate, the world's first photocopy was created.

Refining the design took Carlson several years because he had to develop more sensitive plates and a powder that would stick to paper. In fact, even after he had a good working prototype, it took him two years to find a company that was interested in manufacturing and selling his machine. However, in January, 1947, Haloid Company of Rochester, New York, which was a small photography firm, signed an agreement with Carlson for what he called his "dry printing machine." They were very dubious about whether it would really catch on and told Carlson that they would not pay him for his invention, but only give him a percentage of the profits. The company called the process "Xerography," which meant dry printing. After an almost overnight success, the company gave up all of their other products and renamed the company after their product, "Xerox Corporation."

The early Xerox machines in 1950 needed an operator to actuate the mechanism. However, by 1960 a fully automatic machine produced perfect copies by pressing one button.

Carlson's invention that nobody wanted transformed his lifestyle from a rented bedroom to royalties that paid him several million dollars a year.

Introduction

The first digital ICs contained only four logic gates, so if you wanted to decode a digital signal, you would have to construct a digital decoder circuit on a printed circuit board (PCB) using several logic gate ICs. As semiconductor manufacturers improved their miniaturization techniques, they were able to incorporate larger circuits within a single IC package. Studying the evolving digital electronics industry, semiconductor manufacturers noticed that several circuit functions are always needed in digital electronic systems, so they formed these complete functional circuits within a single IC. They called these digital ICs **combinational logic circuits** since logic gates were combined on a single chip of semiconductor to create functional logic circuits, such as decoders, encoders, multiplexers, demultiplexers, code converters, comparators, and parity generators and checkers. As time has passed, this practice of reducing a frequently needed digital PCB circuit to a single IC has continued, in step with the ability of semiconductor manufacturers to continuously improve their circuit miniaturization techniques.

In this chapter we will take a close look at two types of combinational logic circuits—decoders and encoders.

Combinational Logic Circuits

Digital logic circuits in which the outputs are dependent only on the present input states and the logic path delays.

The term *decode* means to translate coded characters into a more understandable form, or to reverse a previous encoding process. A digital **decoder circuit** is a logic circuit that responds to one specific input word or code, while rejecting all others.

8-1-1   *Basic Decoder Circuits*

Figure 8-1(a) shows how an AND gate and two INVERTERS can be connected to act as a basic decoder. As you know, all inputs to an AND gate need to be 1 for the output to be 1. Referring to the function table for this circuit, you can see that only when the input code is 001 will the AND gate have all of its inputs HIGH and therefore produce a HIGH output.

While some decoders are designed to recognize only one input word combination, other decoders are designed to activate one output for every input word combination. For example, Figure 8-1(b) shows how four AND gates and two INVERTERS can be connected to activate one of four outputs based on the binary input word applied to inputs A and B. The two INVERTERS are included to invert the A and B inputs to create two additional input lines called "NOT A" and "NOT B" (symbolized $\overline{A}$ and $\overline{B}$). Looking at the function table for this circuit, you can see that when inputs A and B are 00_2 (0), only output 0 will be activated or made HIGH. This is because both of the inputs to AND gate 0 are connected to the $\overline{A}$ and $\overline{B}$ input lines, so when A and B are LOW, $\overline{A}$ and $\overline{B}$ are both HIGH, and therefore the output of AND gate 0 will be HIGH. The inputs of the other AND gates are connected to the input lines so that a 01_2 (1) input will make output 1 HIGH, a 10_2 (2) input will make output 2 HIGH, and a 11_2 (3) input will make output 3 HIGH. This

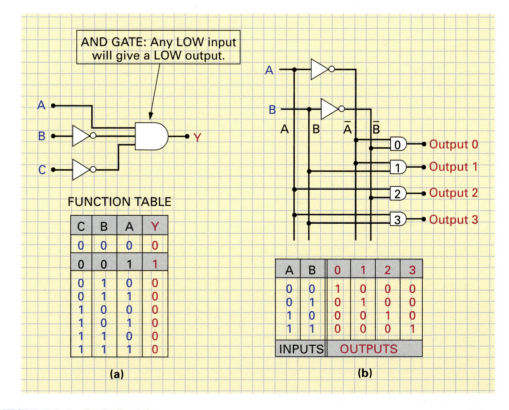

FIGURE 8-1   Basic Decoders.

circuit would be described as a *one-of-four decoder* since an input code is used to select or activate one of four outputs. If the four outputs were connected to light-emitting diodes (LEDs) that were labeled 0 through 3, the decoder would function as a **binary-to-decimal decoder,** since the circuit is translating a binary coded input into a decimal output display.

8-1-2 *Decimal Decoders*

Binary-to-decimal conversion is one of the most common applications for digital decoder circuits.

Application—BCD-to-Decimal Counter Circuit

Figure 8-2 shows how a 7442 can be used in conjunction with a 7490 decade (ten) counter, to produce a BCD-to-decimal counter/decoder circuit. Counters will be discussed in more detail in Chapter 10; for now, however, simply think of the counter as a circuit that counts the number of clock pulses applied to its input (pin 14) and then generates a 4-bit binary equivalent value at its outputs (pins 12, 9, 8, 11). In this circuit, the input clock pulses are continuous, so the output binary count from the 7490 will continually advance or cycle through all ten BCD codes—0000, 0001, 0010, 0011, 0100, 0101, 0110, 0111, 1000, 1001—and then reset to 0000 and repeat the cycle. This incrementing 4-bit BCD count from the 7490 is applied to the 4-bit input of the 7442 decoder.

An input of 0000 to the 7442 will cause it to make its 0 output LOW, an input of 0001 to the 7442 will cause it to make its 1 output LOW, an input of 0010 to the 7442 will cause it to make its 2 output LOW, and so on. Since the LEDs are arranged as a common-anode display, a LOW at any of the 7442 outputs will turn ON the respective LED. As a result, the LED decimal display will turn ON each LED in sequence (0 through 9), and then repeat the cycle.

FIGURE 8-2 Application—BCD-to-Decimal Counter Circuit.

■ **EXAMPLE:**

Imagine you had constructed the circuit in Figure 8-2 in lab, and when you switched on the power the decimal display did not flash sequentially. What tests would you make to try and isolate the problem?

■ *Solution:*

The following tests should help you isolate a problem with the circuit shown in Figure 8-2.

   a. Verify +5 V is applied to the 7490, 7442, and LED anodes using the logic probe.

   b. Verify ground is applied to all necessary IC pins using the logic probe.

   c. Check clock pulse input at pin 14 of the 7490 using the logic probe (dynamic test). You can also use the logic pulser to single pulse the counter input (static test).

   d. Apply the half-split method to the circuit—isolate outputs from 7490 and then test to see if counter is operating as it should. This will determine whether the problem is in the counter half of the circuit or the decoder half of the circuit.

   e. Further isolate to determine whether the problem is internal or external to the suspected IC.

8-1-3   *Hexadecimal Decoders*

Hexadecimal Decoder

A decoder circuit that selects one of sixteen outputs (hexadecimal) depending on the 4-bit binary input applied.

The 74154 is a good example of a **hexadecimal decoder** that will select one of sixteen (hexadecimal) outputs, depending on the 4-bit binary input value.

Application—Sixteen-LED Back-and-Forth Flasher Circuit

Figure 8-3 shows how the 74154 can be used in conjunction with a 74193 up/down counter to produce a sixteen-LED back-and-forth flasher circuit. The circuit in Figure 8-3 operates in the following way. Let us begin by assuming that *B* NAND gate has a HIGH on pin 5, and *A* NAND gate has a LOW on pin 1. These two points are always at opposite logic levels due to the cross-coupling between NAND gates *C* and *D*. With *B* NAND gate pin 5 HIGH, the clock pulses on pin 4 appear at the COUNT-UP input of the counter, and therefore the counter's count will increase in binary (0000, 0001, 0010, and so on toward 1111). This binary count appears at the *ABCD* outputs of the 74193, and is applied to the *ABCD* inputs of the 74154 decoder. The decoder will respond to the binary input and make LOW the equivalent output and consequently turn ON the associated LED (due to the LOW outputs). When the count reaches 1111 (15), the 15 output (pin 17) goes LOW, turning ON the left-most LED. This LOW output at pin 17 of the 74154 also appears at pin 10 of NAND gate *C*, which produces a HIGH output to drive pin 1 of NAND gate *A* and allows the clock pulses through to the COUNT-DOWN input of the 74193. The HIGH output of NAND gate *C* also drives pin 12 of NAND gate *D*, whose other input on pin 13 is also HIGH since the 0 output of the decoder is not active at this time. The two HIGH inputs into NAND gate *D* therefore produce a LOW output, which is applied to NAND gate *B* to block the clock pulses from driving the COUNT-UP input of the 74193. The LOW output of NAND gate *D* is also applied to pin 9 of NAND gate *C* to keep the COUNT-DOWN control line active (HIGH), even after the LOW is no longer present at the 74154's pin 17. The counter, therefore, counts down from 15 to 0, causing the LED's ON/OFF sequence to move from left to right. When the counter reaches a count of 0, the NAND gates will switch the clock pulses once again to the COUNT-UP input of the counter, and the cycle will repeat.

8-1-4   *Display Decoders*

Many systems such as digital watches, calculators, pagers, and cellular phones make use of a multisegment display. Decoders are needed in these systems to decode the binary data into the multisegment data needed to drive the display.

FIGURE 8-3   Application—Sixteen-LED Back-and-Forth Flasher Circuit.

Light-Emitting Diode (LED) Display Decoders

The 7447 IC is one example of a display decoder/driver. Figure 8-4(a) shows how the seven segments of an LED display are labeled—*a, b, c, d, e, f,* and *g*—and Figure 8-4(b) shows how certain ON/OFF segment combinations can be used to display the decimal digits numbers 0 through 9.

Application—Seven-Segment LED Display Decoder/Driver

Figure 8-4(c) shows how a 7447 **BCD-to-seven-segment decoder** IC can be connected to drive a common-anode seven-segment LED display. The function table for this digital IC has been repeated in the upper inset in Figure 8-4(c). Looking at the first line of this table, for example, you can see that if a binary input of 0000 (0) is applied to the *ABCD* inputs, the 7447 will make LOW outputs *abcdef* and make HIGH output *g*. These outputs are applied to the cathodes of the LEDs within the common-anode display (TIL 312), as seen in the lower inset, and since all of the anodes are connected to +5 V, a LOW output will turn ON a segment, and a HIGH output will turn OFF a segment. The 7447, therefore, generates active-LOW outputs, which means that a "LOW" output signal will "activate" its associated output LED. In this example, the active-LOW outputs from the 7447 will turn ON segments *abcdef*, while the HIGH output will turn OFF segment *g*, causing

BCD-to-Seven-Segment Decoder

A decoder circuit that converts a BCD input code into an equivalent seven-segment output code.

(b)

Segments activated
to display digits
0 through 9

Internal Schematic

a
f
b
g
e
c
d

Common Anode Connections

7447 FUNCTION TABLE

INPUTS				Digit Displayed	OUTPUTS						
D	C	B	A		a	b	c	d	e	f	g
L	L	L	L	0	ON	ON	ON	ON	ON	ON	OFF
L	L	L	H	1	OFF	ON	ON	OFF	OFF	OFF	OFF
L	L	H	L	2	ON	ON	OFF	ON	ON	OFF	ON
L	L	H	H	3	ON	ON	ON	ON	OFF	OFF	ON
L	H	L	L	4	OFF	ON	ON	OFF	OFF	ON	ON
L	H	L	H	5	ON	OFF	ON	ON	OFF	ON	ON
L	H	H	L	6	ON	OFF	ON	ON	ON	ON	ON
L	H	H	H	7	ON	ON	ON	OFF	OFF	OFF	OFF
H	L	L	L	8	ON	ON	ON	ON	ON	ON	ON
H	L	L	H	9	ON	ON	ON	OFF	OFF	ON	ON

L = Logic 0
H = Logic 1

LOW Output Turns LED ON
HIGH Output Turns LED OFF

$R_1 - R_7$ are current-limiting resistors of 330 Ω.

+5 V (5)

(1)

+5 V (16)

(14)

7490
Decade Counter

A (12)
B (9)
C (8)
D (11)

(7) A
(1) B
(2) C
(6) D

7447
BCD – 7SEG

(13) R_1 → a
(12) R_2 → b
(11) R_3 → c
(10) R_4 → d
(9) R_5 → e
(15) R_6 → f
(14) R_7 → g

(8)

(10)(2)(3)(6)(7)

SW₁
Count ON/OFF

1 Hz (1 cycle/second)
input from function
generator

+5 V

a (1)
f (2)
(3)
ANODE

a
f   b
e   g   c
d

(14) ANODE
(13) b
(11) g
(10) c
(9) dp (Right)
(8) d

(Left) dp (6)
e (7)

dp   dp

TIL 312

7SEG, Common
Anode Display

Active-Low
Inputs
(Input needs
to be LOW
to activate)

+5 V

a
b
c
d
e
f
g
dp

Anodes All
Have
Common
Connection.

(c)

FIGURE 8-4   Application—Seven-Segment LED Display/Driver.

168

the decimal digit 0 to be displayed. The remaining lines in the function table indicate which segments are made active for the other binary count inputs 0001 (1) through 1001 (9).

■ **EXAMPLE:**

What would be the *abcdefg* HIGH/LOW code generated by a 7447 if 0101 were applied to the input?

■ *Solution:*

Referring to the 7447 function table in Figure 8-4, you can see that when 0101, or LHLH (decimal 5), is applied at the input, the 7447 will turn ON (by generating a LOW output) all of the segments except *b* and *e*. These two segment outputs will therefore be HIGH and the rest will be LOW, so the output code will be ON, OFF, ON, ON, OFF, ON, ON (LHLLHLL).

The circuit shown in Figure 8-4(c) operates in the following way. A 1 Hz clock input signal is applied via a count ON/OFF switch to the input of a 7490 digital counter IC. This counter will produce a 4-bit binary word at its outputs *ABCD* based on the number of input clock pulses received at its input (pin 14). This 4-bit binary word at the output of the 7490 is applied to the 7447 seven-segment decoder/driver, where it will be decoded into a seven-segment code that is applied to a seven-segment common-anode display. When the count ON/OFF switch is closed, the 7490 will count, the 7447 will decode the count, and the display will constantly cycle through the digits 0 through 9. When the count ON/OFF switch is opened, the counter will freeze at its present count, and, therefore, so will the number shown on the display.

The 7447 is an example of a BCD-to-seven-segment decoder/driver that has active-LOW outputs for a common-anode seven-segment display. Also available is a 7448 IC, which is a BCD-to-seven-segment decoder/driver that has active-HIGH outputs for a common-cathode seven-segment display. In most applications, the 7447 is more commonly used because a 7447 can sink more output current from a common-anode display than a 7448 can source current to a common-cathode display. The 7447 and 7448 are also referred to as **code converters,** because they convert a BCD code into a seven-segment code.

The 7447 has three other inputs that should be explained. They are called lamp test input ($\overline{LT}$), blanking input ($\overline{BI}$), and ripple blanking input ($\overline{RBI}$). All three of these input control lines are active-LOW inputs, as indicated by the triangle on the input lines and the NOT bar over the abbreviated letter designations.

When the lamp test input ($\overline{LT}$) is made active (taken LOW), all of the 7447 will be pulled LOW, and therefore all of the seven-segment LEDs will go ON. This feature is good for quickly testing the decoder and all of the LEDs within the seven-segment display. The blanking input ($\overline{BI}$) can be used to blank the display (turn OFF all of the segments), overriding the present binary input being applied. This feature is used when several seven-segment displays are grouped together to form a multidigit display and the most significant displays need to be blanked when they are zero. For example, if we had to display the value "003456," there would be no need to display the first two zeros, so they could be blanked so that the display shows "3456." The third input, the ripple blanking input ($\overline{RBI}$), is an active-LOW control input that allows us to dim or brighten the display without having to adjust the display voltage applied to the LEDs. The ripple blanking input is usually driven by a pulse waveform, as shown in Figure 8-5. The duty cycle, or ratio of the pulse waveform's HIGH time to LOW time, is made variable under the control of a dimmer control. When the pulse waveform is LOW, the display is blanked, and when the pulse waveform is HIGH, the display is ON and controlled by the 4-bit input word applied to the 7447's *ABCD* inputs. By using a pulse frequency that is greater than 25 Hz so no display flicker is visible, the brightness of the display can be varied by changing the pulse waveform's duty cycle, and therefore the display ON-to-OFF ratio, as seen in the examples in the inset in Figure 8-5.

Code Converter

A decoder circuit that converts an input code into a different but equivalent output code.

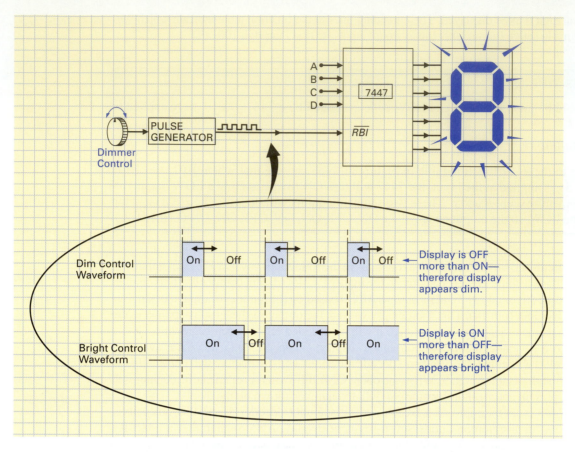

FIGURE 8-5   Ripple Blanking of a Display.

Liquid-Crystal Display (LCD) Decoders

Liquid-crystal displays are also frequently used in electronic systems to display information. Figure 8-6(a) shows a seven-segment LCD. The sealed liquid crystal segments are normally transparent, but when an ac voltage (typically 3 V to 15 V) is applied between a contact and the back electrode or plate, the crystal molecules become disorganized, resulting in a darkening of the segment. In some instances, edge- or back-lighting is used to improve the display's visibility. Displays using LCDs draw much less current than LED displays, which is why you will often see LCDs used in battery-operated devices such as watches and calculators.

Application—Seven-Segment LCD Display Decoder/Driver

Figure 8-6(b) shows how a CMOS 4511 (BCD-to-seven-segment decoder/driver) can be used in conjunction with two CMOS 4070 ICs (quad exclusive-OR gates) to drive a seven-segment liquid crystal display. The XOR gates act as controlled inverters, and the circuit operates as follows. To turn ON a segment, the 4511 generates a HIGH output, which is applied to an XOR gate input. This HIGH input to the XOR will cause the 30 Hz square wave applied to the other input to be inverted due to the action of the XOR, and therefore the 30 Hz being applied to the LCD's segment will be out-of-phase with the 30 Hz being applied to the LCD's backplate. This potential difference between the backplate and a specific segment will cause the respective segment to darken since an electric field will disorganize the segment molecules, as shown in the inset in Figure 8-6(b). On the other hand, when a 4511 output is LOW, there is no inversion of the 30 Hz square wave, so the 30 Hz being applied to the LCD's segment will be in-phase with the 30 Hz being applied to the LCD's backplate. In this instance, there is no potential difference between the backplate and segment, so the molecules remain in their normally organized condition and appear transparent, as shown in the inset in Figure 8-6(b).

FIGURE 8-6 Application—A Liquid-Crystal Display (LCD) Decoder Circuit.

■ **EXAMPLE:**

Would the 4511 BCD-to-seven-segment decoder/driver in Figure 8-6(b) have active-HIGH or active-LOW outputs?

■ *Solution:*

A HIGH output from the 4511 in Figure 8-6(b) will cause the XOR gate to invert the square wave and turn ON the associated segment. A HIGH output therefore activates the segment, so the outputs of the 4511 are active-HIGH.

In most applications CMOS devices are used to drive LCDs because, like the LCD, they also have a very low power consumption, making them ideal for battery-operated systems. The second reason for using CMOS instead of TTL to drive LCD displays is that the LOW output voltage of the TTL gate can be as much as 0.4 V, and this dc level reduces the life span of the LCD display. This condition does not occur with CMOS logic since its worst-case LOW output is 0.1 V.

SELF-TEST EVALUATION POINT FOR SECTION 8-1

Use the following questions to test your understanding of Section 8-1.

1. What would be the 0 through 9 output from a 7442 BCD-to-decimal decoder if the input were 1001?
2. How many outputs would the following decoders have?
   a. BCD-to-decimal
   b. Binary-to-hexadecimal
3. What would be the *abcdefg* logic levels out of a 7447 if 0110 were applied at the input?
4. Why are CMOS decoder/drivers normally used to drive liquid-crystal displays?

8-2   ENCODERS

Encoder Circuit

A circuit that generates specific output codes in response to certain input conditions.

An **encoder circuit** performs the opposite function of a decoder. A decoder is designed to *detect* specific codes, whereas an encoder is designed to *generate* specific codes.

8-2-1   *Basic Encoder Circuits*

Figure 8-7(a) shows how a simple decimal-to-binary encoder circuit can be constructed using three push buttons, three pull-up resistors, and two NAND gates. The pull-up resistors are included to ensure that the input to the NAND gates are normally HIGH. When button 1 is pressed, the upper input of NAND gate A will be pulled LOW, and since any LOW input to a NAND gate will produce a HIGH output, A_0 will be driven HIGH. The two inputs to NAND gate B are unaffected by SW_1, so the two HIGH inputs to this NAND gate will produce a LOW output at A_1. Referring to the truth table in Figure 8-7(a), you can see that the 2 bit code generated when switch 1 is pressed is 01 (binary 1). Studying the circuit and the rest of the function table shown in Figure 8-7(a), you can see that the code 10 (binary 2) will be generated when switch 2 is pressed, and the code 11 (binary 3) will be generated when switch 3 is pressed. This basic decimal-to-binary encoder circuit, which has three inputs and two outputs, would be described as a three-line to two-line encoder.

Figure 8-7(b) expands on the basic encoder circuit discussed in Figure 8-7(a) by including extra push-button switches and NAND gates so we can generate a 4-bit BCD code that is equivalent to the decimal key pressed. This circuit, therefore, is a *decimal-to-BCD encoder* circuit, and it will operate in the same manner as the previously discussed basic encoder. For example, when the decimal 2 key is pressed, it grounds an input to NAND gate B, so only the A_1 output goes HIGH, giving a 0010 (binary 2) output. As another example, when the decimal 7 key is pressed, it grounds inputs to NAND gates A, B, and C, so the A_0, A_1, and A_2 outputs go HIGH, giving a 0111 (binary 7) output.

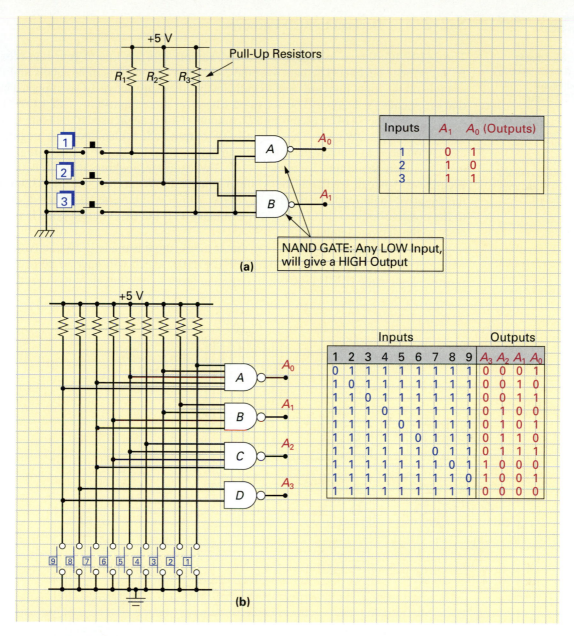

FIGURE 8-7 Basic Encoder Circuits.

■ **EXAMPLE:**

Referring to the encoder circuit in Figure 8-7(b), determine what the output code would be if push buttons 2 and 4 were pressed simultaneously.

■ *Solution:*

If push buttons 2 and 4 in Figure 8-7(b) were pressed simultaneously, NAND gates *B* and *C* would both have a LOW input, so both would generate a HIGH output. The output codes for buttons 2 (0010) and 4 (0100) therefore would be combined, producing a 0110 output code.

Decimal-to-BCD Encoder

An encoder circuit that in response to decimal input conditions generates equivalent BCD output codes.

Negative-Logic

1. An active-LOW signal or code. 2. Digital logic in which the more negative logic level represents a binary 1.

Positive-Logic

1. An active-HIGH signal or code. 2. Digital logic in which the more positive logic level represents a binary 1.

8-2-2 Decimal-to-BCD Encoders

The 74147 IC, shown in Figure 8-8, is an example of a ten-line to four-line (decimal-to-BCD) priority encoder. The bars over the 74147's $\overline{A}_1$ through $\overline{A}_9$ input labels and the small triangle on the input lines are used to indicate that these are active-LOW inputs. This means that the encoder's inputs will be activated whenever these inputs are made LOW. There are also bars over the 74147's $\overline{Q}_0$ through $\overline{Q}_3$ outputs. This is again used to indicate that the output is an active-LOW code. This active-LOW output code is also known as **negative-logic,** and if you need to convert the output to an active-HIGH code, or **positive-logic,** you will need to invert the output from the 74147. For example, looking at the function table in Figure 8-8, you can see that if the $\overline{A}_1$ input is made LOW (active), the 74147 will generate the negative-logic output code HHHL (1110), which will have to be inverted, or reversed, to the positive-logic code LLLH (0001 or binary 1) if the BCD output code is to be equivalent to the activated input $\overline{A}_1$.

As another example, you can see that if the $\overline{A}_9$ input is made LOW (active), the 74147 will generate the negative-logic code LHHL (0110), which will have to be inverted to the positive-logic code HLLH (1001 or binary 9) in order to be equivalent to the activated input

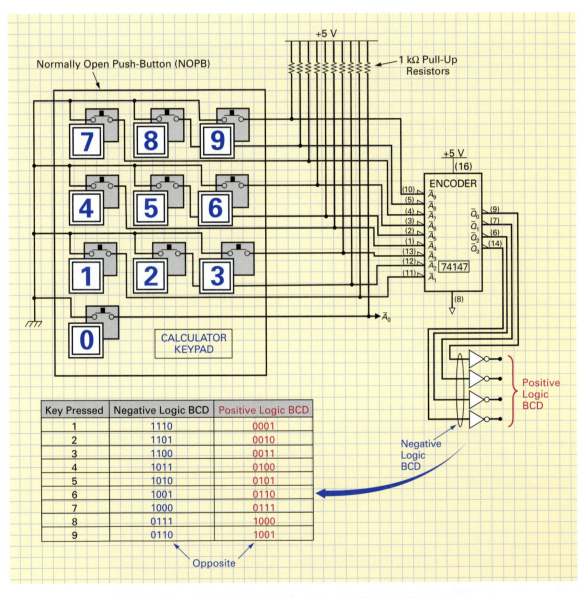

FIGURE 8-8 **Application—A 0-to-9 Keypad Encoder Circuit.**

$\overline{A}_9$. The rest of the function table shows how a LOW on any of the $\overline{A}_1$ through $\overline{A}_9$ inputs will generate an equivalent negative-logic code. To explain one other detail, the 74147 is a **priority encoder,** or high-priority encoder (HPRI), which means that if more than one key is pressed at the same time, the 74147 will only generate the code for the larger digit, or higher priority. For example, if the 9 and 6 keys are pressed at the same time the 74147 will generate a nine output code, responding only to the LOW on the $\overline{A}_9$ input and ignoring the LOW on the $\overline{A}_6$ input. As another example, if input $\overline{A}_8$ is taken LOW, the output code will be LHHH, which is the inverse of 1000, the BCD code for 8. This BCD code of 8, however, will only be present at the output if the $\overline{A}_9$ input is inactive, since the $\overline{A}_9$ input would have a higher priority than the $\overline{A}_8$ input.

Priority Encoder
An encoder circuit that in response to several simultaneous inputs will generate an output code that is equivalent to the highest priority input.

Application—A 0-to-9 Keypad Encoder Circuit

As an application, Figure 8-8 shows how a 74147 can be used to encode a calculator's keypad. Looking first at the push-button switches, which are controlled by the 1 through 9 keypads, you can see that pull-up resistors are included to ensure that the 74147's inputs are normally HIGH. When a key is pressed, the respective input is taken LOW, making that input to the 74147 active. The table shown in the inset in Figure 8-8 shows how this encoder circuit will respond to different inputs. For example, when the 1 key is pressed, the 74147 will receive a LOW at the $\overline{A}_1$ input, so it will generate a 1110 negative-logic output code, which will be inverted by a set of four INVERTER logic gates to the positive-logic code 0001. Therefore, a binary 1 (0001) output code is generated whenever the 1 key is pressed. This circuit, therefore, will generate a positive-logic BCD code at the output of the INVERTER gates that is equivalent to the value of the keypad key pressed.

EXAMPLE:

What would be the negative-logic and positive-logic code generated by the circuit in Figure 8-8 if none of the keys were pressed?

Solution:

Referring to the first line of the function table for the 74147 data sheet shown in Figure 8-7, you can see that when none of the inputs is active (LOW), a 1111 negative-logic output code is generated at the output of the 74147, which, after the inversion, will be a positive-logic code 0000 (0).

SELF-TEST EVALUATION POINT FOR SECTION 8-2

Use the following questions to test your understanding of Section 8-2.

1. How many inputs and outputs would a decimal-to-BCD encoder have?
2. How can the negative BCD logic output of a 74147 be converted to a positive BCD logic output?
3. If the 3, 4, and 7 inputs of a priority encoder were all activated at the same time, what would be the binary output code?
4. Convert the following negative-logic codes to positive-logic codes:
   a. 0010
   b. 1110
   c. LHLL
   d. ON, OFF, ON, OFF

A digital **multiplexer** (MUX, pronounced "mucks"), or selector, is a circuit that is controlled to switch one of several inputs through to one output. The most basic of all multiplexers is the rotary switch, shown in Figure 8-9(a), which will switch any of the six inputs through to the one output based on the switch position selected. In most high-speed applications, the multiplexer will need to be electronically rather than mechanically controlled. **Analog multiplexer** circuits use relays or transistors to electronically switch analog signals, while **digital multiplexer** circuits use logic gates to switch one of several digital input signals through to one output. Figure 8-9(b) illustrates a simple digital multiplexer designed to switch one of its two inputs through to the output (a one-of-two multiplexer). The logic level applied to the SELECT input determines whether AND gate A or B is enabled and therefore which of the data inputs (D_0 or D_1) is switched to the output. If SELECT = 0, AND gate A is enabled while AND gate B is disabled, and therefore output Y will follow input D_0 ($Y = D_0$). On the other hand, if SELECT = 1, AND gate B is enabled while AND gate A is disabled, and output Y will follow input D_1 ($Y = D_1$).

8-3-1   *One-of-Eight Data Multiplexer/Selector*

The 74151 IC shown in Figure 8-10 is an eight-input data multiplexer, or selector. The binary value applied to the three DATA SELECT control inputs (ABC) will determine which one of the eight inputs is switched through to the output. For example, when $ABC = 000$ (binary 0) data input D_0 is switched through to the Y output, or when $ABC = 110$ (binary 6) data input D_6 is switched through to the Y output. An eight-input MUX will therefore need three data select input control lines so that eight binary combinations can be obtained ($2^3 = 8$), and each of these combinations can be used to select one of the eight data inputs. The 74151 has a direct output Y (pin 5) and a complemented, or inverted, output W (pin 6). The active-LOW strobe enable or gate enable ($\overline{G}$) is used to either enable ($\overline{G} = 0$) or disable ($\overline{G} = 1$), as detailed in the function table.

Application—Parallel-to-Serial Data Conversion

Figure 8-10(a) shows how a 74151 can be used as a **parallel-to-serial converter.** Binary data can be transferred either in parallel form or serial form between two digital circuits. The circuit in Figure 8-10 shows the differences between these two data formats by showing how we can convert an 8-bit parallel-data input into an 8-bit serial-data output. Two 4-bit storage registers hold the 8-bit parallel word that has to be converted to serial form. This 8-bit parallel binary word (0110 1001) is applied to the data inputs of the 8-bit multiplexer.

FIGURE 8-9 **Basic Multiplexers.**

FIGURE 8-10 Application—Parallel-to-Serial Data Conversion.

A clock or timing signal is applied to the input of a number 0 through 7 counter whose three outputs (Q_0, Q_1, and Q_2) are applied to the three multiplexer data select inputs (A, B, and C). Referring to the waveforms in Figure 8-10(b), you can see that each positive-going edge of the clock signal causes the counter to advance by one count. As the 3-bit counter is incremented through each of its eight output states (000, 001, 010, 011, 100, 101, 110, and 111), the ABC data-select MUX inputs are sequenced in order, causing each of the eight data inputs (D_0 through D_7) to be selected one after the other. As a result, the 8-bit parallel binary word at the input of the 74151 is switched, one bit at a time, producing an 8-bit serial binary word at the output.

Parallel data transfer is generally not used for the long-distance transmission of binary data (for example, over a telephone line) since it requires several data lines connected in parallel. Parallel data is therefore converted to serial data before it is transmitted, since serial data transmission only requires a single data line. The serial data transfer of binary digits is, however, a lot slower than parallel data transmission. To explain this, let us assume that the time interval allotted to each bit in the example in Figure 8-10 is one microsecond. It will therefore take eight microseconds for the 8-bit word to be transmitted in serial form; however, all of the binary bits in a parallel word are transferred simultaneously in one microsecond over a group of parallel lines. The difference between these two data transfer methods therefore seems to come down to the standard speed-versus-cost trade-off. Serial data transfer needs only one transmitter circuit, one receiver circuit, and one data line to transfer any number of binary bits. Parallel data transfer, on the other hand, requires a separate transmitter, receiver, and line for each bit of the word. The transmission and reception of parallel data is therefore more expensive; however, when high-speed data transfer is required, the data handling time for a parallel transfer of data is extremely fast compared to the data handling time needed for a serial transfer of data.

EXAMPLE:

Which input would be switched through to the Y output if the following logic levels were applied to the 74151?

    a. $A = 0, B = 1, C = 0, D_0 = 1, D_2 = 1, D_3 = 0, D_4 = 1, D_5 = 1, D_6 = 1, D_7 = 0,$
        $G = 1$

    b. $A = 1, B = 1, C = 1, D_0 = 1, D_2 = 1, D_3 = 0, D_4 = 1, D_5 = 1, D_6 = 1, D_7 = 0,$
        $G = 0$

■ *Solution:*

    a. Y = LOW because the gate enable or strobe is not active.

    b. $Y = D_7 = 0$ because select inputs $ABC = 111$, or decimal 7.

Application—A Logic-Function Generator Circuit

Logic-Function Generator

A circuit that will generate the desired logic level at the output for each input combination.

Multiplexers are also very useful as **logic-function generators.** As an example, Figure 8-11(a) shows a truth table and sum-of-products expression for a desired logic-function. By connecting the multiplexer's data inputs either HIGH or LOW, as shown in Figure 8-11(b), we can generate the desired HIGH or LOW output at Y for each of the ABC input combinations. Using a MUX instead of discrete logic gates will greatly reduce the number of ICs used and therefore the circuit cost, size, and power consumption. For example, to implement the sum-of-products expression in Figure 8-11 using discrete logic gates would require four three-input AND gates, one four-input OR gate, and three INVERTERS—a total of four ICs compared to one 74151 IC.

8-3-2   *Four-of-Eight Data Multiplexer/Selector*

Figure 8-12 shows the 74157 IC, which is an eight-input four-output multiplexer. This multiplexer will switch either the A 4-bit word (1A, 2A, 3A, and 4A) or the B 4-bit word (1B, 2B, 3B, and 4B) through to the 4-bit output (1Y, 2Y, 3Y, and 4Y). The active-LOW gate enable input ($\overline{G}$) is used to either enable ($\overline{G} = 0$) or disable ($\overline{G} = 1$) all of the logic gates within the IC. When the gate-enable line is active, the Y outputs will follow whichever inputs have been selected by the $\overline{A}/B$ select control input. For example, when $\overline{A}/B = 0$, the Y outputs will follow whatever is applied to the A inputs, whereas when $\overline{A}/B = 1$, the Y outputs will follow whatever is applied to the B inputs.

C	B	A	Input Selected	Desired Output (Y)
0	0	0	D_0	0
0	0	1	D_1	1 → $\overline{C}\overline{B}A$
0	1	0	D_2	1 → $\overline{C}B\overline{A}$
0	1	1	D_3	0
1	0	0	D_4	0
1	0	1	D_5	1 → $C\overline{B}A$
1	1	0	D_6	1 → $CB\overline{A}$
1	1	1	D_7	0

(a)

$$Y = \overline{C}\overline{B}A + \overline{C}B\overline{A} + C\overline{B}A + CB\overline{A}$$

$$Y = \overline{C}\overline{B}A + \overline{C}B\overline{A} + C\overline{B}A + CB\overline{A}$$

(b)

FIGURE 8-11 Application—A Logic-Function Generator Circuit.

Application—A Multiplexed Display

Most calculators have an $\boxed{X \leftrightarrow M}$ key that is used to select whether the calculator's seven-segment display shows the current value being operated on (X) or the value stored in memory (M). Each time the $\boxed{X \leftrightarrow M}$ key is pressed, the display switches between these two values. Figure 8-12 shows how this function could be implemented using a 74157, 7447, and a seven-segment display. The calculator's $\boxed{X \leftrightarrow M}$ key controls the $\overline{A}/B$ input of the 74157. When $\overline{A}/B = 0$, the contents of the X storage register (which is applied to the 74157's A inputs) is switched through to the Y outputs ($Y = 0110$), decoded by the 7447, and displayed on the seven-segment display (display will show decimal 6). On the other hand, when $\overline{A}/B = 1$, the contents of the M storage register (which is applied to the 74157's B inputs) is switched through to the Y outputs ($Y = 1001$), decoded by the 7447, and displayed on the seven-segment display (display will show decimal 9).

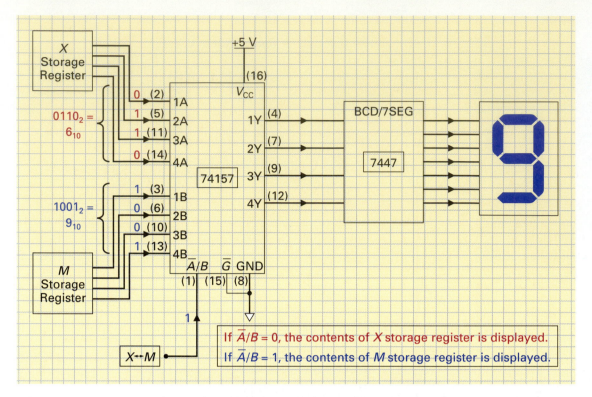

FIGURE 8-12   Application—A Multiplexed Display.

A **multiplexed display** uses one display to present two or more different pieces of information. This "timesharing" of the decoder driver, display, and interconnect wiring greatly reduces the number of ICs needed, power consumption, size, and printed circuit board interconnections. The big disadvantage of using a single multiplexed display instead of two separate displays is that you cannot see both values at the same time.

Multiplexed Display

A display circuit that timeshares the display or the display drive circuitry.

■ **EXAMPLE:**

Which input would be switched through to the output if the following logic levels were applied to the 74157?

a. $\overline{A}/B = 0$, $\overline{G} = 0$, $A = 1001$, $B = 0110$.

b. $\overline{A}/B = 1$, $\overline{G} = 0$, $A = 1001$, $B = 0110$.

■ *Solution:*

a. $Y = A = 1001$. Because the $\overline{G}$ input is LOW, the $\overline{A}/B$ input is LOW and so the A inputs are selected.

b. $Y = A = 0110$. Because the $\overline{G}$ input is LOW, $\overline{A}/B$ input is HIGH and so the B inputs are selected.

SELF-TEST EVALUATION POINT FOR SECTION 8-3

Use the following questions to test your understanding of Section 8-3.

1. A multiplexer is a circuit that through a control input will switch _____ input(s) through to _____ output(s).
   a. one, one of several
   b. one of several, one
   c. one, one
   d. one of several, one of several

2. How many data selects inputs would a one-of-sixteen multiplexer need?

3. True or false: multiplexers are often used for serial-to-parallel data conversion.

4. Which data transmission method is considered the fastest: serial or parallel?

5. Which data transmission method requires more hardware: serial or parallel?

6. Would the 74151 or 74157 best be described as a 4-bit word multiplexer?

8-4   DEMULTIPLEXERS

A digital multiplexer (MUX), or selector, is a circuit that is controlled to switch one of several inputs through to one output. On the other hand, a digital **demultiplexer** (DMUX, pronounced "demucks"), or data distributor, operates in exactly the opposite way since it is a circuit that is controlled to switch a single input through to one of several outputs. The demultiplexer is also known as a *data distributor* since the data applied to the single input is distributed to one of the many outputs.

Figure 8-13(a) shows how a rotary switch can be made to act as a mechanically controlled demultiplexer. In this circuit the switch is controlled to connect a single input through to one of several outputs. Like the multiplexer, mechanical devices are too slow for high-speed applications, and therefore most demultiplexers are also electronically rather than mechanically controlled.

Analog demultiplexers will switch a single analog signal input through to one of several outputs, while **digital demultiplexers** will switch a single digital signal input through to one of several outputs. Figure 8-13(b) shows a simple digital demultiplexer circuit designed to switch its single input through to one of two outputs. This one-line to two-line demultiplexer is controlled by the SELECT INPUT which will determine which AND gate is enabled and which AND gate is disabled, as described in the function table.

Demultiplexer (DMUX)
Also called a data distributor, it is a circuit that is controlled to switch a single input through to one of several outputs.

Analog Demultiplexer
An analog signal demultiplexer circuit.

Digital Demultiplexer
A digital signal demultiplexer circuit.

FIGURE 8-13   Basic Demultiplexers/Data Distributors.

In the following sections you will be introduced to three typical demultiplexer ICs. Two of these ICs were discussed previously in the decoder section and they are reintroduced in this section since most decoder ICs can be made to function as either decoders or demultiplexers.

8-4-1   *A One-Line to Eight-Line Demultiplexer*

The 7442 BCD-to-decimal decoder can be used as a demultiplexer, as shown in Figure 8-14. The internal schematic of the 7442 IC is shown in Figure 8-14(a). As a decoder, any one of

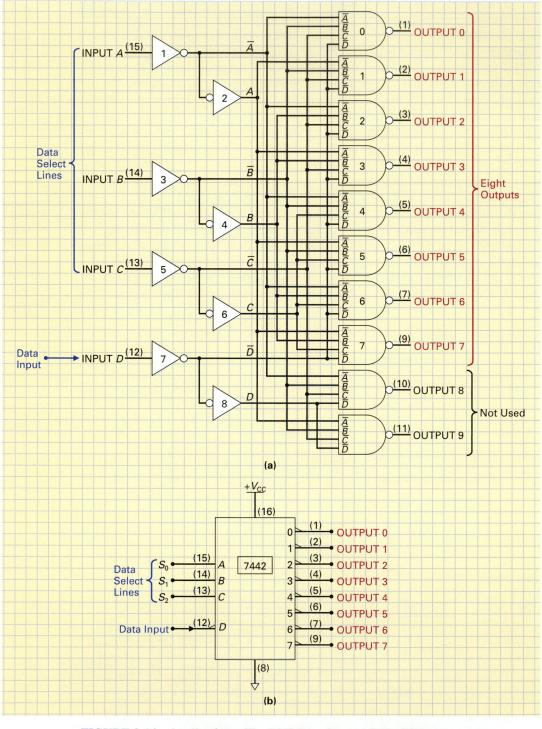

(a)

(b)

FIGURE 8-14   Application—The 7442 Decoder as a Demultiplexer.

the 7442's ten outputs could be driven LOW by applying an equivalent *BCD* input code to the *ABCD* inputs. As a demultiplexer, the 7442 can be made to function as a one-line to eight-line data distributor. To operate in this way, inputs *A, B,* and *C* are used to select one of the internal logic gates 0 through 7 (three select lines will give a total of eight combinations). For example, if 010 is applied to inputs ABC, internal gate 2 will be enabled. The serial data to be distributed is applied to input D, and will be applied to all of the lower inputs of NAND gates 0 through 7. INVERTER 7 will invert the input data, but this process will be reversed by the inversion of the NAND gates. The output of NAND gate 2 will therefore be under the control of the *D* input, so a HIGH input on *D* will appear as a HIGH on output 2 and a LOW input on *D* will appear as a LOW on output 2. When a 7442 IC is used as a demultiplexer, its logic symbol is drawn differently from that of a decoder logic symbol, as shown in Figure 8-14(b).

■   **EXAMPLE:**

The *D* data input would be switched through to which output if the following logic levels were applied to the 7442?

   a. *CBA* = 011   b. *CBA* = 101

■   *Solution:*

   a. Output 3   b. Output 5

8-4-2   *A One-Line to Sixteen-Line Demultiplexer*

Figure 8-15 shows how a 74154 decoder IC can be made to function as a one-line to sixteen-line demultiplexer. With this IC all four of the decoder OUTPUT SELECT inputs are used to select one of the sixteen available outputs ($2^4 = 16$), and the DATA input is applied to one

FIGURE 8-15   Application—The 74154 as a Four-line to Sixteen-line Demultiplexer.

of the gate-enable inputs. When the DATA input is LOW, all of the 74154's internal gates are enabled, so the selected output (based on the OUTPUT SELECT code applied) is driven LOW. On the other hand, when the DATA input is HIGH, all of the 74154's internal gates are disabled, so the selected output is HIGH.

EXAMPLE:

The D input would be switched through to which output if the following logic levels were applied to the 74154?

a. $S = 1011$ b. $S = 1110$

Solution:

a. Output 11   b. Output 14

8-4-3   *A Three-Line to Eight-Line Decoder/Demultiplexer*

Like the 7442 and 74154 ICs, the 74138 shown in Figure 8-16 is listed as a decoder/demultiplexer in the IC data books. The 74138 has three inputs. The G_1 input is active-HIGH, while the $\overline{G}_{2A}$ and the $\overline{G}_{2B}$ inputs are both active-LOW. All three of these gate-enable inputs must be active in order for the IC to be enabled. The three SELECT inputs (ABC) are used to select one of the eight ($2^3 = 8$) active-LOW outputs (Y_0 through Y_7). For example, when $CBA = 101$ (decimal 5), the Y_5 output is made active (LOW) while all the other outputs remain inactive (HIGH).

EXAMPLE:

What would be the outputs from a 74138 for the following input conditions?

a. $CBA = 011, G_1 = $ LOW, $\overline{G}_2 = $ LOW

b. $CBA = 101, G_1 = $ HIGH, $\overline{G}_2 = $ LOW

Solution:

a. All outputs will be HIGH because G_1 is not active.

b. All outputs will be HIGH, except for Y_5 (101), which will be LOW.

Application—A Basic Home Security System

Figure 8-16 shows how a 74151 eight-line to one-line multiplexer (MUX), and a 74138 one-line to eight-line demultiplexer (DMUX) could be connected to form a basic home security system. A binary 0 (000) to 7 (111) counter drives the ABC select or control inputs of both the MUX and DMUX. As the counter counts from 0 to 7, it "selects" in turn each of the 74151 inputs and switches the HIGH or LOW data through to the Y output, as seen in the function table in the inset in Figure 8-16. For example, when the counter is at a count of 011 (3), the HIGH or LOW data present at the three-input of the 74151 (pin 1) is switched through to the Y output. In this example circuit, the three-input of the 74151 is connected to a window switch. If the window is open, its associated switch will also be open and the input to the 74151 will be HIGH due to the pull-up resistors. On the other hand, if the window and its associated switch are closed, ground or a LOW logic level will be applied to the input of the 74151.

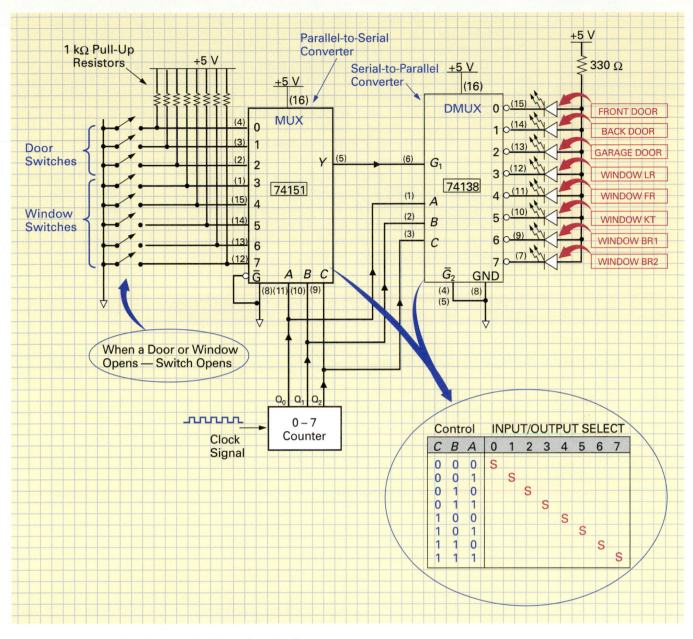

FIGURE 8-16 Application—Basic Home Security System.

The 74138 DMUX in this circuit application operates in exactly the opposite way of the 74151. For example, as we mentioned earlier, when the counter is at a count of 011 (3), the HIGH or LOW data present at the three-input of the 74151 (pin 1) is switched through to the Y output. This HIGH or LOW data is applied to the G_1 input of the 74138, which is also receiving a 011 (3) select input from the counter, so the data at G_1 is switched through to the three-output of the 74138 (pin 12). Since the 74138 has active-LOW outputs, there will be an inversion between input and output. This inversion between the 74138's input and outputs is needed since we are driving a common-anode LED display whose inputs, as we know, are also active-LOW (LEDs are activated by a LOW). To explain the operation from input to output, if the window connected to the three-input is opened, the switch will open and give a HIGH to the three-input of the 74151. The 74151 will switch this HIGH data through to the Y output, where it will be applied to the G_1 input of the 74138, which will invert the HIGH to a LOW and then switch it to the three-output, where it will turn ON the living room (LR) window LED. This circuit, therefore, first senses the condition of switch 0 and then transmits its data to LED

0, and then senses the condition of switch 1 and then transmits its data to LED 1, and then senses the condition of switch 2 and then transmits its data to LED 2, and so on until it reaches 7 (111). The counter then resets and the process is repeated.

The 74151 functions as a parallel-to-serial converter since it converts the parallel 8-bit word applied to its inputs into an 8-bit serial data stream. Conversely, the 74138 functions as a **serial-to-parallel converter,** since it converts an 8-bit serial data stream applied to its input into a parallel 8-bit output word. One question you may have about this circuit is, why don't we simply connect the switches straight to the LEDs and bypass the 74151 and 74138? The answer lies in the basic difference between serial data transmission and parallel data transmission. With this circuit, the 74151 can be placed in close proximity to all the windows and doors, and then only four wires (three for the counter and one for data) need to be run to the 74138 and its display. If the switches were connected directly to the display, eight wires would be needed to connect the switch sensors to the LED display. The advantage of serial data transmission, therefore, is that only one data line is needed. On the other hand, the advantage of parallel data transmission is speed, since a parallel connection would mean that all of the data is transmitted at one time.

Serial-to-Parallel Converter

A circuit that will convert serial data into a parallel format.

Use the following questions to test your understanding of Section 8-4.

1. A demultiplexer or data distributor is a circuit that through a control input will switch _____ input(s) through to _____ output(s).
   a. one, one of several
   c. one of several, one
   b. one, one
   d. one of several, one of several

2. A/an _____ can also be used as a demultiplexer. (encoder/decoder)

3. Which 3-bit code must be applied to the *CBA* inputs of a 74138 to select the Y_4 output?

4. The main difference between a decoder and a demultiplexer is that a decoder has only SELECT inputs, while a demultiplexer has both SELECT inputs and a DATA input. (true/false)

8-5   COMPARATORS

Comparator

A circuit that compares two inputs to determine whether or not the two inputs are equal.

A digital **comparator** compares two binary input words to see if they are equal. The exclusive NOR (XNOR) gate is a comparator logic gate since it compares the two binary input digits applied, as seen in Figure 8-17(a). When the two input bits are the same, the output of the XNOR gate is a 1; however when the input bits are different, the output of the XNOR gate is 0. Let us now see how we could use a group of XNOR gates to create a 4-bit binary comparator circuit.

8-5-1   *A 4-Bit Binary Comparator*

Figure 8-17(b) shows a 4-bit binary comparator circuit that will compare two 4-bit words being applied by two storage registers. Four XNOR gates are used to compare WORD A (A_0, A_1, A_2, and A_3) with WORD B (B_0, B_1, B_2, and B_3). Logic gate XNOR 0 compares A_0 and B_0, and if they are the same, Y_0 is a 1. Logic gate XNOR 1 compares A_1 and B_1, and if they are the same, Y_1 is a 1. In turn, XNOR 2 compares A_2 and B_2, and XNOR 3 compares A_3 and B_3, producing a 1 output if the bits are equal ($=$) and a 0 output if the bits are not equal ($\neq$). If WORD A is identical to WORD B, all of the XNOR outputs will be HIGH, and therefore the output of the AND gate will also be HIGH. If WORD A differs from WORD B in any way, the AND gate will receive a LOW input and will produce a LOW output. Therefore, if WORD A equals WORD B, the output $Y = 1$, whereas if WORD A is not equal to WORD B, the output $Y = 0$.

A	B	Output (Y)	
0	0	1	A = B
0	1	0	A ≠ B
1	0	0	A ≠ B
1	1	1	A = B

(= equal, ≠ not equal)

(a)

$A = B, Y = 1$
$A \neq B, Y = 0$

(b)

FIGURE 8-17   Basic Comparators.

8-5-2   *A 4-Bit Magnitude Comparator*

A wide variety of digital comparator ICs are available that contain a complete internal binary comparator circuit. Let us now examine the operation and application of a frequently used digital comparator IC, the TTL 7485, which is shown in Figure 8-18. The 7485 compares the 4-bit word inputs (WORD *A* and WORD *B*) and provides an active-HIGH output that indicates whether the two input words are equal ($A = B$, pin 6). In addition, the 7485 provides two other active-HIGH outputs indicating whether WORD *A* is greater than WORD *B* ($A > B$, pin 5), and whether WORD *A* is less than WORD *B* ($A < B$, pin 7). The greater-than and less-than magnitude outputs account for why this IC is called a magnitude comparator.

Application—A Photocopier Control Circuit

Figure 8-18 shows how a 7485 comparator could be used in a photocopier to control the number of copies made. A storage register (*A*) holds the 4-bit binary word 0010 (2), indicating the number of copies that have been made so far. The storage register (*B*) holds the 4-bit binary word 1000 (8), indicating the number of copies requested. The 7485 compares these two inputs and generates three output control signals. If the 4-bit input *A* is less than the 4-bit input *B*, the 7485 pin 7 output will be HIGH, signaling that the copier should continue copying. When input *A* is equal to input *B*, the 7485 pin 6 output will be HIGH, signaling that the requested number of copies have been made and the copier should stop copying.

FIGURE 8-18 Application—Comparator Control of a Photocopier.

SELF-TEST EVALUATION POINT FOR SECTION 8-5

Use the following questions to test your understanding of Section 8-5.

1. Which logic gate type is considered a basic comparator?
2. What are the three outputs generally present at the output of a magnitude comparator IC, such as the 7485?
3. Why would a comparator IC have cascading inputs?
4. What would be the logic levels at the three outputs of a 7485 if WORD $A = 1010$ and WORD $B = 1011$?

8-6   PARITY GENERATORS AND CHECKERS

The main objective of a digital electronic system is to manage the flow of digital information or data. In all digital systems, therefore, it is no surprise that a large volume of binary data is continually being transferred from one point to another. The accuracy of this data is susceptible to electromagnetic noise, interference, and circuit fluctuations that can cause a binary 1 to be misinterpreted as a binary 0, or a binary 0 to be misinterpreted as a binary 1. Although the chance of a data transfer error in a two-state system is very remote, it can occur, and if it does, it is important that some checking system is in place so that a data error can be detected. The **parity check** system uses circuitry to check whether the odd or even number of binary 1s present in the transmitted code matches the odd or even number of binary 1s present in the received code.

Parity Check

An odd-even check that makes the total number of 1s in a binary word either an odd or even number.

188   CHAPTER 8 / COMBINATIONAL LOGIC CIRCUITS

8-6-1   *Even or Odd Parity*

Figure 8-19(a) shows a simplified 4-bit parallel data transmission system that has a **parity generator** circuit and a **parity checker** circuit included for data error detection. The parity generator circuit monitors the 4-bit code being transmitted from the keyboard storage register and generates a **parity bit** based on the total number of binary 1s present in the 4-bit parallel word. The table in Figure 8-19(b) shows what EVEN-PARITY BIT (P_E) or ODD-PARITY BIT (P_O) would be generated for each of the possible sixteen 4-bit codes from the keyboard. For example, let us assume that we wish to transmit an even-parity check bit. The P_E will be made a 1 or 0 so that the number of binary 1s in the final transmitted 5-bit code (D_0, D_1, D_2, D_3, and P_E) will always equal an even number (which means that there are zero, two, or four binary 1s in the transmitted 5-bit code). If, on the other hand, we wish to transmit an odd-parity check bit, the P_O will be made a 1 or 0 so that the number of binary 1s in the final transmitted 5-bit code (D_0, D_1, D_2, D_3, and P_O) will always equal an odd number (which means that there are one, three, or five binary 1s in the transmitted 5-bit code).

Referring to the circuit in Figure 8-19(a), you can see that XOR gates are used in both the parity generator and parity checker circuits. In this example circuit, an even-parity check system is being used, and therefore an even-parity bit is generated by XOR *A* and transmitted along with the 4-bit code from keyboard to computer. Within the computer, the 4-bit code is stored in a register, and at the same time it is monitored by XOR gates *B* and *C,* which will produce an active-HIGH PARITY ERROR output. This HIGH error signal could be used to turn on a DATA ERROR LED, signal the computer to display a keyboard data transmission error, or signal the computer to reject the transmitted code and signal the keyboard to retransmit the code for the key pressed.

8-6-2   *A 9-Bit Parity Generator/Checker*

Parity generator and checker ICs, such as the 74180 shown in Figure 8-20, are available for use in parity check applications. The 74180, like all parity generators and checkers, assumes that if an error were to occur, it would more than likely occur in only one of the bits within the transmitted word. If an error were to occur in two bits of the transmitted word, it is possible that an incorrect word will pass the parity test (if a 1 is degraded by noise to a 0, and at the same time a 0 is upgraded by noise to a 1). Multi-bit errors, however, are extremely rare due to the reliability and accuracy of two-state digital electronic systems.

The 74180 is able to function as either a parity generator or as a parity checker. The best way to fully understand the operation of this IC is to see it in an application.

Application—8-Bit Parallel Data Transmission System with Even-Parity Error Detection

Figure 8-20 shows how two 74180 ICs can be used in an 8-bit parallel data transmission system with even-parity error detection. The parity generator (74180 *A*) has its EVEN control input LOW and its ODD control input HIGH. To explain all of the possibilities, Figure 8-21 lists each condition.

a. In the first condition, the 8-bit code to be transmitted from the keyboard contains an even number of 1s, and therefore the parity generator (74180 *A*) generates a LOW, even-parity bit output. The parity checker IC (74180 *B*) receives an even number of 1s at its *A* through *H* inputs, and the same control inputs as the parity generator IC (EVEN input = LOW, ODD input = HIGH), so the even output is LOW, indicating no transmission error has occurred.

b. In the second condition, the 8-bit code to be transmitted from the keyboard contains an even number of 1s, and therefore the parity generator (74180 *A*) generates a LOW, even-parity bit output. In this condition, however, a bit error occurs due to electrical noise so an

Parity Generator

A circuit that generates the parity bit that is added to a binary word so that a parity check can later be performed.

Parity Checker

A circuit that tests the parity bit that has been added to a binary word so that a parity check can be performed.

Parity Bit

A binary digit that is added to a binary word to make the sum of the bits either odd or even.

FIGURE 8-19 Simplified 4-Bit Parallel Data Transmission System with Even-Parity Error Detection.

FIGURE 8-20 Application—Simplified 8-Bit Parallel Data Transmission System with Even-Parity Error Detection.

ODD number of 1s is received at the computer. The parity checker IC (74180 *B*) receives an ODD number of 1s at its *A* through *H* inputs, and the same control inputs as the parity generator IC (EVEN input = LOW, ODD input = HIGH), so the even output is HIGH, indicating a transmission error has occurred.

   c. In the third condition, the 8-bit code to be transmitted from the keyboard contains an odd number of 1s, and therefore the parity generator (74180 *A*) generates a HIGH, even-parity bit output. The parity checker IC (74180 *B*) receives an odd number of 1s at its *A* through *H* inputs, and opposite control inputs to the parity generator IC (EVEN input = HIGH, ODD input = LOW), so the even output is LOW, indicating no transmission error has occurred.

   d. In the fourth condition, the 8-bit code to be transmitted from the keyboard contains an odd number of 1s, and therefore the parity generator (74180 *A*) generates a HIGH, even-parity bit output. In this condition, however, a bit error occurs due to electrical noise so an even number of 1s is received at the computer. The parity checker IC (74180 *B*) receives an

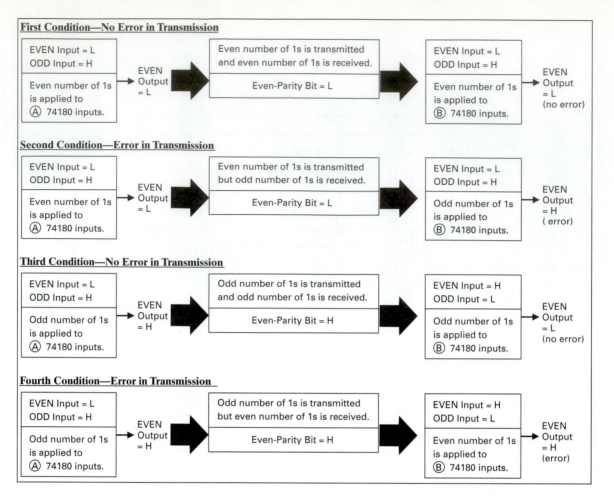

Figure 8-21
```
First Condition—No Error in Transmission
```
| EVEN Input = L
ODD Input = H

Even number of 1s is applied to Ⓐ 74180 inputs. | → EVEN Output = L → | Even number of 1s is transmitted and even number of 1s is received.

Even-Parity Bit = L | → | EVEN Input = L
ODD Input = H

Even number of 1s is applied to Ⓑ 74180 inputs. | → EVEN Output = L (no error) |

```
Second Condition—Error in Transmission
```
| EVEN Input = L
ODD Input = H

Even number of 1s is applied to Ⓐ 74180 inputs. | → EVEN Output = L → | Even number of 1s is transmitted but odd number of 1s is received.

Even-Parity Bit = L | → | EVEN Input = L
ODD Input = H

Odd number of 1s is applied to Ⓑ 74180 inputs. | → EVEN Output = H (error) |

```
Third Condition—No Error in Transmission
```
| EVEN Input = L
ODD Input = H

Odd number of 1s is applied to Ⓐ 74180 inputs. | → EVEN Output = H → | Odd number of 1s is transmitted and odd number of 1s is received.

Even-Parity Bit = H | → | EVEN Input = H
ODD Input = L

Odd number of 1s is applied to Ⓑ 74180 inputs. | → EVEN Output = L (no error) |

```
Fourth Condition—Error in Transmission
```
| EVEN Input = L
ODD Input = H

Odd number of 1s is applied to Ⓐ 74180 inputs. | → EVEN Output = H → | Odd number of 1s is transmitted but even number of 1s is received.

Even-Parity Bit = H | → | EVEN Input = H
ODD Input = L

Even number of 1s is applied to Ⓑ 74180 inputs. | → EVEN Output = H (error) |

FIGURE 8-21   **Four Basic Conditions of 74180 Even-Parity Circuit.**

even number of 1s at its *A* through *H* inputs, and the opposite control inputs to the parity generator IC (EVEN input = HIGH, ODD input = LOW), so the even output is HIGH, indicating a transmission error has occurred.

SELF-TEST EVALUATION POINT FOR SECTION 8-6

Use the following questions to test your understanding of Section 8-6.

1. Include an odd-parity bit for the following 4-bit gray codes.

DECIMAL	GRAY CODE	ODD PARITY BIT
0	0000	?
1	0001	?
2	0011	?
3	0010	?
4	0110	?
5	0111	?

2. The parity system can only be used to detect single-bit errors. (true/false)

3. Is the 74180 used as a parity generator or as a parity checker?

4. If an even-parity bit were added to the ASCII codes being generated by a keyboard, what would be the size of the words being transmitted?

To be an effective electronics technician or troubleshooter, you must have a thorough knowledge of electronics, test equipment, troubleshooting techniques, and equipment repair. In most cases a technician is required to quickly locate the problem within an electronic system and then make the repair. To review, the procedure for fixing a failure can be broken down into three steps.

Step 1:   DIAGNOSE
The first step is to determine whether a circuit problem really exists, or if it is simply an operator error. To carry out this step, a technician must collect as much information as possible about the system, circuit, and components used, and then diagnose the problem.

Step 2:   ISOLATE
The second step is to apply a logical and sequential reasoning process to isolate the problem. In this step, a technician will operate, observe, test, and apply troubleshooting techniques in order to isolate the malfunction.

Step 3:   REPAIR
The third and final step is to make the actual repair, and then final test the circuit.

As far as troubleshooting is concerned, practice really does make perfect. Since this chapter has been devoted to additional combinational logic circuits, let us first examine the operation of a typical combinational logic circuit and then apply our three-step troubleshooting process to this circuit.

8-7-1   *A Combinational Logic Circuit*

Figure 8-22 shows a 2-digit counter with a multiplexed display circuit. The 555 timer A functions as a square-wave oscillator with its frequency controlled by R_2. The clock signal generated by this 555 timer serves as a multiplexing clock signal and is applied to both the 74157 multiplexer and the 1458 dual op-amp. The A op-amp of the 1458 is connected to operate as a comparator, while the B op-amp is connected to operate as an INVERTER. These two op-amps provide the two complementary, or opposite, waveforms shown at the top of Figure 8-22. Since both the units and 10s seven-segment displays are common-anode types, only when the anode pin (14) is made HIGH will the display be activated. Since these anode control signals are out-of-phase with one another, the displays will alternate between ON and OFF, with the units ON and the 10s OFF during one half-cycle, and then the units OFF and the 10s ON during the following half-cycle.

As mentioned previously, the multiplexer control signal from 555 timer A is also applied to the $\overline{A}/B$ switching control input of the 74157 multiplexer, making this multiplexer synchronized with the ON/OFF switching of the seven-segment displays. When the $\overline{A}/B$ multiplexer control line is HIGH, the 74157 switches the 4-bit B-input word from the units counter through to the 74157 Y-outputs. This unit count is decoded by the seven-segment decoder (7447) and the seven-segment code is applied to both of the displays; however, only the units display is enabled at this time and therefore the units count will be displayed on the units display. When the $\overline{A}/B$ multiplexer control line is LOW, the 74157 switches the 4-bit A-input word from the 10s counter through to the 74157 Y-outputs. This 10s count is decoded by the seven-segment decoder (7447), and the seven-segment code is applied to both of the displays; however, only the 10s display is enabled at this time, and therefore the 10s count will be displayed on the 10s display.

The 555 timer B provides a clock signal for the counters. By varying this timer's clock frequency (adjusting R_5), you can vary the rate at which the counters count. Switch 1 is used to clear both counters to 0 when the normally open push-button switch is pressed.

FIGURE 8-22 Troubleshooting Combinational Logic Circuits—A 2-Digit Counter with a Multiplexed Display Circuit.

194

Let us now apply our three-step troubleshooting procedure to this combinational logic circuit so that we can practice troubleshooting procedures and methods.

Step 1: Diagnose

It is extremely important that you first understand the operation of a circuit and how all of the devices within it are supposed to work so that you are able to determine whether or not a circuit malfunction really exists. If you were preparing to troubleshoot the circuit in Figure 8-22, your first step should be to read through the circuit description and review the operation of each integrated circuit until you feel completely confident with the correct operation of the entire circuit. The circuit description, or theory of operation, for an electronic circuit can generally be found in a service or technical manual, along with troubleshooting guides. As far as the circuit's ICs are concerned, manufacturer's digital data books contain a full description of the IC's operation, characteristics, and pin allocation. Referring to all of this documentation before you begin troubleshooting will generally speed up and simplify the isolation step. Many technicians bypass this data collection step and proceed directly to the isolation step. If you are completely familiar with the circuit's operation, this shortcut would not hurt your performance. However, if you are not completely familiar with the circuit, keep in mind the following expression and let it act as a brake to stop you from racing past the problem: *Less haste, more speed.*

Once you are fully familiar with the operation of the circuit, you will easily be able to diagnose the problem as either an *operator error* or a *circuit malfunction.* Distinguishing an operator error from an actual circuit malfunction is an important first step, and a wrong diagnosis can waste a lot of time and effort. For example, the following could be interpreted as circuit malfunctions, when in fact they are simply operator errors:

Example 1.
Symptom: If switch 1 is pressed, both displays go to 0.
Diagnosis: Operator error—This is normal since switch 1 is a display reset switch.

Example 2.
Symptom: Only one display is on at a time. The circuit seems to be switching between the units and tens.
Diagnosis: Operator error—The 555 timer A clock frequency is set too low. Adjust R_2 until the display switching is so fast that the eye sees both displays as being constantly ON.

Example 3.
Symptom: The circuit is malfunctioning because the display is counting up in half-seconds instead of seconds.
Diagnosis: Operator error—There is not a circuit malfunction; the 555 timer B clock frequency is set too high. Adjust R_5 until the display counts up in seconds.

Once you have determined that the problem is not an operator error, but is in fact a circuit malfunction, proceed to step 2 and isolate the circuit failure.

Step 2: Isolate

No matter what circuit or system failure has occurred, you should always follow a logical and sequential troubleshooting procedure. Let us review some of the isolating techniques and apply them to our example circuit in Figure 8-22.

a. Use a cause-and-effect troubleshooting process, which means study the effects you are getting from the faulty circuit and then logically reason out what could be the cause.

b. Check first for obvious errors before leaping into a detailed testing procedure. Is the power OFF or not connected to the circuit? Are there wiring errors? Are all of the ICs correctly oriented?

c. Using a logic probe or voltmeter, test that power and ground are connected to the circuit and are present at all points requiring power and ground. If the whole circuit, or a large section of the circuit, is not operating, the problem is normally power. Using a multimeter, check that all of the dc voltages for the circuit are present at all IC pins that should have power or a HIGH input and are within tolerance. Secondly, check that 0 V or ground is connected to all IC ground pins and all inputs that should be tied LOW.

d. Use your senses to check for broken wires, loose connectors, overheating or smoking components, pins not making contact, and so on.

e. Test the clock signals using the logic probe or the oscilloscope. Although clock signals can be easily checked with a logic probe, the oscilloscope is best for checking the signal's amplitude, frequency, pulse width, and so on. The oscilloscope will also display timing problems and noise spikes (glitches), which could be false-triggering a circuit into operation at the wrong time.

f. Perform a static test on the circuit. With our circuit example in Figure 8-22, you could static test the circuit by disconnecting the clock input to the units counter, and then use a logic pulser's SINGLE PULSE feature to clock the counter in single steps. A logic probe could then be used to test that valid logic levels are appearing at the outputs of the units and tens counters. For example, after resetting the counter and then applying nineteen clock-pulse inputs from the logic pulser, the logic probe should indicate the following logic levels at the counter outputs:

	TENS	UNITS
Counter	0001	1001_{BCD}
Display	1	9_{10}

By also disconnecting the clock output from 555 timer A and then switching the multiplexer control line first HIGH and then LOW, you can trace the 4-bit values through the 74157 multiplexer and 7447 decoder to the displays. Holding the circuit stationary at different stages in its operation (static testing) and using the logic pulser to single-step the circuit and the logic probe to detect logic levels at the inputs and outputs during each step will enable you to isolate any timing problems within the circuit.

g. With a dynamic test, the circuit is tested while it is operating at its normal clock frequency, and therefore all of the inputs and outputs are continually changing. Although a logic probe can detect a pulse waveform, the oscilloscope will display more signal detail, and since it can display two signals at a time, it is ideal for making signal comparisons and looking for timing errors. A *logic analyzer* is a type of oscilloscope that can display typically eight to sixteen digital signals at one time.

In some instances you will discover that some circuits will operate normally when undergoing a static test and yet fail a dynamic test. This effect usually points to a timing problem involving the clock signals and/or the propagation delay times of the ICs used within the circuit.

h. "Noise" due to electromagnetic interference (EMI) can false-trigger an IC, such as the counters in Figure 8-22. This problem can be overcome by not leaving any of the IC's inputs unconnected and therefore floating. Connect unneeded active-HIGH inputs to ground and active-LOW inputs to $+V_{CC}$.

i. Apply the half-split method of troubleshooting first to a circuit and then to a section of a circuit to help speed up the isolation process. With our circuit example in Figure 8-22, a good midpoint check would be to first static-test the inputs to the seven-segment displays. If the units control signal or the tens control signal to the seven-segment displays is not as it should be, then the problem is more than likely in the upper section of the circuit (555 timer A and the 1458). If, on the other hand, the seven-segment code from the 7447 is not as it should be, then the problem more than likely lies in the lower section of the circuit (555 timer B, the counters, the 74157, and the 7447).

Remember that a load problem can make a source appear at fault. If an output is incorrect, it would be safer to disconnect the source from the rest of the circuit and then recheck the output. If the output is still incorrect, the problem is definitely within the source; however, if the output is correct, the load is probably shorting the line either HIGH or LOW.

j. Substitution can be used to help speed up your troubleshooting process. Once the problem is localized to an area containing only a few ICs, substitute suspect ICs with known-working ICs (one at a time) to see if the problem can be quickly remedied.

Step 3: Repair

Once the fault has been found, the final step is to repair the circuit, which could involve simply removing an excess piece of wire, resoldering a broken connection, reconnecting a connector, or adjusting the power supply voltage or clock frequency. In most instances, however, the repair will involve the replacement of a faulty component. For a circuit that has been constructed on a prototyping board or bread board, the removal and replacement of a component is simple; however, when a printed circuit board is involved, you should make a note of the component's orientation and observe good soldering and desoldering techniques. Also be sure to handle any MOS ICs with care to prevent any damage due to static discharge.

When the circuit has been repaired, always perform a final test to see that the circuit and the system are now fully operational.

8-7-2   *Sample Problems*

Once you have constructed a circuit like the 2-digit counter with a multiplexed display shown in Figure 8-22, introduce a few errors to see what effect or symptoms they produce. Then logically reason out why a particular error or cause has a particular effect on the circuit. Never short any two points together unless you have carefully thought out the consequences. It is, however, generally safe to open a path and see the results. Here are some problems for our example circuit in Figure 8-22.

a. Disconnect the output of 555 timer *A* from the circuit.

b. Disconnect the output of 555 timer *A* from the 1458 input.

c. Open the pin 14 connection to the tens display.

d. Disconnect an input to the 7447.

e. Disconnect the output of 555 timer *B* from the counter.

f. Open the connection between 555 timer *A* and the 74157 $\overline{A}/B$ input.

g. Disconnect power to the 7447.

h. Disconnect one of the 7447 outputs from the 7447 and connect it to ground.

SELF-TEST EVALUATION POINT FOR SECTION 8-7

Use the following questions to test your understanding of Section 8-7.

1. What are the three basic troubleshooting steps?

2. What is a static test?

3. What is the half-split troubleshooting technique?

4. What symptom would you get from the circuit in Figure 8-22 if the units and tens control signals from the 1458 were in-phase with one another?

REVIEW QUESTIONS

Multiple-Choice Questions

1. A _____ is controlled to switch a single input through to one of several outputs.
   a. multiplexer   c. demultiplexer
   b. comparator   d. encoder

2. A _____ is controlled to switch one of several inputs through to one output.
   a. multiplexer   c. demultiplexer
   b. comparator   d. encoder

3. What code conversion is performed by a 7448 IC?
   a. BCD to gray   c. BCD to seven-segment
   b. Hex to BCD   d. Gray to seven-segment

4. _____ are designed to detect specific codes, while _____ are designed to generate specific codes.
   a. Code converters, decoders
   b. Decoders, encoders
   c. Encoders, code converters
   d. Encoders, decoders

5. A decimal-to-BCD encoder will have _____ inputs and _____ outputs.
   a. ten, four   c. four, ten
   b. three, ten   d. ten, three

6. A _____ translates coded characters into a more understandable form.
   a. decoder   c. multiplexer
   b. encoder   d. code converter

7. How many inputs and outputs would a BCD-to-decimal decoder have?
   a. three inputs, sixteen outputs
   b. ten inputs, four outputs
   c. ten inputs, one output
   d. four inputs, ten outputs

8. A common-cathode seven-segment display would need active-_____ inputs, while a common-anode seven-segment display would need active-_____ inputs.
   a. HIGH, LOW   c. LOW, HIGH
   b. LOW, LOW   d. HIGH, HIGH

9. Multiplexers can be used for _____ data conversion, while demultiplexers can be used for _____ data conversion.
   a. parallel-to-serial, serial-to-parallel
   b. parallel-to-serial, parallel-to-serial
   c. serial-to-parallel, serial-to-parallel
   d. serial-to-parallel, parallel-to-serial

10. Demultiplexers are often referred to as _____.
    a. encoders   c. comparators
    b. data selectors   d. data distributors

11. Multiplexers are often referred to as _____.
    a. encoders   c. comparators
    b. data selectors   d. data distributors

12. _____ ICs can also be used as _____ ICs.
    a. Encoder, decoder   c. Decoder, demultiplexer
    b. Decoder, multiplexer   d. Demultiplexer, encoder

13. A digital _____ is a circuit used to determine whether two parallel binary input words are equal or unequal.
    a. multiplexer   c. decoder
    b. comparator   d. parity generator

14. What would the even-parity bit be for the code 0011011?
    a. HIGH   c. Either 1 or 0
    b. LOW   d. None of the above

15. That would be the odd-parity bit for the code 0011011?
    a. HIGH   c. Either 1 or 0
    b. LOW   d. None of the above

16. The parity check system can be used to detect single- or multiple-bit errors.
    a. True   b. False

17. Which basic logic gate functions as an even-parity generator?
    a. XNOR   c. NOR
    b. NAND   d. XOR

18. What are the three steps in the three-step troubleshooting procedure?
    a. Isolate, diagnose, repair   c. Diagnose, isolate, repair
    b. Repair, diagnose, isolate   d. Isolate, repair, diagnose

19. Disconnecting the clock signal to a circuit and then point-to-point testing all the logic levels throughout the circuit is an example of _____ testing.
    a. cause-and-effect   c. half-split
    b. dynamic   d. static

20. Running a circuit at its normal operating frequency and testing different points throughout the circuit is an example of _____ testing.
    a. cause-and-effect   c. half-split
    b. dynamic   d. static

Practice Problems

To practice your circuit recognition and operation, refer to Figure 8-23 and answer the following six questions.

21. What is the function of the 7442 in this circuit?

22. What is the purpose of R_1 through R_4?

23. Do the LEDs in this circuit form a common-anode or common-cathode display?

24. Are the outputs from the 7442 active-LOW or active-HIGH?

25. What would happen to the output display if the following input switches were pressed simultaneously?
    a. 3 and 1
    b. 0, 2, and 3
    c. 2 and 1

26. If the number 4 LED were ON, what binary logic levels would you expect to find on the *ABCD* inputs of the 7442, if you were to check using the logic probe?

To practice your circuit recognition and operation, refer to Figure 8-24 and answer the following six questions.

27. What is the function of the 7448 in this circuit?

28. Is the seven-segment display in this circuit a common-anode or common-cathode display?

29. Are the outputs of the 7448 in this circuit active-LOW or active-HIGH?

30. What is the purpose of the push-button switch in this circuit?

31. What would be displayed if the start/stop push button were held down permanently?

32. If decimal 8 were displayed on the seven-segment display, what binary logic levels would you expect to find on the 7448's input and output pins if you were to check using a logic probe?

To practice your circuit recognition and operation, refer to Figure 8-25 and answer the following three questions.

33. Is this circuit a decoder or encoder?

34. Briefly describe the operation of this circuit.

35. Determine the *A, B,* and *C* outputs for each of the rotary switch positions and insert the results in the function table alongside the circuit.

To practice your circuit recognition and operation, refer to Figure 8-26 and answer the following two questions.

36. Is the circuit operating as a serial-to-parallel or parallel-to-serial data transmission circuit?

37. What binary data will appear at the output as the counter cycles from 0 through 7?

FIGURE 8-23 Application—A Decimal Decoder Circuit.

FIGURE 8-24 Application—A 0–9 Second Timer Circuit.

To practice your circuit recognition and operation, refer to Figure 8-27 and answer the following five questions.

38. Briefly describe the operation of the circuit, including how the display will show the binary value of whichever push-button key is pressed.

39. If push button 4 is pressed, the output of the 74151 will output a LOW when the counter has reached a count of _____, and this LOW output from the 74151 will cause the NAND gate to _____ (block/pass) the clock pulses to the 74193. The binary display will therefore display _____.

40. The display in this circuit is a common-_____ display, and therefore a LED ON indicates a (LOW/HIGH) logic level.

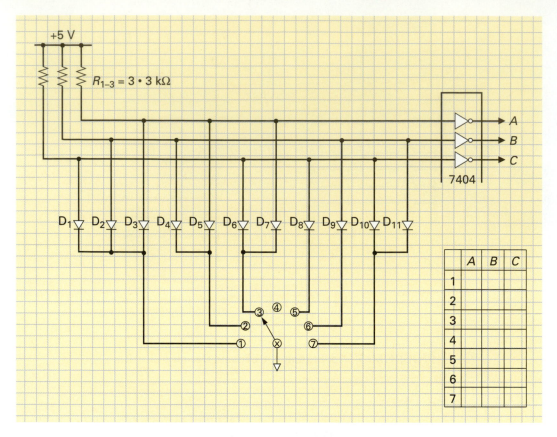

FIGURE 8-25   Application—A Switch Encoder Circuit.

FIGURE 8-26   Application—8-Bit Serial Word Generator.

FIGURE 8-27 Application—0 through 7 Keyboard Encoder Circuit.

Web Site Questions

Go to the web site http://www.prenhall.com/cook, select the textbook *Practical Digital Electronics,* select this chapter, and then follow the instructions when answering the multiple choice practice problems.

Flip-Flops and Timers

William Shockley

The Persistor

In the late 1930s, experimenters had successfully demonstrated that semiconductors could act as rectifiers and be used in place of the then-dominant vacuum tube. The only problem was that nobody could work out how to control semiconductors to make them predictable so they could function like the vacuum tube as an amplifier or a switch. It was not until the beginning of World War II that researchers and physicists began seriously investigating semiconductors in hopes that they could be used to create superior components for radar applications.

Convinced that semiconductor devices would be playing a major role in the future of electronics, AT&T's Bell Laboratories launched an extensive research program. Their goal was to develop a component to replace the vacuum-tube amplifier and mechanical-relay switch that were being used extensively throughout the nation's telephone system. The three men leading the assault were Walter Brattain, a sixteen-year Bell experimenter, John Bardeen, a young theoretician, and the team leader, William Shockley.

Shockley had been fascinated by semiconductors for the past ten years and had a talent for simplifying a research problem to its basic elements and then directing the project into its next avenue of investigation. He had a dominant, intensely serious, and competitive personality and on many occasions forged ahead without waiting to evaluate all of the details of the present situation. On one occasion, after several months of extensive research on a semiconductor amplifier, one of Shockley's ideas that worked perfectly in theory failed completely when tested. Shockley forged off in another direction, but Bardeen and Brattain became fascinated by the reasons for the failure.

On December 23, 1947, after three years of research and at a cost of $1 million, Brattain and Bardeen conducted an experiment on their newly constructed semiconductor prototype. Applying an audio signal to the input of the device, a replica was viewed on an oscilloscope at the output which was fifty times larger in size—they had achieved amplification. The input signal seemed to control the resistance between the output terminals, and this controlled the amount of current flowing through the resistive section of the device. Since the input controlled the amount of current transferring through the resistive section of the device, the component was called a *transresistor,* a name which was later shortened to *transistor.* Six months later, the device was released on the market, but its unpredictable behavior and cost of eight dollars, compared to its seventy-five-cent vacuum-tube counterpart, resulted in it not making much of an impression.

To perhaps make up for his lack of involvement in the final design, Shockley immediately began work on refining the design and removing what he called "the mysterious witchcraft" that the device seemed to possess. In a matter of days, Shockley had much of the theory outlined, but perfecting the design was going to be a long and arduous task. One colleague working with Shockley jokingly referred to the device as a "persistor" because of its stubbornness to work. Shockley's persistence, however, eventually won out, and in 1951 the first reliable commercial transistor was released. It could do everything that its vacuum-tube counterpart could do and it was smaller, had no fragile glass envelope, did not need a heating filament to warm it up, and only consumed a fraction of the power.

In 1956, Shockley, Brattain, and Bardeen were all recognized by the world's scientific community when they shared the Nobel Prize in physics. Today this point in history is recognized as the most significant milestone in the history of electronics. Every electronic system manufactured today has an abundance of transistors within its circuitry—not bad for a device that was originally ridiculed by many as "a flash in the pan."

Introduction

In the previous chapter we discussed how logic gates could be used in combination to decode, encode, select, distribute, and check binary data. In this chapter you will see how logic gates can be used in combination to form a memory circuit that is able to store binary information. This memory circuit is called a flip-flop because it can be flipped into its set condition in which it stores a binary 1 or flopped into its reset condition in which it stores a binary 0.

There are three basic types of flip-flop circuits available:

the Set-Reset (S-R) flip-flop
the Data-type (D-type) flip-flop
the J-K flip-flop

The flip-flop is a digital logic circuit that is capable of storing a single bit of data. It is able to store either a binary 1 or a binary 0 because of the circuit's two stable operating states—SET and RESET. Once the flip-flop has been flipped into its set condition (in which it stores a binary 1) or flopped into its reset condition (in which it stores a binary 0), the output of the circuit remains latched or locked in this state as long as power is applied to the circuit. This latching or holding action accounts for why the SET or RESET (*S-R*) flip-flop is also known as a SET or RESET latch. Let us now examine this circuit in detail.

9-1-1 *NOR* S-R *Latch and NAND* S-R *Latch*

Figure 9-1 shows how set-reset flip-flop circuits can be constructed using logic gates. Figure 9-1(a) shows how two NOR gates could be connected to form a set-reset NOR latch, or flip-flop. Looking at the truth table for the NOR latch in Figure 9-1(b), you can see that when

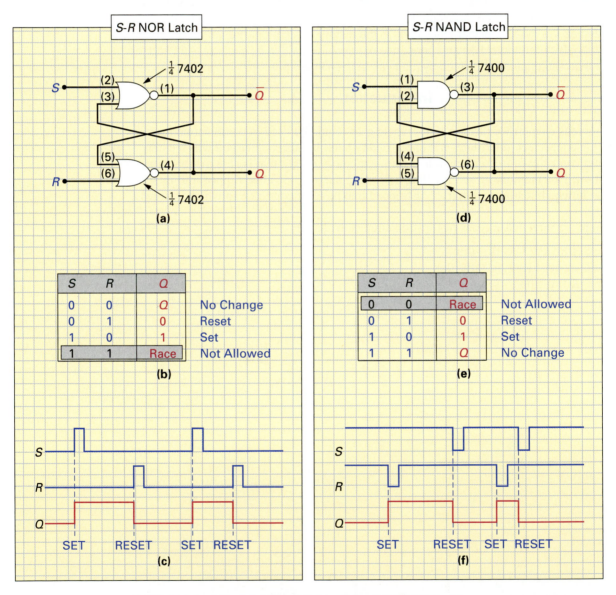

FIGURE 9-1 The *S-R* NOR Latch and the *S-R* NAND Latch.

both the S and R inputs are LOW, there will be a **NO CHANGE condition** in the Q's output logic level. This means that if the Q output is HIGH it will stay HIGH, and similarly, if the Q output is LOW it will stay LOW. In the second line of the truth table, you can see that a HIGH on the R input will RESET the Q output to 0, while the third line of the truth table shows how a HIGH on the S input will SET the Q output to 1. The last line of the truth table shows that when the S and the R inputs are both HIGH, the circuit will race and the Q output will be unpredictable. The S-R latch is only operated in the NO CHANGE, RESET, and SET conditions. The *RACE* condition is not normally used because the output cannot be controlled or predicted.

The timing diagram in Figure 9-1(c) shows how the Q output can be either SET HIGH by momentarily pulsing the S input HIGH or RESET LOW by momentarily pulsing the R input HIGH. Let us study a few examples to see how this circuit operates.

EXAMPLE:

Remembering that any HIGH into a NOR gate will give a LOW output, and that only when both inputs are LOW will the output be HIGH, study Figure 9-2, which shows a variety of input-output conditions for the set-reset NOR latch. Determine the Q output for each of the examples in Figure 9-2.

Solution:

By considering the S and R inputs to the NOR gates and the Q and $\overline{Q}$ outputs that are cross-coupled to the other NOR gate inputs, we can determine the final NOR gate outputs.

a. In Figure 9-2(a), NOR gate X inputs = 0 and 0, therefore $\overline{Q}$ will be 1. NOR gate Y inputs = 0 and 1, therefore Q will be 0. In this instance, the Q and $\overline{Q}$ outputs will not change ($Q = 0$, $\overline{Q} = 1$). This ties in with the truth table given in Figure 9-1(b), which states that when $S = 0$ and $R = 0$, there will be no change in Q ($Q = Q$).

b. Moving to the example in Figure 9-2(b), you can see that the only change that has occurred is that the S input is now HIGH. This HIGH input to NOR gate X will produce a LOW output at $\overline{Q}$, which will be coupled to the input of NOR gate Y. With both inputs LOW,

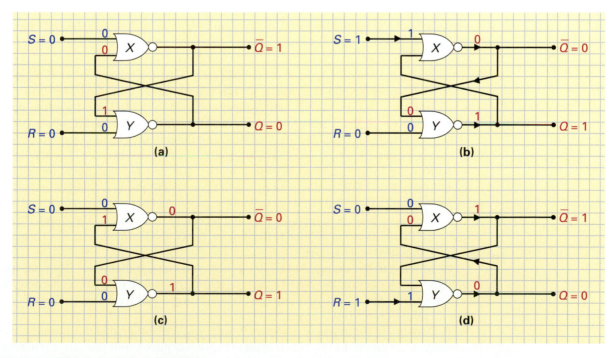

FIGURE 9-2   NOR Latch Input Conditions.

the output of NOR gate Y will go HIGH, and Q will therefore change to a 1 and $\overline{Q}$ to a 0. In this example, a SET condition was applied to the input to the NOR latch and the Q output was therefore SET to 1.

    c. Moving to the example in Figure 9-2(c), you can see that the Q output is now SET HIGH. The S input, which was taken HIGH momentarily, has now gone LOW and therefore the NOR latch inputs are in the NO CHANGE condition. NOR gate X has inputs of 0 and 1, and therefore $\overline{Q} = 0$ (no change). NOR gate Y has inputs 0 and 0, and therefore $Q = 1$ (no change).

    d. Moving to the example in Figure 9-2(d), you can see that the R input is now being pulsed HIGH. This HIGH input to NOR gate Y will produce a LOW output at Q, which will be cross-coupled back to the input of NOR gate X. With two LOW inputs to NOR gate X, its $\overline{Q}$ output will be HIGH. This HIGH level at $\overline{Q}$ will be coupled to the input of NOR gate Y and will keep Q at 0 even when R returns to a LOW. This RESET input condition ($S = 0$, $R = 1$) will therefore reset the Q output to 0.

    From these examples you may have noticed that it is the cross-coupling that keeps the circuit outputs latched in either the SET or RESET condition.

    An S-R latch can also be constructed using two NAND gates, as shown in Figure 9-1(d). Referring to the truth table in Figure 9-1(e), you can see that the NO CHANGE and RACE conditions for a NAND latch are opposite that of the NOR latch. With the NAND latch, a LOW on both inputs will cause the circuit to race and result in an unpredictable output, while a HIGH on both inputs will result in no change at the outputs. Since the NO CHANGE condition occurs when both inputs are HIGH, the S and R inputs will normally be HIGH, as seen in the timing diagram in Figure 9-1(f). To SET the Q output to a 1, therefore, the R input must be pulsed LOW so that $S = 1$ and $R = 0$ (SET condition). On the other hand, to reset the Q output to a 0, the S input must be pulsed LOW so that $S = 0$ and $R = 1$ (RESET condition). Let us study a few examples to see how this circuit operates.

■ EXAMPLE:

Remembering that any 0 into a NAND gate gives a 1 output, and only when both inputs are 1 will the output be 0, study Figure 9-3, which shows a variety of input-output conditions for the set-reset NAND latch. Determine the Q output for each of the examples in Figure 9-3.

■ *Solution:*

By considering the S and R inputs to the NAND gates and the Q and $\overline{Q}$ outputs that are cross-coupled to the other NAND gate inputs, we can determine the final NAND gate outputs.

    a. In Figure 9-3(a), NAND gate X inputs $= 1$ and 0 and $\overline{Q}$ will therefore be 1, and NAND Y inputs $= 1$ and 1 and Q will therefore be 0. With both S and R inputs HIGH, the NAND latch is in the NO CHANGE condition, and the outputs Q and $\overline{Q}$ therefore remain in their last condition.

    b. Moving to the example in Figure 9-3(b), you can see that the R input has been brought LOW. This LOW to the input of NAND gate Y will produce a HIGH output, which will change the Q to a 1. The HIGH Q will be cross-coupled back to the input of NAND gate X, which will now have both inputs HIGH and therefore produce a LOW output. In this example, a SET condition was applied to the input of the NAND latch ($S = 1$, $R = 0$), and therefore the Q output was set HIGH.

    c. Moving to the example in Figure 9-3(c), you can see that the R input has returned to a HIGH level, causing the NAND latch to return to a NO CHANGE condition. NAND gate X inputs $= 1$ and 1, and therefore $\overline{Q} = 0$ (no change), and NAND gate Y inputs $= 0$ and 1, and therefore $Q = 1$ (no change).

    d. Moving to the example in Figure 9-3(d), you will notice that the S input is now being momentarily pulsed LOW. This 0 applied to NAND gate X will produce a HIGH output, and therefore $\overline{Q} = 1$. This HIGH at $\overline{Q}$ will be cross-coupled to the input of NAND gate Y,

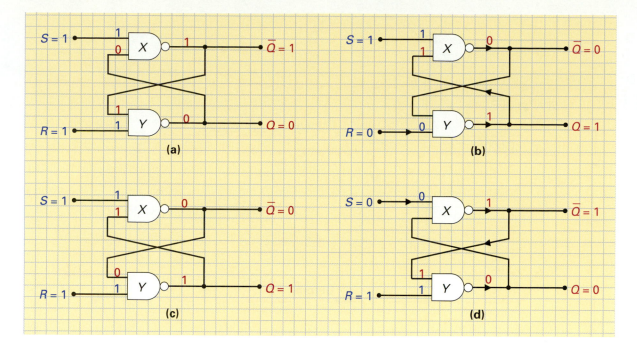

FIGURE 9-3 NAND Latch Input Conditions.

and since both of its inputs are now HIGH, it will produce a LOW output at Q. This RESET input condition ($S = 0$, $R = 1$) will therefore reset Q to a 0.

Like the NOR latch, it is the cross-coupling between the NAND gates that keeps the circuit latched into either the SET or RESET condition.

Application—Contact-Bounce Eliminator Circuit

Set-reset flip-flops or latches are often used as **switch debouncers,** as shown in Figure 9-4. Whenever a switch is thrown into a new position, the moving and stationary contacts of the switch tend to bounce, causing the switch to make and break for a few milliseconds before finally settling in the new position. This effect is illustrated in Figure 9-4(a) and (b), which show the before and after conditions. In Figure 9-4(a), the ON/OFF is in the OFF position, and due to the action of the pull-up resistor R_1, the input to the INVERTER gate is HIGH and the final output is therefore LOW. In Figure 9-4(b), the switch has been thrown from the OFF to the ON position, and due to the bouncing of the switch's moving contact (pole) on the stationary ON contact, the output vibrates HIGH and LOW several times before it settles down permanently in the ON position. If the output of this circuit were connected to the input of a counter, the counter would erroneously assume that several OFF-to-ON transitions had occurred instead of just one. It is therefore vital in any digital system that contact bounce be eliminated.

Figure 9-4(c) shows how a S-R NAND latch can be used as a *contact-bounce eliminator.* With the switch in the OFF position, the S input is LOW and the R input is HIGH. This RESET input condition ($S = 0$, $R = 1$) resets the Q output to 0, as seen in the timing diagram in Figure 9-4(c) at time t_0. When the switch is thrown to the ON position (time t_1), the S input is pulled permanently HIGH and the R input will bounce between LOW and HIGH for a few milliseconds. The first time that the switch's pole makes contact and the R input goes LOW, the latch will be SET since $S = 1$ and $R = 0$. Subsequent bounces will have no effect on the SET or HIGH Q output, since the alternating R input is only causing the latch to switch between its SET ($S = 1$, $R = 0$) and NO CHANGE ($S = 1$, $R = 1$) input conditions.

Similarly, when the switch is thrown into its OFF position, the R input will be pulled permanently HIGH and the S input will bounce between LOW and HIGH for a few milliseconds.

Switch Debouncer

A circuit designed to eliminate the contact bounce that occurs whenever a switch is thrown into a new position.

FIGURE 9-4   Application—The *S-R* Latch as a Contact-Bounce Eliminator.

The first time that the switch's pole makes contact and the S input goes LOW (t_2), the latch will be RESET ($S = 0$, $R = 1$). Subsequent bounces will have no effect on the RESET or LOW Q output, since the alternating S input is only causing the latch to switch between its RESET ($S = 0, R = 1$) and NO CHANGE ($S = 1, R = 1$) input conditions.

9-1-2   *Level-Triggered S-R Flip-Flops*

A typical digital electronic system contains many thousands of flip-flops. To coordinate the overall operation of a digital system, a clock signal is applied to each flip-flop to ensure that each device is **triggered** into operation at the right time. A clock signal, therefore, controls when a flip-flop is enabled or disabled and when its outputs change state.

Figure 9-5(a) illustrates how a *S-R* NAND latch can be controlled by the HIGH or positive level of a clock signal. NAND gates A and B act as controlled switches, only allowing the S and R inputs through to the inputs of the latch when the clock signal is HIGH.

Triggered

A device or circuit that initiates an action in response to a trigger pulse.

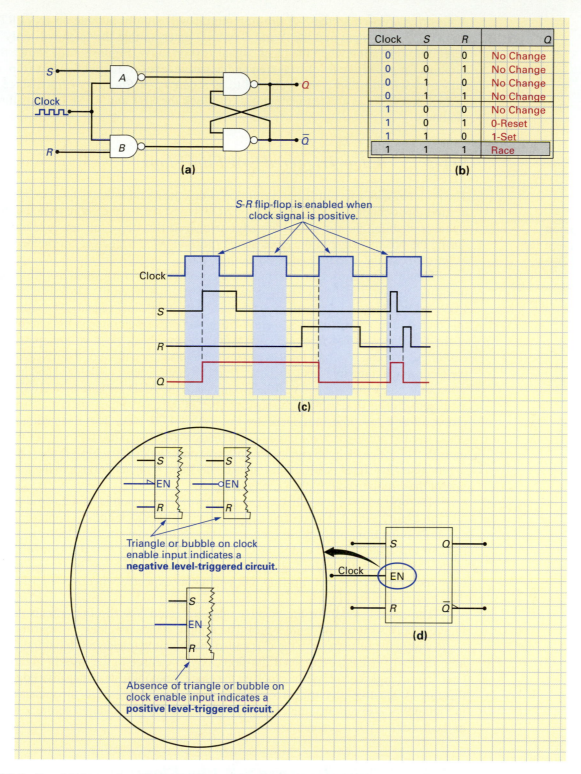

Clock	S	R	Q
0	0	0	No Change
0	0	1	No Change
0	1	0	No Change
0	1	1	No Change
1	0	0	No Change
1	0	1	0-Reset
1	1	0	1-Set
1	1	1	Race

(b)

S-R flip-flop is enabled when clock signal is positive.

(c)

Triangle or bubble on clock enable input indicates a **negative level-triggered circuit**.

Absence of triangle or bubble on clock enable input indicates a **positive level-triggered circuit**.

(d)

FIGURE 9-5   Level-Triggered or Gated Set-Reset Flip-Flop.

This operation is summarized in the truth table in Figure 9-5(b). Whenever the clock signal is 0, the latch is in the NO CHANGE condition; however, when the clock signal is 1, the *S-R* latch functions.

Referring to the waveforms in Figure 9-5(c), you can see that this *S-R* latch is only enabled when the clock signal is positive. This circuit is normally referred to as a *positive level-triggered S-R flip-flop* since a positive clock level will trigger the circuit to operate.

Level-Triggered

A device or circuit that initiates an action in response to a trigger's logic level.

A *negative level-triggered S-R flip-flop* circuit will simply include an INVERTER between the clock input and the *A* and *B* NAND gates so that the latch will only be triggered into operation when the clock signal is 0 or LOW. Figure 9-5(d) shows the logic symbols for a positive level-triggered and negative level-triggered *S-R* flip-flop. A level-triggered *S-R* flip-flop is also referred to as a *gated S-R flip-flop or latch* since NAND gates *A* and *B* act as a gate to either pass or block the SET and RESET inputs through to the latch.

9-1-3   *Edge-Triggered* S-R *Flip-Flops*

Edge-Triggered

A device or circuit that initiates an action in response to a trigger's positive or negative edge.

The problem with a level-triggered flip-flop is that the *S* and *R* inputs have to be held in the desired input condition (SET, RESET, or NO CHANGE) for the entire time that the clock signal is enabling the flip-flop. With **edge-triggered** flip-flops, the device is only enabled during the positive edge of the clock signal *(positive edge-triggering)* or the negative edge of the clock signal *(negative edge-triggering)*.

Figure 9-6(a) shows how the previously discussed level-triggered *S-R* flip-flop can be modified so that it will only be triggered into operation when the clock is changing from LOW to HIGH (positive transition or positive edge of the clock signal). The modification to this circuit involves including a *positive-transition pulse generator* in the internal circuitry of the *S-R* flip-flop. The inset in Figure 9-6(a) shows a basic positive edge pulse generator circuit. As you can see from the waveforms in the inset, one input of the AND gate receives the clock signal directly (input *Y*) while the other input receives a delayed and inverted version of the clock input (input *X*). The two inputs to the AND gate are both HIGH for only a short period of time, starting at the positive edge of the clock input and lasting as long as the propagation delay time introduced by the INVERTER gate. The final output from this positive-edge pulse-generator circuit is one narrow positive pulse for every LOW to HIGH transition of the clock input signal. This positive pulse will enable NAND gates *A* and *B* during the time this signal is HIGH and connect the *S* and *R* inputs through to the *S-R* latch section of the circuit to control the *Q* and $\overline{Q}$ outputs.

The operation of a positive edge-triggered *S-R* flip-flop is summarized by the truth table in Figure 9-6(b). The up and down arrows in the clock column represent the rising (positive transition) and falling (negative transition) edges of the clock. The first three rows of this table indicate that if the clock is at a LOW level (0), HIGH level (1), or on a negative edge (↓), the output *Q* will remain unchanged. The last four rows of this truth table indicate how the *S-R* flip-flop will operate as expected whenever a positive edge of the clock is applied (↑).

The logic symbol for the positive edge-triggered *S-R* flip-flop circuit discussed in Figure 9-6(a) is shown in Figure 9-6(c). The *C* input stands for clock, or control, and the triangle within the block next to the *C* is called the dynamic (changing) input indicator and is used to identify an edge-triggered flip-flop. As you can see in the inset in Figure 9-6(c), a bubble or triangle outside of the block on the clock input indicates a negative edge-triggered flip-flop. A negative edge-triggered flip-flop circuit would be exactly the same as the circuit in Figure 9-6(a) except that the clock input would have to be inverted before it was applied to the transition pulse generator.

9-1-4   *Pulse-Triggered* S-R *Flip-Flops*

Pulse-Triggered

A device or circuit that initiates an action in response to a complete trigger pulse.

Up to this point, we have seen how set-reset flip-flops can be controlled by the level or edge of a clock signal. A **pulse-triggered** *S-R* flip-flop is a level-clocked flip-flop; however, both the HIGH and LOW levels of the input clock are needed before the *Q* outputs will reflect the *S* and *R* input condition. With this circuit, therefore, a complete cycle of the input clock (a complete clock pulse) is needed to trigger the circuit into operation, which is why this circuit is called a pulse-triggered flip-flop.

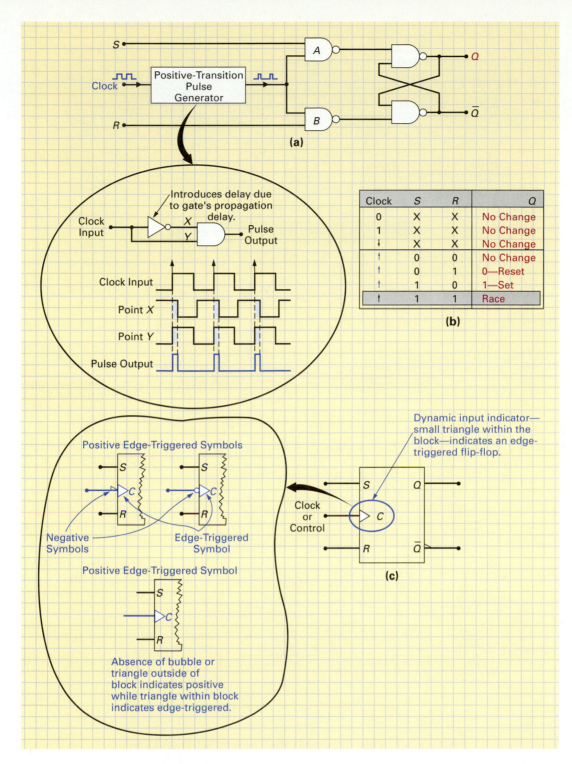

FIGURE 9-6 **Edge-Triggered Set-Reset Flip-Flops.**

Clock	S	R	Q
0	X	X	No Change
1	X	X	No Change
↓	X	X	No Change
↑	0	0	No Change
↑	0	1	0—Reset
↑	1	0	1—Set
↑	1	1	Race

(b)

Figure 9-7(a) shows a simplified circuit of a pulse-triggered *S-R* flip-flop. Pulse-triggered circuits are also referred to as **master-slave flip-flops** since they contain two clocked *S-R* latches called the master latch and the slave latch. The first-stage master latch receives the clock directly *(C),* while the second-stage slave latch receives an inverted version of the clock ($\overline{C}$). Referring to the *C* and $\overline{C}$ waveforms in Figure 9-7(c), you can see that the master section is enabled while the clock *(C)* is positive. The slave section, on the other

Master-Slave Flip-Flop

A two-latch flip-flop circuit in which the master latch is loaded on the leading edge of the clock pulse, and the slave latch is loaded on the trailing edge of the clock pulse.

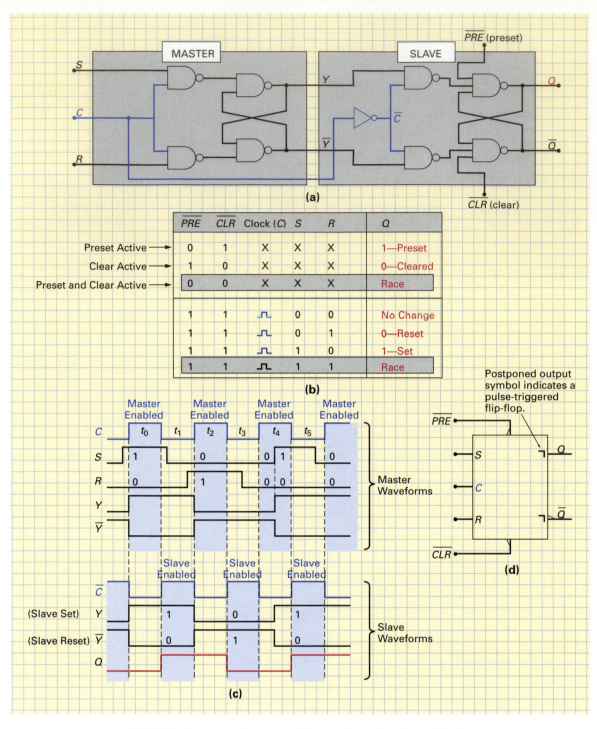

Preset Active →

Clear Active →

Preset and Clear Active →

$\overline{PRE}$	$\overline{CLR}$	Clock (C)	S	R	Q
0	1	X	X	X	1—Preset
1	0	X	X	X	0—Cleared
0	0	X	X	X	Race
1	1	⊓	0	0	No Change
1	1	⊓	0	1	0—Reset
1	1	⊓	1	0	1—Set
1	1	⊓	1	1	Race

(b)

Postponed output symbol indicates a pulse-triggered flip-flop.

FIGURE 9-7   Pulse-Triggered (Master-Slave) Set-Reset Flip-Flop.

hand, is enabled when the clock is negative, because it is at this time that the inverted clock ($\overline{C}$) is positive. There are therefore two distinct steps involved before the Q and $\overline{Q}$ outputs reflect the input condition applied to the SET and RESET inputs.

Step 1:   During the HIGH level of the clock input, the master SET-RESET latch is enabled and its output Y is either SET, RESET, or left unchanged (NO CHANGE).

Step 2:   During the LOW level of the clock input, the slave SET-RESET latch is enabled and its output Q will follow whatever logic level is present on the Y input.

This two-step action is sometimes referred to as *cocking and triggering*. The master latch is cocked during the positive level of the clock and the slave latch is triggered during the negative level of the clock.

The truth table in Figure 9-7(b) summarizes the operation of the pulse-triggered (master-slave) *S-R* flip-flop. The last four lines of the truth table reflect the normal operating conditions (NO CHANGE, RESET, SET, and RACE) of a *S-R* flip-flop. A complete clock pulse is shown in the clock column for these conditions since both the HIGH and LOW levels of the clock input are needed to enable both sections of the flip-flop. The first three lines of the truth table describe how the active-LOW **PRESET** ($\overline{PRE}$) and active-LOW **CLEAR** ($\overline{CLR}$) input functions. When power is first applied to a digital system, flip-flops may start up in either the SET ($Q = 1$) or RESET ($Q = 0$) condition. This could be both dangerous and damaging if these Q outputs were controlling external devices. For this reason, the direct PRESET (synonymous with SET) and direct CLEAR (synonymous with RESET) inputs are generally always available with most commercially available flip-flop ICs. They are called direct inputs because they do not need the clock signal to be active in order to PRESET the Q to a 1 or CLEAR the Q to a 0. One application for these direct PRESET and CLEAR inputs would be to control the outputs of any flip-flops controlling machinery in an automated manufacturing plant so that when power is first applied, all devices start off in their OFF state.

Returning to the truth table in Figure 9-7(b), you can see that when the CLEAR is HIGH and the PRESET is LOW (active), the output is PRESET to 1 regardless of the C, S, and R inputs. In the second line of the table, the PRESET is HIGH and the CLEAR input is LOW (active), causing the Q to be cleared to 0, regardless of the C, S, and R inputs. These active-LOW inputs are therefore normally pulled HIGH to make them inactive. Referring to the circuit in Figure 9-7(a) you can see that the PRESET and CLEAR inputs control the final NAND gates of the slave latch. As with any NAND gate, any LOW input will produce a HIGH output, and therefore a LOW PRESET will produce a HIGH Q (LOW $\overline{Q}$), while a LOW CLEAR will produce a LOW Q (HIGH $\overline{Q}$). Bringing both the PRESET and CLEAR LOW at the same time will produce a RACE condition, as summarized in the truth table in Figure 9-7(b), since both Q and $\overline{Q}$ will try to go HIGH.

Figure 9-7(c) shows a timing diagram for a pulse-triggered (master-slave) set-reset flip-flop. Referring to the clock input (C), let us step through times t_0 through t_5 and see how this circuit responds to a variety of input conditions.

At time t_0, the master latch section is enabled by the positive level of C. Since the S input is HIGH and the R input is LOW at this time (SET condition), the Y output of the master latch will be SET to 1 ($\overline{Y} = 0$).

At time t_1, the master latch section is disabled by a LOW C, while the slave latch section is enabled by a HIGH $\overline{C}$. Since the Y and $\overline{Y}$ inputs to the slave latch section are in the SET condition ($Y = 1$, $\overline{Y} = 0$), the Q output is SET to 1. This action demonstrates how the slave section simply switches its inputs through to the outputs when enabled by the clock (if $Y = 1$ and $\overline{Y} = 0$, $Q = 1$ and $\overline{Q} = 0$ when $\overline{C}$ is HIGH). It can therefore be said that the second latch is a slave to the master latch.

At time t_2, the master latch section is enabled by a HIGH C. Since its SET input is LOW and its RESET input is HIGH at this time, the Y output will be reset to 0.

At time t_3, the master latch section is disabled and the slave latch section is enabled. The reset output of the master latch is clocked into the enabled slave section and transferred through to the output, resetting Q LOW (if $Y = 0$ and $\overline{Y} = 1$, $Q = 0$ and $\overline{Q} = 1$ when $\overline{C}$ is HIGH).

At time t_4, the master-slave *S-R* flip-flop is a level-clocked circuit, which means that the inputs have to remain constant for the entire time that the clock is enabling the latch. At the beginning of time t_4, the S and R inputs are both LOW and therefore Y remains in its last state, which was RESET ($Y = 0$). Half-way through time t_4, the S input goes HIGH and the Y output is therefore SET HIGH.

At time t_5, the master latch is disabled while the slave latch is enabled, connecting the HIGH Y input through to the Q output.

The logic symbol for a pulse-triggered master-slave flip-flop is shown in Figure 9-7(d). The active-LOW PRESET ($\overline{PRE}$) and active-LOW CLEAR ($\overline{CLR}$) inputs are included, with the traditional triangle being used to represent the fact that they are activated by a LOW logic level.

Preset

A control input that is independent of the clock and is used to set the output HIGH.

Clear

A control input that is independent of the clock and is used to reset the output LOW.

FIGURE 9-8 Application—A Memory Latch/Unlatch Circuit.

The *postponed output symbol* (⌐) at the Q and $\overline{Q}$ outputs indicates that these outputs do not change state until the clock input has fallen from a HIGH to a LOW level. This is due to the slave section, which, as you know, is only enabled—and the output updated—when the clock signal drops LOW. If the postponed output were turned the other way, it would indicate that the master latch would be enabled by a LOW clock input and the slave latch would be enabled by a HIGH clock input. The output with this type of S-R master-slave flip-flop would be updated whenever the clock rose to a HIGH level.

Application–A Memory Latch/Unlatch Circuit

The inset in Figure 9-8 shows the function table for a 74L71, which is a pulse-triggered set-reset master-slave flip-flop IC with three ANDed SET inputs and three ANDed RESET inputs. By ANDing the SET and RESET inputs in this way, the IC has more control versatility, since all inputs will have to be HIGH for the control line to be HIGH and any LOW input will pull the corresponding control line LOW. The first three lines of the function table show how the active-LOW PRESET and CLEAR inputs will override the outputs, and the last four lines of the function table show the NO CHANGE, SET, RESET, and RACE input conditions.

■ **EXAMPLE:**

What would be the Q output from the 7471 for the following input conditions?

a. $S_1 = 1$, $S_2 = 1$, $S_3 = 1$, $R_1 = 0$, $R_2 = 0$, $R_3 = 0$, CLK = complete pulse, PRE = 0, CLR = 1.

b. $S_1 = 1$, $S_2 = 1$, $S_3 = 0$, $R_1 = 0$, $R_2 = 0$, $R_3 = 0$, CLK = complete pulse, PRE = 1, CLR = 1.

■ *Solution:*

a. Q would be HIGH since PRESET is active.

b. Q would not change, since $S = 0$ (due to S_3) and $R = 0$.

To understand the operation of this IC better, let us see how it could be used in an application.

The circuit in Figure 9-8 shows how a pulse-triggered set-reset master-slave flip-flop can be used to latch or unlatch a control output. The output Q is connected to +5 V via an LED and current-limiting resistor (R_4), so a HIGH output (SET) will turn OFF the LED and a LOW output (RESET) will turn ON the LED.

All of the SET inputs are tied together and connected to the LED OFF push-button switch, SW_1. The RESET inputs are also all tied together and connected to the LED ON push-button switch, SW_2. The SET and RESET inputs are normally both LOW (NO CHANGE) due to the pull-down resistors R_1 and R_2; however, when a push-button switch is pressed, the corresponding input is switched HIGH. For example, if SW_1 were momentarily pressed, the Q output would be SET HIGH after one clock pulse and the LED would turn OFF. On the other hand, if SW_2 were momentarily pressed, the Q output would be RESET LOW after one clock pulse and the LED would turn ON. A momentary press of SW_2, therefore, will lock or ON the LED, while a momentary press of SW_1 will lock or OFF the LED.

As described earlier, a flip-flop's output may start out either HIGH or LOW when power is first applied to a circuit. To ensure that a flip-flop's output always starts off in the desired condition, a start circuit is normally included and connected to either the PRESET or CLEAR input. In Figure 9-8, R_3 and C_1 form a *slow-start circuit* that is connected to the flip-flop's PRESET input to ensure that the LED is initially OFF when the circuit is powered up. This circuit operates in the following way. When power is first applied to the circuit, there is no charge on C_1 and therefore the PRESET input to the 7471 will be active, so the Q output will be PRESET to a 1, causing the LED to turn OFF. After a small delay (equal to five time constants of R_3 and C_1), C_1 will have charged to 5 V and the PRESET input will be inactive (HIGH). This slow-start circuit therefore delays the PRESET input from going immediately HIGH to ensure that the Q output starts off SET or HIGH, causing the LED to be OFF when power is first applied. If you wished the Q output to be 0 at power-up, you would simply connect the PRESET input to +5 V and connect the CLEAR input to the slow-start circuit.

SELF-TEST EVALUATION POINT FOR SECTION 9-1

Use the following questions to test your understanding of Section 9-1.

1. Once triggered, the SET-RESET flip-flop's output will remain locked in either the SET condition or RESET condition, which is why a flip-flop is also known as a _____.

2. List the three different ways a flip-flop can be triggered.

3. If $S = 0$ and $R = 0$, the NOR latch would be in the _____ input condition and the NAND latch would be in the _____ input conditions.

4. Which form of triggering makes use of a
   a. transition pulse generator?   b. master latch and slave latch?

5. What is a contact-bounce eliminator circuit?

6. What is a slow-start circuit?

D-Type Flip-Flop or Latch

A data-type flip-flop circuit that can latch or store a binary 1 or 0.

The **D-type flip-flop** or **D-type latch** is basically a SET-RESET flip-flop with a small circuit modification. This modification was introduced so that the data-type flip-flop could RACE like the *S-R* flip-flop.

The basic D-type flip-flop logic circuit is shown in Figure 9-9(a). The modification to the circuit is the inclusion of an INVERTER gate, which ensures that the *R* and *S* inputs to the NAND latch are never at the same logic level. The single data input bit (D) appears at the *S* input of the latch, and its complement ($\overline{D}$) appears at the *R* input of the latch. Therefore, when the *D* input is 1, the *Q* output is SET to 1, and when the *D* input is 0, the *Q* output is RESET to 0, as summarized in the truth table in Figure 9-9(b). The logic symbol for this unclocked D-type flip-flop is shown in Figure 9-9(c). As can be seen in the timing waveforms in Figure 9-9(d), the *Q* output is either SET or RESET as soon as the *D* input goes HIGH or LOW.

To coordinate the overall action of a digital system, devices are triggered into operation at specific times. This synchronization is controlled by a master timing signal called a clock. Like the *S-R* flip-flop, the D-type flip-flop can be either level-triggered, edge-triggered, or pulse-triggered. Let us now examine each of these three types in more detail.

9-2-1 *Level-Triggered D-Type Flip-Flops*

Figure 9-10(a) shows the basic logic circuit for a *level-triggered* or *gated D-type flip-flop*. A LOW clock level will disable the input gates *A* and *B* and prevent the latch from changing states, as indicated in the first line of the truth table in Figure 9-10(b). When the clock level is HIGH, however, gates *A* and *B* are enabled and the *D* input controls the *Q* output, as seen in the second and third lines of the truth table. In the timing diagram in Figure 9-10(c), the operation of the level-clocked or level-triggered D-type flip-flop is further reinforced. This circuit's operation can be summarized by saying that when the clock input is HIGH, the *Q* output equals the *D* input.

The logic symbol for both the positive level-triggered and negative level-triggered D-type flip-flop is shown in Figure 9-10(d).

FIGURE 9-9 **The Basic Data-Type (D-type) Flip-Flop or Latch.**

FIGURE 9-10 Level-Triggered or Gated D-Type Flip-Flop.

The 7475 IC is shown in Figure 9-11. This device contains four level-triggered D-type latches. Referring to the logic symbol, you can see that the four data inputs are labeled 1D, 2D, 3D, and 4D. There are only two enable or control inputs (*C*); pin 13 will enable the first and second latches, and pin 4 will enable the third and fourth latches.

Application—A Decimal Counting Unit with Freeze Control

The four latches of a 7475 are ideally suited for use as temporary storage of binary information between a processing unit and an input-output, or indicator, unit. Figure 9-11 shows how a 7475 can be connected to operate in such an application. The two enable inputs for the 7475 (pins 4 and 13) are tied together and connected to switch 2 (SW_2). When this switch is placed in the LATCH position, all four of the 7475 D-type latches are enabled and the BCD count from the 7490 will pass through the 7475 to the 7447 and finally be displayed on the common-anode seven-segment display. When SW_2 is placed in the DISABLE position, however, all four of the 7475 D-type latches are disabled, but the previously latched BCD code will remain at the output and, therefore, on the display, regardless of the BCD

FIGURE 9-11 Application—Decimal Counting Unit with Freeze Control.

code variations at the *D* inputs. The effect is to freeze the display when SW_2 disables the 7475 but allow the display to reflect the 7490 count when SW_2 enables the 7475.

■ **EXAMPLE:**

What would be the *Q* output from the 7475 for the following input conditions?

$$1D = 1, 2D = 0, 1C, 2C = H, 3D = 1, 4D = 1, 3C, 4C = L$$

■ *Solution:*

Outputs 1Q = 1 and 2Q = 0, since the enables (1C and 2C = HIGH) for these two flip-flops are active. On the other hand, 3Q = NO CHANGE and 4Q = NO CHANGE, since these two flip-flops are disabled (3C and 4C = LOW).

9-2-2 *Edge-Triggered D-Type Flip-Flops*

To be useful in fast-acting digital circuits, a flip-flop needs to respond to the edge of a clock signal rather than the level of a clock. The advantage of edge-triggered devices is that the inputs do not need to be held stable for the entire time of the HIGH or LOW level of the clock.

Figure 9-12(a) shows how an edge-triggered D-type flip-flop incorporates the same positive transition pulse generator as the edge-triggered *S-R* flip-flop. Referring to the truth table in Figure 9-12(b), you can see that the active-LOW PRESET and CLEAR inputs are not dependent on the clock input or data input and can be used to PRESET or CLEAR the *Q* output. The last three lines of the truth table summarize how a positive edge-triggered D-type flip-flop will operate. If the clock input is at a HIGH level, LOW level, or making a negative transition from LOW to HIGH (↓), the D-type flip-flop will not be enabled and therefore NO

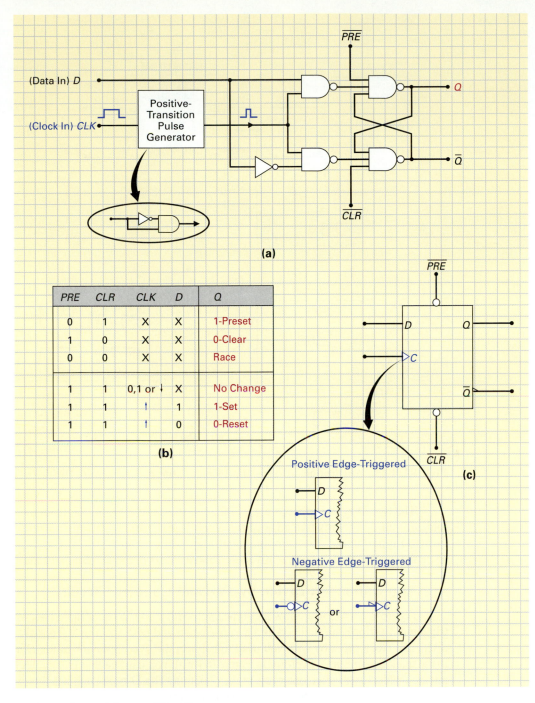

(a)

PRE	CLR	CLK	D	Q
0	1	X	X	1-Preset
1	0	X	X	0-Clear
0	0	X	X	Race
1	1	0,1 or ↓	X	No Change
1	1	↑	1	1-Set
1	1	↑	0	0-Reset

(b)

Positive Edge-Triggered

Negative Edge-Triggered

(c)

FIGURE 9-12 Edge-Triggered D-type Flip-Flop.

CHANGE will occur at the output. On a positive transition of the clock (↑), however, the Q output will follow the D input.

The logic symbol for a positive edge-triggered D-type flip-flop is shown in Figure 9-12(c). The inset illustrates how positive edge-triggered and negative edge-triggered symbols will differ.

Application—A Storage Register Circuit

Figure 9-13(a) shows how the two D-type flip-flops within a 7474 IC can be used as a 2-bit storage register. In this example, a combinational logic circuit produces a Y and Z data output, which is only present for a short space of time.

FIGURE 9-13 Application—D-type Flip-Flop. (a) Storage Register. (b) Divide-by-Two Counter.

The two D-type flip-flops are included to store the 2-bit output from the combinational logic circuit on the positive edge of the clock signal. Once this information is safely stored in the D-type latches, the combinational logic circuit can begin working on another task, since the 2-bit word is permanently stored and appears at the Q outputs of the D-type latches.

Application—A Divide-by-Two and Counter Circuit

Figure 9-13(b) shows how a D-type latch with a 74175 (quadruple D-type positive edge-triggered flip-flop) can function as a *divide-by-two* and *counter circuit.* To make a D-type latch divide the clock input by two, you need to connect the $\overline{Q}$ output back to the D input. Referring to the thick lines in the waveform in Figure 9-13(b), you can see that two cycles of the clock input are needed to produce one cycle at the Q output. The D-type latch has therefore divided the clock input by two by making use of the phase reversal between the D input and the $\overline{Q}$.

Referring to the binary values within the thick lines of the waveforms, you may have noticed that the divide-by-two D-type latch is also counting down in binary from 3 to 0. For example, at time t_0, the binary count present on the clock input and the $\overline{Q}$ output is 11_2, or 3_{10}. At time t_1, the count has dropped to 10_2, or 2_{10}, and then at time t_2 the count has decreased to 01_2, or 1_{10}. For each clock pulse following, the count decreases by 1, starting at the maximum 2-bit count of 3_{10} (11_2), then decreasing to 0 (00), and then repeating the cycle.

9-2-3   *Pulse-Triggered D-Type Flip-Flops*

The logic circuit, truth table, and logic symbol for a pulse-triggered (master-slave) D-type flip-flop is shown in Figure 9-14. Like the pulse-triggered *S-R* flip-flop, the pulse-triggered D-type flip-flop requires a complete clock pulse input before the Q output will reflect the D input.

FIGURE 9-14 Pulse-Triggered (Master-Slave) D-Type Flip-Flop.

$\overline{PRE}$	$\overline{CLR}$	Clock	D	Q
0	1	X	X	1—Preset
1	0	X	X	0—Cleared
0	0	X	X	Race
1	1	⊓	0	0—Reset
1	1	⊓	1	1—Set

(b)

(c)

EXAMPLE:

What would be the *Q* output for the pulse-triggered D-type flip-flop shown in Figure 9-15, assuming the *D* and *C* inputs shown in the waveforms?

Solution:

All pulse-triggered flip-flops are level-clocked devices, which means that the master section is enabled for the entire time that the *C* input is HIGH and the slave section is enabled during the LOW input of *C*. The *D* input is therefore transferred through the master section when the clock is HIGH and then transferred through the slave section, and to the *Q* output, when the clock input is LOW, as shown in Figure 9-15.

FIGURE 9-15 Input/Output Waveforms for a Pulse-Triggered D-Type Flip-Flop.

Use the following questions to test your understanding of Section 9-2.

1. The "D" in D-type stands for _____.
2. Can the D-type flip-flop RACE?
3. What are the input and output differences between an *S-R* flip-flop and a D-type flip-flop?
4. A D-type flip-flop could be used as a/an _____.
   a. binary counter   **c.** storage register
   b. frequency divider   **d.** all of the above

9-3   *J-K* FLIP-FLOPS

J-K Flip-Flop

A circuit that will operate in the same way as a set-reset flip-flop, except that it will toggle its output instead of RACE when both inputs are HIGH.

Toggle

An input combination or condition that causes the output to toggle or switch its output to the opposite logic level.

With the previously discussed *S-R* flip-flop and D-type flip-flop, the letters "S," "R," and "D" were abbreviations for SET, RESET, and DATA. With the ***J-K* flip-flop,** the letters "J" and "K" are not abbreviations, they are arbitrarily chosen letters. As you will see in the following section, the *J-K* flip-flop operates in almost exactly the same way as the *S-R* flip-flop in that there will be NO CHANGE at the output when the *J* and *K* inputs are both LOW, the output will be SET HIGH when the *J* input (set input) is HIGH, and the output will be RESET LOW when the *K* input (reset input) is HIGH. The distinctive difference with the *J-K* flip-flop is that it will not RACE when both its SET (*J*) and RESET (*K*) inputs are HIGH. When this input condition is applied (*J* = 1 and *K* = 1), the *J-K* flip-flop will **toggle,** which means that it will simply switch or reverse the present logic level at its *Q* output. This toggle feature of the *J-K* flip-flop is achieved by modifying the basic *S-R* flip-flop's internal logic circuit to include two cross-coupled feedback lines between the output and the input. This circuit modification means, however, that the *J-K* flip-flop cannot be level-triggered; it can only be edge-triggered or pulse-triggered.

9-3-1   *Edge-Triggered* J-K *Flip-Flops*

Figure 9-16(a) shows the internal logic circuit for an edge-triggered *J-K* flip-flop circuit, and Figure 9-16(b) shows this circuit's function table. The first three lines of the function table show that if the clock input line is LOW, HIGH, or making a transition from LOW to HIGH (a positive edge), there will be no change at the *Q* output regardless of the *J* and *K* inputs. This is because this *J-K* flip-flop is a negative edge-triggered device, which means it will only "wake up" and perform the operation applied to the *J* and *K* inputs when the clock input is making a transition from HIGH to LOW (a negative edge). The negative-transition pulse generator will generate a positive output pulse every time the clock pulse input drops from a HIGH to a LOW level. In the last four lines of the function table, you can see that when a negative edge is applied to the clock input, the *J-K* flip-flop will react to its *J* (SET) and *K* (RESET) inputs. For example, when negative-edge-triggered, the *J-K* flip-flop will operate in almost the same way as a *S-R* flip-flop, since it will not change its *Q* output when *J* and *K* are LOW, RESET its *Q* output when the reset input (*K*) is HIGH, and SET its *Q* output when the set input (*J*) is HIGH. In the last line of the function table, you can see why the *J-K* flip-flop is an improvement over the *S-R* flip-flop. When both the *S* and *R* inputs of a *S-R* flip-flop are HIGH, it will race and generate an unpredictable *Q* output, which is why this input condition is never used. When both the *J* and *K* inputs of a *J-K* flip-flop are made HIGH, on the other hand, the *Q* output will toggle, or switch, to the opposite state. This means that if *Q* is HIGH, it will switch to a LOW, and if *Q* is LOW, it will switch to a HIGH. The function table shows that if the *J-K* flip-flop is in the toggle condition and the clock input makes a transition from HIGH to LOW, the *Q* output will switch to the opposite logic level, or the logic level of the $\overline{Q}$ output.

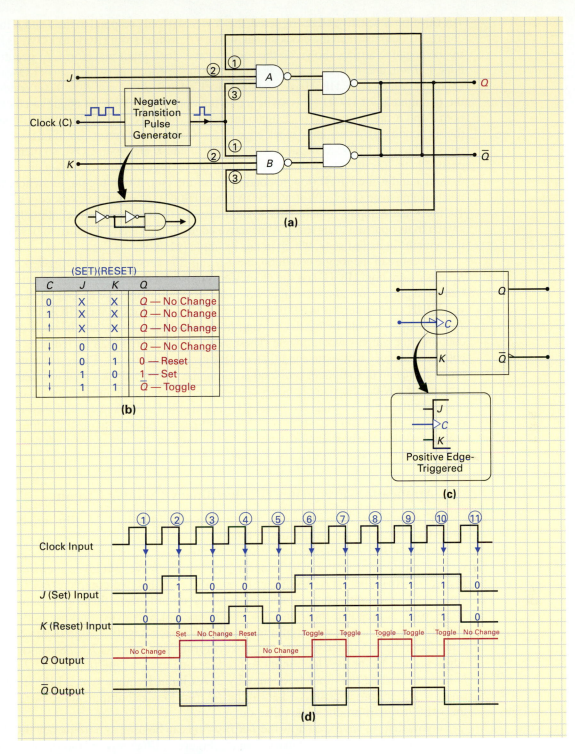

FIGURE 9-16 Edge-Triggered *J-K* Flip-Flops.

Figure 9-16(c) shows the logic symbol for a negative edge-triggered *J-K* flip-flop. A full triangle at the clock input, inside the rectangular block, is used to indicate that this device is edge-triggered, and the smaller right triangle (or bubble) outside the rectangular block is used to indicate that this input is active-LOW, or negative edge-triggered. The inset in Figure 9-16(c) shows the logic symbol for a positive edge-triggered *J-K* flip-flop.

To reinforce your understanding of the *J-K* flip-flop, Figure 9-16(d) shows how a negative edge-triggered *J-K* flip-flop will respond to a variety of input combinations. Since this

J-K flip-flop is negative edge-triggered, the device will only respond to the *J* and *K* inputs and change its outputs when the clock signal makes a transition from HIGH to LOW, as indicated by the negative arrows shown on the square-wave clock input waveform. At negative clock edge 1, $J = 0$ and $K = 0$, so there is no change in the *Q* output (which stays LOW), and the $\overline{Q}$ output (which stays HIGH). At negative clock edge 2, $J = 1$ and $K = 0$ (set condition), so the *Q* output is set HIGH ($\overline{Q}$ is switched to the opposite, LOW). At negative clock edge 3, both *J* and *K* are logic 0, so once again the output will not change (*Q* will remain latched in the set condition). At negative clock edge 4, $J = 0$ and $K = 1$ (reset condition), so the *Q* output will be reset LOW. At negative edge 5, both *J* and *K* are again LOW, so there will be no change in the *Q* output (*Q* will remain latched in the reset condition). At negative clock edge 6, both *J* and *K* are HIGH (toggle condition), so the output will toggle, or switch, to its opposite state (since *Q* is LOW it will be switched HIGH). At negative clock edges 7, 8, 9, and 10, the *J* and *K* inputs remain HIGH, and therefore the *Q* output will continually toggle or switch to its opposite logic level. At negative edge 11, both *J* and *K* return to 0, so the *Q* output remains in its last state, which in this example is HIGH.

Because of the toggle condition, the *J-K* flip-flop cannot be level-triggered. If the *J* and *K* inputs were both HIGH, and the clock were connected directly to the input gates without a transition pulse generator, the output would toggle continuously for as long as the clock input was HIGH. To explain this in a little more detail, when the clock is at its HIGH level, the HIGH *J* and *K* inputs would be passed through to the flip-flop outputs, and because of the cross-coupling, the outputs would toggle. These new outputs would then be fed back to the input gates, producing a new input condition, and the cycle would repeat. The result would be a continual change in the output or oscillations during the time the clock signal was active. For this reason, the *J-K* flip-flop is only either edge-triggered (so that the input gates are only enabled momentarily) or pulse-triggered (master is first enabled while slave—and therefore the output—is disabled). With edge-triggered *J-K* flip-flops, the pulse from the transition generator is always less than the propagation delay of the flip-flop, so that output oscillation will not occur when the flip-flop is in the toggle condition.

Application–Frequency Divider/Binary Counter Circuit

Figure 9-17(a) shows how a 555 timer and a 74LS76A *J-K* flip-flop can be connected to form a frequency divider or binary counter circuit. The 555 timer, which will be discussed in detail in the second section of this chapter, is connected to operate as an astable multivibrator and will generate the square-wave output waveform shown in Figure 9-17(b). The 74LS76A digital IC actually contains two complete *J-K* flip-flop circuits, and since only one is needed for this circuit, the flip-flop is labeled "½ of 74LS76A." Notice that the *J* and *K* inputs of the 74LS76A flip-flop are both connected HIGH, so the *Q* output will continually toggle for each negative edge of the clock signal input from the 555 timer, as seen in the waveforms in Figure 9-17(b). Since two negative edges are needed at the clock input to produce one cycle at the output, the *J-K* flip-flop is in fact dividing the input frequency by 2. An input frequency of 2 kHz from the 555 timer will therefore appear as 1 kHz at the *J-K* flip-flop's *Q* output, since two pulses in to the 74LS76A will produce one pulse out.

Now that we have seen how the *J-K* flip-flop can be used as a frequency divider, let us see how it can function as a binary counter. Figure 9-17(c) repeats the output waveforms from the 555 timer and the 74LS76A *Q* output. If these two outputs were connected to two LEDs, the LEDs would count up in binary, as shown in the table in Figure 9-17(c). To explain this in more detail, at time t_0, both the 555 and 74LS76A *Q* output are LOW, so our LED binary display shows a count of 00 (binary 0). At time t_1, the binary display will be driven by a HIGH from the 555 timer output and a LOW from the 74LS76A *Q* output and therefore will display 01 (binary 1). At time t_2, the display will show 10 (binary 2), and at time t_3 the display will show 11 (binary 3). Combined, the 555 timer and the divided-by-two output of the 74LS76A can be used to generate a 2-bit word that will count from binary 0 (00) to binary 3 (11) and then continuously repeat the count.

FIGURE 9-17   Application—The Edge-Triggered *J-K* Flip-Flop Acting as a Frequency Divider and Binary Counter.

Figure 9-18(a) shows how two 74LS76A ICs can be connected to produce a 4-bit binary up-counter. This and other counters will be discussed in more detail in Chapter 10; however, for now let us understand how this circuit operates. Since the *J* and *K* inputs of all the flip-flops in this circuit are tied HIGH, the flip-flops will permanently be in the toggle-input condition. The timing diagram for this circuit is shown in Figure 9-18(b). In this circuit, the clock input triggers flip-flop 1 (FF1), the output of FF1 (1Q) triggers FF2, the output of FF2

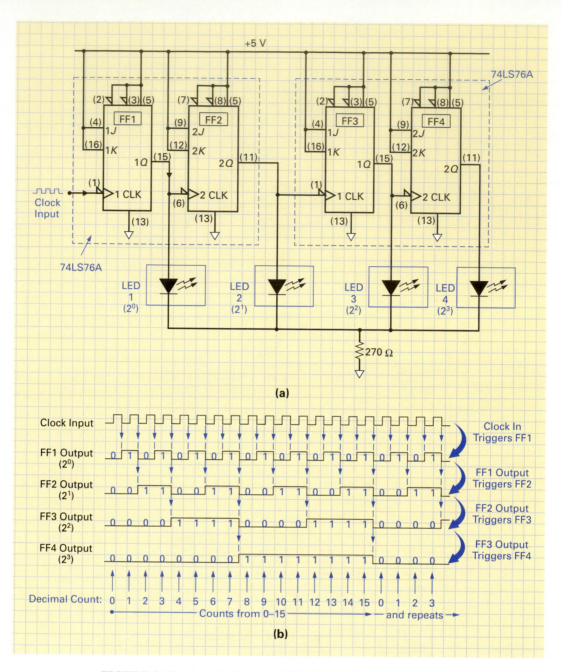

FIGURE 9-18 **Application—A 4-Bit Binary Counter.**

triggers FF3, and the output of FF3 triggers FF4. Since all of the flip-flops are negative edge-triggered, each negative edge of the clock input will toggle FF1, each negative edge of the FF1 output will toggle FF2, each negative edge of the FF2 output will toggle FF3, and each negative edge of the FF3 output will toggle FF4. Four active-HIGH light-emitting diodes are connected to display the logic level at each of the outputs. As far as the binary count is concerned, the display is in its reverse order, since LED1 is the display's LSB (2^0) and LED4 is the display's MSB (2^3). As a result, we will see a 1248 binary display instead of the customary 8421 binary display.

Referring to the decimal count line in Figure 9-18(b), you can see that at first all outputs are LOW and therefore all LEDs are OFF (a binary count of 0000 or decimal 0). The count then advances by 1 for every clock pulse input until it reaches its maximum 4-bit count with all LEDs ON (a binary count of 1111 or decimal 15).

9-3-2 *Pulse-Triggered J-K Flip-Flops*

The basic circuit for a pulse-triggered (master-slave) *J-K* flip-flop is shown in Figure 9-19(a). The circuit is almost identical to the master-slave *S-R* flip-flop except for the distinctive *J–K* cross-coupled feedback connections from the Q and $\overline{Q}$ slave outputs back to the master input gates. Figure 9-19(b) summarizes the operation of the pulse-triggered *J–K* flip-flop.

FIGURE 9-19 **Pulse-Triggered (Master-Slave) *J-K* Flip-Flop.**

The operation of the PRESET and CLEAR inputs are not included in this table, since they will operate in exactly the same way as any other flip-flop. The logic symbol for this flip-flop is shown in Figure 9-19(c). The postponed output symbol (¬) is used to indicate that the Q and $\overline{Q}$ outputs will only change when the clock input (C) falls from a HIGH to a LOW level (which is when the slave is enabled).

The timing diagram in Figure 9-19(d) serves as a visual summary of the master-slave J-K flip-flop's operation. When the clock input (C) goes HIGH, the master section is enabled (ME). If $J = 1$ and $K = 0$, the Y output will be SET to 1 (S), whereas if $J = 0$ and $K = 1$, the Y output will be RESET to 0 (R). On the other hand, if $J = 0$ and $K = 0$, there will be NO CHANGE at the output (NC), whereas if $J = 1$ and $K = 1$, the output will toggle (T) based on the feedback control inputs from the Q and $\overline{Q}$ outputs. When the clock input (C) goes LOW and disables the master section, the inverted clock input ($\overline{C}$) goes HIGH and enables the slave section. Looking at the thick Y and Q output waveforms in Figure 9-19(d), you can see that the Q output is a replica of the Y output except that the Q lags Y by half a clock pulse. This is because the Q output will only follow the Y output when the slave section is enabled, and this occurs when the clock input (C) falls from a HIGH to a LOW (¬).

Application–Frequency Divider

Figure 9-20(a) shows how the two pulse-triggered J-K flip-flops within a 74107 could be connected as a divide-by-four circuit. Referring to the waveforms in Figure 9-20(b), you can see that two complete cycles of the clock input are needed to produce one complete cycle at the output of flip-flop 1 (FF₁ therefore divides by 2). Comparing the clock input to the output of FF₂, you can see that four complete cycles of the clock input are needed to produce one complete cycle at the output of FF₂ (FF₁ divides by 2 and then FF₂ divides by 2, resulting in a final divide-by-four output).

FIGURE 9-20   Application—Divide-by-Four Circuit.

Use the following questions to test your understanding of Section 9-3.

1. Which of the following input conditions will set HIGH the Q output of a negative edge-triggered J-K flip-flop?
   a. $J = 1, K = 1, C = \uparrow, \overline{PRE} = 1, \overline{CLR} = 1$.
   b. $J = 1, K = 0, C = \downarrow, \overline{PRE} = 1, \overline{CLR} = 0$.
   c. $J = 1, K = 1, C = \downarrow, \overline{PRE} = 1, \overline{CLR} = 1$.
   d. $J = 1, K = 0, C = \downarrow, \overline{PRE} = 1, \overline{CLR} = 1$.

2. When both the inputs of a J-K flip-flop are _____, the Q output will _____ or switch to the opposite state.

3. If the clock input to a J-K flip-flop were a 126 kHz square wave and $J = 1, K = 1, \overline{PRE} = 1, \overline{CLR} = 1$, what would the Q output be?
   a. A 126 kHz square wave.   c. A 31.5 kHz square wave.
   b. A 63 kHz square wave.   d. A 252 kHz square wave.

4. The _____ flip-flop can perform all the functions of the _____ flip-flop and _____ flip-flop.
   a. S-R, J-K, D-type   c. J-K, D-type, S-R
   b. D-type, J-K, S-R d. S-R, D-type, J-K

9-4   DIGITAL TIMER AND CONTROL CIRCUITS

Timing is everything in digital logic circuits. To control the timing of digital circuits, a clock signal is distributed throughout the digital system. This square-wave clock signal is generated by a clock oscillator, and its sharp positive (leading) and negative (trailing) edges are used to control the sequence of operations in a digital circuit.

The S-R, D-type, and J-K flip-flops are all examples of bistable multivibrators, since they have two (bi) stable states (SET and RESET). In this section we will discuss the astable, or unstable, multivibrator, which has no stable states and is commonly used as a clock oscillator. The third type of multivibrator is the monostable multivibrator which has only one (mono) stable state and when triggered will generate a rectangular pulse of a fixed duration.

Logic Gate Astable Multivibrator Circuits

The astable multivibrator can also be constructed using logic gates, as seen in Figure 9-21. In Figure 9-21(a), a Schmitt-trigger INVERTER is connected to operate as a clock oscillator. When power is first applied to this circuit, the capacitor will have no charge and this LOW input is inverted by the NOT gate, giving a HIGH output (seen as red in the illustration). The capacitor (C) will begin to charge via the resistor (R), and the increasing positive charge across the capacitor will be felt at the input of the INVERTER. After a time (which is dependent on the values of R and C), the capacitor charge will be large enough to apply a valid HIGH to the INVERTER input. This HIGH input will cause the output of the INVERTER to go LOW, so the capacitor will begin to discharge (seen as blue in illustration). When the capacitor's charge falls to a valid LOW logic level, the INVERTER will generate a HIGH output, and the cycle will repeat.

Figure 9-21(b) shows another basic astable multivibrator circuit using two INVERTERS. The operation of this circuit is similar to the one in Figure 9-21(a). A continual capacitor charge and discharge causes the astable or free-running multivibrator to switch back and forth between its two unstable states, producing a repeating rectangular wave at the output (two conditions are shown in the illustration as red and blue).

FIGURE 9-21 Application—Logic-Gate Astable Multivibrators.

Logic-Gate Monostable Multivibrator Circuits

Like the astable multivibrator, the monostable multivibrator can be constructed using logic gates, as seen in Figure 9-22(a). When the trigger input is LOW and the Q output is LOW, the output from the NOR gate is HIGH, and therefore the output from the INVERTER is LOW, keeping the circuit in its stable state. When the trigger input pulses HIGH, it causes the output of the NOR gate to go LOW. This HIGH-to-LOW transition is coupled through the capacitor to the INVERTER, which produces a HIGH Q output, as can be seen in the timing diagram below the circuit. This HIGH Q output is fed back to the NOR gate's other input and keeps the NOR gate output LOW even after the trigger pulse has ended. A LOW NOR gate output will produce a potential difference across the resistor-capacitor network, and therefore the capacitor will begin to charge. After a time, dependent on the values of R and C, the charge on C is large enough for the INVERTER to recognize it as a valid HIGH input, and therefore it generates a LOW Q output ending the pulse.

The typical block logic symbols for a monostable multivibrator, or one-shot, are shown in Figure 9-22(b). Some logic symbols group the entire circuit within a block, while others block only the logic gates and show the time-determining capacitor and resistor separately.

Today, one-shot circuits are very rarely constructed with discrete transistors or logic gates since integrated circuits are available. These one-shot ICs are classified as being either *nonretriggerable one-shots* or *retriggerable one-shots*. The difference between these two types is best described by comparing the waveforms in Figure 9-23.

FIGURE 9-22 **Application—Logic-Gate Monostable (One-Shot) Multivibrator.**

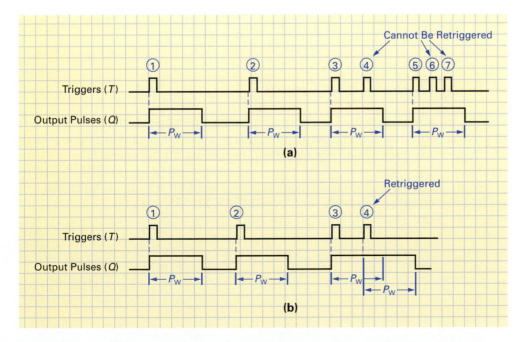

FIGURE 9-23 **One-Shot IC Types. (a) Nonretriggerable Action. (b) Retriggerable Action.**

Figure 9-23(a) shows the action of the nonretriggerable one-shot. When triggered, an output pulse of a certain width (P_w) is generated, as shown when triggers 1 and 2 are applied. If another trigger is applied before the output pulse has ended, it will be ignored. For example, trigger 3 starts an output pulse, but trigger 4 is ignored since it occurs before the output pulse has *timed out*. As another example, trigger 5 *fires* the one-shot, while triggers 6 and 7 are ignored because the one-shot is in its unstable state. Once triggered, therefore, this type of one-shot cannot be retriggered until its output pulse has ended (the one-shot is nonretriggerable).

Figure 9-23(b) shows the action of the retriggerable one-shot. When triggered, an output pulse is produced of a certain pulse width (P_w), as shown when triggers 1 and 2 are applied. The difference with this type of one-shot can be seen with triggers 3 and 4. Trigger 3 fires the one-shot, so it begins to produce an output pulse; however, before this pulse has timed out, trigger 4 retriggers the one-shot. The output pulse will now be extended for a time equal to a pulse width starting at the time trigger 4 occurred. Once triggered, therefore, this type of one-shot can be retriggered (the one-shot is retriggerable). The 74121 is an example of a nonretriggerable one-shot IC while the 74122 is an example of a retriggerable one-shot.

Applications–Pulse Generator, Sequencer and Delay Circuits

Figure 9-24(a) shows how a 74121 can be connected to function as a variable output pulse width generator. The normally open push-button (NOPB) switch is used to trigger the one-shot, and the external 100 kΩ variable resistor is used to adjust the output pulse width. The $\overline{Q}$ output is used to drive an output LED instead of the Q output, since a common-anode display is being used for a higher current and therefore a higher light-level display.

Figure 9-24(b) shows how three one-shots can be used to generate a sequence of timing pulses. Since all inputs are negative edge-triggered, the trailing edge of each output is used to trigger the next one-shot, as can be seen in the associated waveforms in Figure 9-24(b).

Figure 9-24(c) shows how two one-shots can be used to introduce a delay. An input trigger fires one-shot *Y*, which introduces a certain time delay, and its trailing edge triggers one-shot *Z*, which then produces an output pulse. The result is that the input pulse is delayed from appearing at the output for a time equal to the pulse width time of the *Y* one-shot. This time delay function is shown in the accompanying waveforms in Figure 9-24(c).

9-4-1   *The 555 Timer Circuit*

One of the most frequently used low-cost integrated circuit timers is the **555 timer.** It is a highly versatile timer that can be made to function as an astable multivibrator, monostable multivibrator, frequency divider, or modulator, depending on the connection of external components.

Nearly all the IC manufacturers produce a version of the 555 timer, which can be labeled in different ways, such as SN72 555, MC14 555, SE 555, and so on.

FIGURE 9-24 Application—One-Shot. (a) Pulse Generation. (b) Timing and Sequencing. (c) Time Delay.

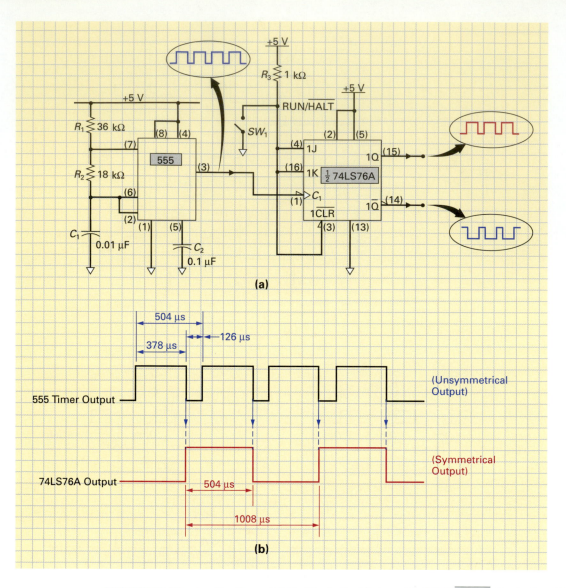

FIGURE 9-25 Application—1 kHz Clock Oscillator with RUN/$\overline{\text{HALT}}$ Control and Low-Level Start.

Application–Clock Oscillator Circuit

Figure 9-25 shows how a 555 timer and a *J-K* flip-flop could be connected to form a clock oscillator circuit.

EXAMPLE:

Figure 9-25(a) shows a typical clock-oscillator circuit using a 555 timer and a *J-K* flip-flop. Perform the following:

a. Calculate the output frequency from the 555 timer.

b. Determine the HIGH time and LOW time of the 555 timer's output.

c. Determine the output frequency from the *J-K* flip-flop.

d. Sketch the 555 timer output and the 74LS76A's *Q* output in time relation.

e. Explain the purpose of the RUN/$\overline{\text{HALT}}$ switch.

Solution:

a.

$$f = \frac{1.43}{(R_1 + 2R_2) \times C_1}$$

$$= \frac{1.43}{[36 \text{ k}\Omega + (2 \times 18 \text{ k}\Omega)] \times 0.01 \text{ }\mu\text{F}} = 2 \text{ kHz}$$

b. The positive half-cycle will last for

$$t_p = 0.7 \times C \times (R_A + R_B)$$
$$= 0.7 \times 0.01 \text{ }\mu\text{F} \times (36 \text{ k}\Omega + 18 \text{ k}\Omega) = 378 \text{ }\mu\text{s}$$

The negative half-cycle will last for

$$t_p = 0.7 \times C \times R_B$$
$$= 0.7 \times 0.01 \text{ }\mu\text{F} \times 18 \text{ k}\Omega = 126 \text{ }\mu\text{s}$$

c. Since the J and K inputs of the flip-flop are normally tied HIGH, the output will toggle and therefore divide the input frequency by 2. Since we know that the 555 timer's output is 2 kHz, the J-K flip-flop's output will be 1 kHz (2 kHz ÷ 2 = 1 kHz).

d. The 555 timer output and the 74LS76A's Q output in time relation are shown in Figure 9-25(b). The 555 timer output drives the clock input (C) of the negative edge-triggered 74LS76A J-K flip-flop. Each negative edge of the unsymmetrical 2 kHz clock input will cause the output to toggle, or change state, producing a symmetrical 1 kHz output.

e. With SW_1 open, the RUN/$\overline{\text{HALT}}$ line will be HIGH, so the J and K inputs of the flip-flop will also be in HIGH (toggle) condition. If SW_1 is closed, the RUN/$\overline{\text{HALT}}$ line will be LOW, forcing the J-K flip-flop into the NO CHANGE condition. This will freeze the flip-flop's outputs and therefore halt or stop the clock signal division. The RUN/$\overline{\text{HALT}}$ line is also connected to the active-LOW clear input of the J-K flip-flop. This will ensure that when we switch from HALT to the RUN mode, the clock output at Q will always start LOW (in the HALT mode Q is reset LOW, and therefore when we switch to the RUN mode, Q will start LOW and then switch HIGH). Having both a Q and $\overline{Q}$ output means that this circuit can be used to supply either a single-phase clock signal (Q only) or a two-phase clock signal (Q and $\overline{Q}$).

SELF-TEST EVALUATION POINT FOR SECTION 9-4

Use the following questions to test your understanding of Section 9-4.

1. Which of the multivibrator circuits has only one stable state?
2. Which multivibrator is also called a set-reset flip-flop?
3. In what applications are one-shots used?
4. The 555 timer derived its number identification from _____.

REVIEW QUESTIONS

Multiple-Choice Questions

1. What is another name for a flip-flop circuit?
   a. Astable multivibrator
   b. Monostable multivibrator
   c. Bistable multivibrator
   d. Horstable multivibrator

2. Which two multivibrator types require trigger inputs?
   a. Astable and monostable
   b. Monostable and bistable
   c. Bistable and astable
   d. Both a. and c. are true

3. What input condition does the *J-K* flip-flop have that the *S-R* and D-type flip-flops don't have?
   a. SET   c. NO CHANGE
   b. RESET   d. TOGGLE

4. What would be the output from a negative edge-triggered *J-K* flip-flop if $J = 1$, $K = 0$, C = LOW to HIGH, and $Q = 0$?
   a. NO CHANGE   c. $Q = 0$
   b. $Q = 1$ d. Both a. and c. are true

5. A four-stage *J-K* flip-flop can be made to function as a _____ binary counter.
   a. 0-to-7   c. 0-to-15
   b. 0-to-3   d. 0-to-16

6. If $J = 0$ and $K = 1$, a *J-K* flip-flop when clocked is said to be in the _____ condition.
   a. SET   c. NO CHANGE
   b. RESET   d. TOGGLE

7. If $J = 1$, $K = 1$, and the input clock frequency is 2 MHz, what would be the Q output frequency?
   a. 2 MHz   c. 0.5 MHz
   b. 1 MHz   d. 4 MHz

8. Which triggering system is used to control a master-slave latch?
   a. Level-triggering   c. Pulse-triggering
   b. Edge-triggering   d. Either a. or c.

9. The *J-K* flip-flop responds to only _____.
   a. Level-triggering   c. Pulse-triggering
   b. Edge-triggering   d. Both b. and c.

10. Which multivibrator type is generally used as a clock oscillator?
    a. Astable multivibrator
    b. Monostable multivibrator
    c. Bistable multivibrator
    d. Tristable multivibrator

11. Which multivibrator type is used as a pulse generator?
    a. Astable multivibrator
    b. Monostable multivibrator
    c. Bistable multivibrator
    d. Tristable multivibrator

12. Monostable or one-shot multivibrators are generally used as a
    a. pulse generator.   c. delay.
    b. timer and sequencer.   d. all of the above.

13. Which multivibrator type has two stable states?
    a. Astable multivibrator
    b. Monostable multivibrator
    c. Bistable multivibrator
    d. Tristable multivibrator

14. Which multivibrator is also called a set-reset flip-flop?
    a. Astable   c. Bistable
    b. Monostable   d. Schmitt

15. How many *J-K* flip-flop stages will be needed to achieve a divide-by-eight frequency division?
    a. Two   c. Four
    b. Three   d. Five

Practice Problems

To practice your circuit recognition and operation, refer to Figure 9-26 and answer the following six questions.

16. What type of flip-flop circuit is being used in this application?

17. If the LOAD control line were LOW, what would be applied to the *D* inputs of the flip-flops?

18. If the LOAD control line were HIGH, what would be applied to the *D* inputs of the flip-flops?

19. What type of triggering is used in this circuit?

20. Describe the basic steps involved in loading a 4-bit word into this storage register circuit.

21. How is the $\overline{CLR}$ control line used in this circuit?

To practice your circuit recognition and operation, refer to Figure 9-27 and answer the following three questions.

22. The logic gates *A* and *B* form what type of flip-flop circuit?

23. What will be the output at point *X* when the switch is in the OFF position and in the ON position?

24. What will be the output at point *Y* when the switch is in the OFF position and in the ON position?

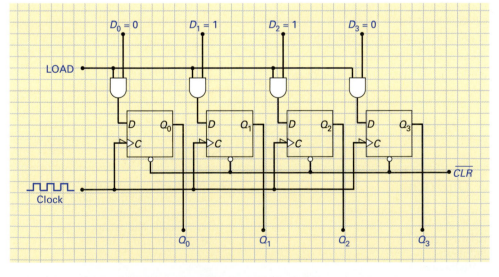

FIGURE 9-26   Application—A 4-Bit Storage Register.

To practice your circuit recognition and operation, refer to Figure 9-28 and answer the following three questions.

25. What type of flip-flop is being used in the application in this circuit?

26. What will be the output frequency at
   a. Q_A?   **b.** Q_B?

27. Sketch the input/output waveforms for this circuit.

To practice your circuit recognition and operation, refer to Figure 9-29 and answer the following four questions.

28. What type of flip-flop is being used in this circuit?

29. What type of triggering is being used in this circuit?

30. Describe how this circuit will operate when inputs A and B are out-of-phase with one another.

31. Describe how this circuit will operate when inputs A and B are in-phase with one another.

FIGURE 9-27   **Application—ON/OFF Clock-Control Circuit.**

FIGURE 9-28   **Application—Clock-Division Circuit.**

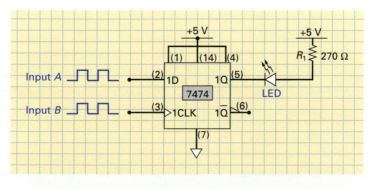

FIGURE 9-29   **Application—A Phase-Detector Circuit.**

FIGURE 9-30 Application—Divide-by-Two Clock Circuit with ON/OFF Control.

FIGURE 9-31 A *J-K* Flip-Flop Timing Example.

FIGURE 9-32 A *J-K* Flip-Flop Timing Example.

To practice your circuit recognition and operation, refer to Figure 9-30 and answer the following three questions.

32. How is the *J-K* flip-flop in Figure 9-30 triggered?

33. Identify the circuit formed by NAND gates *A* and *B*.

34. What will be the output frequency at *Q* when (a) the switch is in the ON position? (b) the switch is on the OFF position?

To practice your circuit recognition and operation, refer to Figure 9-31 and answer the following three questions.

35. Is this *J-K* flip-flop edge-triggered or pulse-triggered?

36. Sketch the *Q* output waveform for the input timing waveforms shown.

37. Sketch the $\overline{Q}$ output waveform for the input timing waveforms shown.

To practice your circuit recognition and operation, refer to Figure 9-32 and answer the following three questions.

38. Is this *J-K* flip-flop edge-triggered or pulse-triggered?

39. Sketch the *Q* output waveform for the input timing waveforms shown.

40. Sketch the $\overline{Q}$ output waveform for the input timing waveforms shown.

To practice your circuit recognition and operation, refer to Figure 9-33 and answer the following four questions.

FIGURE 9-33 Application—Tone Generator.

FIGURE 9-34 Application—Divider Circuits.

41. The 555 timer is connected to operate as a/an _____ multivibrator.

42. Calculate the frequency range of the 555 timer if R_1 were adjusted between the two extremes of 1 Ω and 100 kΩ.

43. Is the 74121 nonretriggerable or retriggerable?

44. Calculate the pulse length range of the 74121 one-shot if R_3 were adjusted between the two extremes of 1 Ω and 100 kΩ.

To practice your circuit recognition and operation, refer to Figure 9-34 and answer the following four questions.

45. Which type of *J-K* flip-flops are being used?

46. Which of the *J-K* flip-flop circuits will function as a divide-by-two circuit?

47. Which of the *J-K* flip-flop circuits will function as a divide-by-three circuit?

48. Which of the *J-K* flip-flop circuits will function as a divide-by-four circuit?

Web Site Questions

Go to the web site http://www.prenhall.com/cook, select the textbook *Practical Digital Electronics,* select this chapter, and then follow the instructions when answering the multiple choice practice problems.

Sequential Logic Circuits

Trash

In 1976 a Texas-based chain of stores known as Tandy Radio Shack, which up to that time had only been involved with electronic kits and parts for citizen band (CB) radios and stereo equipment, sensed a sudden demand among its hobbyists for a personal computer. In July of that year the company recruited Steven Leininger, who had been working for National Semiconductor since graduating from college. He seemed to possess the right qualifications in that he was an electronics engineer and a computer buff. In a double feat, Leininger designed both the computer's architecture and its built-in software and used at the heart of his computer the brand-new Z80 microprocessor. This new microprocessor from the newly formed company, Zilog, resembled Intel's 8080 microprocessor but was exceedingly more powerful. The reason it resembled Intel's processor was because Zilog was founded by a group of disgruntled Intel engineers; a situation that would later lead to some legal problems—but then, that's another story.

On February 2, 1977, after Leininger had not left the plant for the last two weeks and had been working almost completely around the clock, the TRS-80-Model 1 was demonstrated faultlessly to company owner Charles Tandy. Blowing smoke from his ever-present large cigar, Tandy said that it was "quite a nice little gadget." An executive then asked, "How many should be built—one thousand, two thousand?" The company's financial controller answered by saying, "We've got 3,500 company-owned stores. I think we can build that many. If nothing else, we can use them in the back for accounting." For Leininger, his baby had not received the welcome he expected from the general management, who did not share his confidence in its ability or attraction.

Not even Leininger was prepared for the TRS-80 (nicknamed Trash 80) splash that was going to take place. Going on sale in September of 1977, 10,000 orders flooded in almost overnight and the demand remained so strong that the company's production plant took over a year to catch up with the orders. A year later, in 1978, Tandy Computers took an impressive lead in the personal computer race, leaving behind their chief competitor, Commodore, and its PET computer.

Introduction

Flip-flops are memory elements that are the basic building blocks we will use to construct sequential logic circuits, which can perform such functions as storing, sequencing, and counting. To help define the term **sequential logic circuits,** let us compare them to combinational logic circuits, which perform such functions as decoding, encoding, and comparing. The output from a combinational logic circuit is determined by the present state of the inputs, whereas the output of a sequential logic circuit, because of its memory ability, is determined by the previous state of the inputs.

Like combinational logic circuits, there is a large variety of sequential logic circuits; however, two types predominate in digital systems: registers and counters. Both of these circuit types have been briefly discussed in previous applications; however, in this chapter we will examine them in more detail.

Sequential Logic Circuits

Logic circuits in which outputs are dependent on the input states, delays encountered in the logic path, the presence of a discrete timing interval, and the previous state of the logic circuit.

Register

A group of flip-flops or
memory elements that
store and shift a group of
bits or binary word.

Buffer Register

A temporary data storage
circuit able to store a
digital word.

A **register** is a group of flip-flops or memory elements that work together to store and shift a group of bits or a binary word. The most basic type of register is called the **buffer register,** and its function is simply to store a binary word. Other registers, such as the shift register, modify the stored word by shifting the stored bits to the left or to the right.

The buffer register simply stores a digital word. Figure 10-1(a) shows how a buffer register could be constructed using positive edge-triggered D-type flip-flops. This buffer register has actually been constructed using the four D-type flip-flops within a 74175 IC, which was covered in Chapter 9. The 4-bit data input word (1D, 2D, 3D, and 4D) will be stored in the register and appear at the Q output (Q_1, Q_2, Q_3, and Q_4) when the first positive clock edge occurs at the 74175 clock input (pin 9). The 4-bit Q output word drives a 4-bit LED display, which will display the contents of the 4-bit buffer register. Referring to the waveforms in

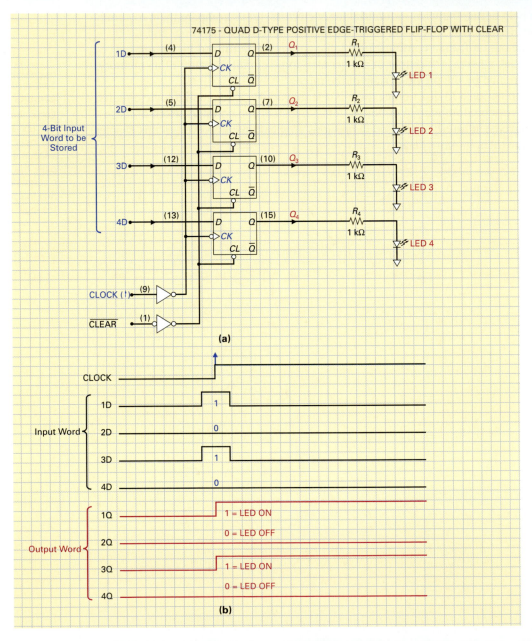

FIGURE 10-1   A Buffer Register. (a) 74175 Parallel-In, Parallel-Out Register. (b) Timing Waveforms.

Figure 10-1(b), you can see that the 4-bit data input, which is only momentarily present on the input data lines (1D, 2D, 3D, and 4D), is stored, or retained, in the buffer register when a rising clock edge occurs. This data is permanently present at the output, and therefore displayed on the LED display.

Since a 4-bit parallel word is delivered to the input of the register, and a 4-bit parallel word is present at the output, buffer registers are often referred to as **parallel-in, parallel-out (PIPO) registers.** The active-LOW clear input of the 74175 can be used to clear all of the flip-flops (erase the stored word) and therefore turn OFF all of the LEDs.

Parallel-In, Parallel-Out (PIPO) Register

A register that will parallel-load an input word and parallel-output its stored contents.

SELF-TEST EVALUATION POINT FOR SECTION 10-1

Use the following questions to test your understanding of Section 10-1.

1. Buffer registers are also referred to as _____ registers.
   a. SISO   b. PISO   c. PIPO   d. SIPO

2. What is the main function of a buffer register?

10-2   SHIFT REGISTERS

A **shift register** is a storage register that will move or shift the bits of the stored word either to the left or to the right. The three basic types of shift registers are

   a. serial-in, serial-out (SISO) shift registers

   b. serial-in, parallel-out (SIPO) shift registers

   c. parallel-in, serial-out (PISO) shift registers

Each of these shift register classifications is shown in Figure 10-2. The names given to these three types actually describe how data is entered, or inputted, into the shift register

Shift Register

A temporary data storage circuit able to shift or move the stored word either left or right.

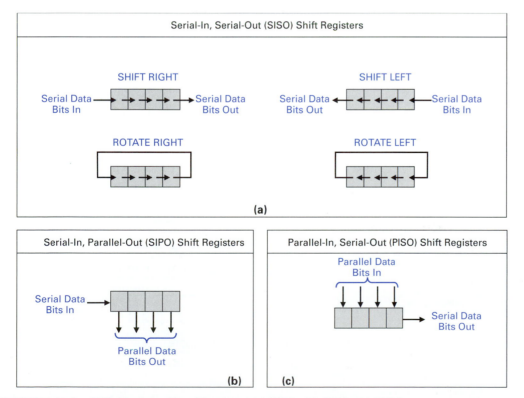

FIGURE 10-2   **Shift Register Classifications. (a) SISO. (b) SIPO. (c) PISO.**

for storage, and how data exits, or is outputted from, the shift register. To understand these operations in more detail, let us take a close look at each of these three shift registers in the following sections.

10-2-1   *Serial-In, Serial-Out (SISO) Shift Registers*

To begin with, Figure 10-3(a) shows how a shift register operates. In this example, the initial condition shows that the 4-bit word 0110 is stored within the 4-bit shift register,

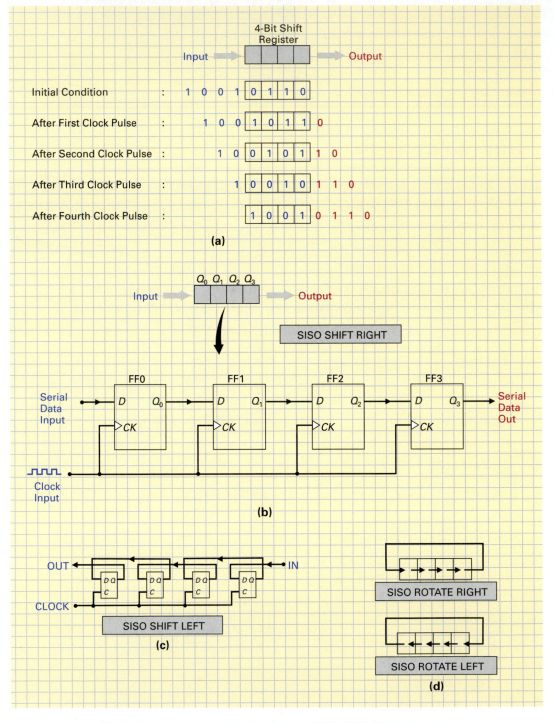

FIGURE 10-3   Serial-In, Serial-Out (SISO) Shift Registers.

FIGURE 10-4   **Test Circuit for a 7491A Shift Register.**

while the external serial word 1001 is being applied to the shift register's input. After the first clock pulse, the data stored within the shift register is shifted one position to the right, and the first bit of the applied serial word is shifted into the first position of the shift register. After the second clock pulse, two bits of the stored 0110 word have been shifted out of the shift register, while two bits of the applied 1001 word have been shifted into the shift register. After the third clock pulse, three shift-right operations have occurred. After four clock pulses, the originally stored 0110 word has been completely shifted out of the register, while the applied 1001 input word has been completely shifted into the register and is now being stored.

Now that the basic operation of the shift register is understood, let us see how we can use flip-flops to construct a shift register circuit. Figure 10-3(b) shows how a 4-bit shift register circuit can be constructed using four D-type flip-flops. The serial data input is applied to the D input of flip-flop 0 (FF0). The output from FF0 (Q_0) is applied to the D input of FF1, the output of FF1 (Q_1) is applied to the D input of FF2, the output of FF2 (Q_2) is applied to the D input of FF3, and the output of FF3 (Q_3) is the final serial data output of the 4-bit shift register. The clock input is applied simultaneously to all of the D-type flip-flops, with each positive edge of the clock causing the stored 4-bit word to shift one position to the right. With this **serial-in, serial-out shift register,** data is serially fed into the register for storage under the control of the clock (1 bit is stored per clock pulse, so four clock pulses will be needed to shift in a serially applied 4-bit word). To extract a stored 4-bit word from the shift register, four clock pulses will have to be applied to serially shift the stored word out of the register.

In summary, the circuit in Figure 10-3(b) shows how four D-type flip-flops should be connected to form a *SISO shift-right shift register.* Figure 10-3(c) shows how the *D* and *Q* connections can be altered to form a *SISO shift-left shift register.* In some applications, the serial data outputs of the circuits shown in Figure 10-3(b) and (c) are connected directly back to the serial data inputs so that the outputted data is immediately inputted. These SISO operations are called *SISO rotate-right* and *SISO rotate-left* and are shown in Figure 10-3(d).

The 7491A shown in Figure 10-4 is an 8-bit SISO shift-right shift register IC. *S-R* flip-flops are used to form an 8-bit SISO shift register, which is positive edge-triggered and has a gated *A* and *B* data input. Since the *A* and *B* data inputs are ANDed, both inputs must be HIGH to shift in a binary 1, while either input can be LOW to shift in a binary 0.

> **Serial-In, Serial-Out (SISO) Register**
>
> A register that will serial-load an input word, and serial-output its stored contents.

■ **EXAMPLE:**

Referring to the circuit shown in Figure 10-4, describe how you would

a. input or store the word "11001010" within the 7491A SISO shift register.

b. output the stored 8-bit word so that it can be displayed on the LED.

This test circuit operates as follows. SW_1 controls the data input to the IC, while SW_2 controls the dock input to the IC. If SW_1 is in the upper position, a HIGH will be applied to the IC's serial data input, whereas if SW_1 is in the lower position, a LOW will be applied to the IC's serial data input. To shift-right the data stored within the 7491A, and therefore load the input data applied by SW_1, SW_2 should be momentarily moved into its upper position and then placed back in its lower position. Switching SW_2 up and then down in this way will apply a positive edge to the clock input of the 7491A and therefore all of the S-R flip-flops within the 7491A.

a. To store the word "11001010" within the 7491A SISO shift register, first apply a LOW (LSB) from SW_1 and then clock the IC using SW_2. Then apply a HIGH and clock, then a LOW and clock, and then a HIGH and clock, and so on until all eight bits are inputted.

b. To display the previously stored 8-bit word on the LED, simply clock the 7491A eight times, and each bit of the word 11001010 will be shifted right and appear in turn at the 7491A output (pin 13). Since the 8-bit word's LSB was the first to be applied to the input of the shift register, it will be the first to appear at the output of the shift register.

10-2-2   *Serial-In, Parallel-Out (SIPO) Shift Registers*

Serial-In, Parallel-Out (SIPO) Register

A register that will serial-load an input word and parallel-output its stored contents.

Figure 10-5 shows the second shift register type, which is called a **serial-in, parallel-out (SIPO) shift register.** As can be seen in the simplified diagram in Figure 10-5(a), data is entered into this shift register type in serial form, and the stored data is available at the output as a parallel word.

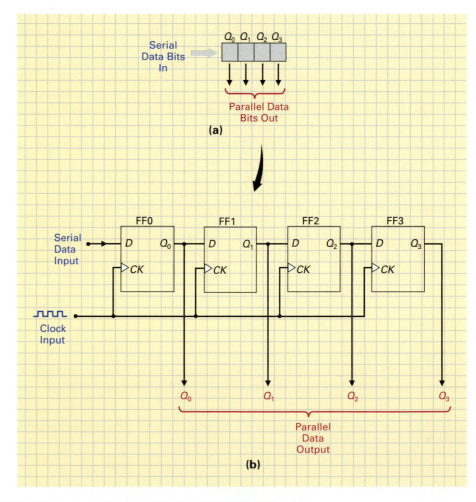

FIGURE 10-5   **Serial-In, Parallel-Out (SIPO) Shift Registers.**

Figure 10-5(b) shows how a 4-bit SIPO shift register can be constructed using D-type flip-flops. To input data into this shift register, a 4-bit serial word is applied at the serial data input and shifted in under the control of the clock input (one shift-right/clock pulse). To input or store a 4-bit serial word into this shift register, therefore, would require four clock pulses. The data stored within the shift register is available at the four Q outputs (Q_0, Q_1, Q_2, and Q_3) as a 4-bit parallel data output.

Figure 10-6 features the 74164, which is an 8-bit SIPO shift-right shift register IC.

Application—Serial-to-Parallel Data Converter

Figure 10-6(a) shows how a 74164 can be used to convert a serial data bitstream into a parallel data word output. Referring to the waveforms in Figure 10-6(b), you can see that one bit of serial data is loaded into the 74164 for each positive transition of the clock. The clock signal is also fed to a divide-by-eight circuit which divides the clock input by eight and loads the 8-bit data from the 74164 into an 8-bit buffer register (made up of two 74175s) at 8-bit intervals.

FIGURE 10-6   Application—Serial-to-Parallel Data Converter. (a) Circuit. (b) Waveforms.

10-2-3   *Parallel-In, Serial-Out (PISO) Shift Registers*

The simplified diagram in Figure 10-7(a) shows the basic operation of a **parallel-in, serial-out shift register.** With this shift register type, the data bits are entered simultaneously, or in parallel, into the register, and then the binary word is shifted out of the register bit by bit as a serial data stream.

Figure 10-7(b) shows how a 4-bit PISO shift register can be constructed using D-type flip-flops. The circuit is controlled by the SHIFT/LOAD control input. When the SHIFT/LOAD control line is LOW, all of the shaded AND gates are enabled, due to the inversion of this control signal by the shaded INVERTER. These enabled AND gates connect the 4-bit parallel input word on the data input lines (D_0, D_1, D_2, and D_3) to the data inputs of the flip-flops. When a clock pulse occurs, this 4-bit parallel word will be latched simultaneously into the four flip-flops and appear at the Q outputs (Q_0, Q_1, Q_2, and Q_3). When the SHIFT/LOAD control line is HIGH, all of the unshaded AND gates are enabled. These enabled AND gates connect the Q_0 output to the D input of FF_1, the Q_1 output to the D input of FF_2, and the Q_2 output to the D input of FF_3. In this mode, therefore, the data stored in the shift register will be shifted to the right by one position for every clock pulse applied.

The 74165 IC can function as either an 8-bit parallel-in (inputs *A-H*), serial-out (output Q_H), or an 8-bit serial-in (input SER) shift register. A LOW on the $SH/\overline{LD}$ (pin 1) control line will enable all of the NAND gates and allow simultaneous loading of the eight S-R flip-flops. A LOW on any of the parallel *A* through *H* inputs will RESET the respective flip-flop, while a HIGH on any of the parallel *A* through *H* inputs will SET the respective flip-flop. The loading of the register is called an *asynchronous operation* since the loading operation will not be in step with the occurrence of an active clock signal (the loading operation and clock are not synchronized). A HIGH on the $SH/\overline{LD}$ line will disable all of the NAND gates and shift

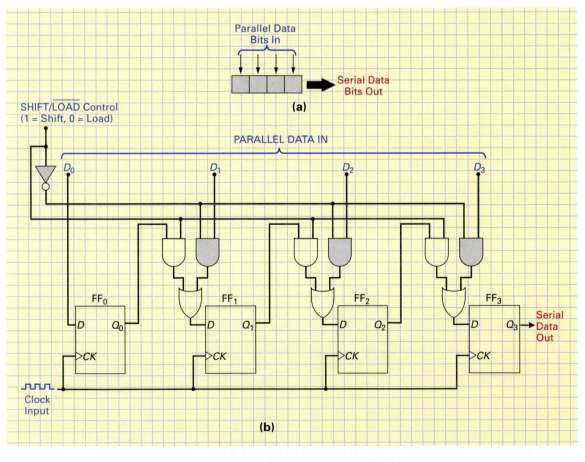

FIGURE 10-7   **Parallel-In, Serial-Out (PISO) Shift Registers.**

the stored 8-bit word to the right at a rate of one position for every one clock pulse. The shifting of data within the shift register is called a *synchronous operation* since the shifting of data will be in step or in sync with the occurrence of an active clock signal (the shifting operation and clock are synchronized). The clock input for this IC can be inhibited, if desired, by applying a HIGH to the *CLK INH* (pin 15) input.

Application—Parallel-to-Serial Data Converter

Figure 10-8 shows how a 74165 can be connected to convert an 8-bit parallel data input into an 8-bit serial data output. In this example, switches have been used to supply an 8-bit parallel input of 11010110 to the *A* through *H* inputs of the 74165. Moving SW_1 to the LOAD position will make pin 1 of the 74165 ($SH/\overline{LD}$) LOW, so the 8-bit word applied to the parallel inputs *A* through *H* will be loaded into the 74165's eight internal flip-flops. Since the LSB *H* input is LOW, the output LED will initially be OFF since Q_H is LOW. The 74165 is positive edge-triggered, and therefore as SW_2 is toggled, the stored bits will be shifted to the output (in the order *H, G, F, E, D, C, B,* and then *A*) at a rate of one bit per HIGH/LOW switching of SW_2.

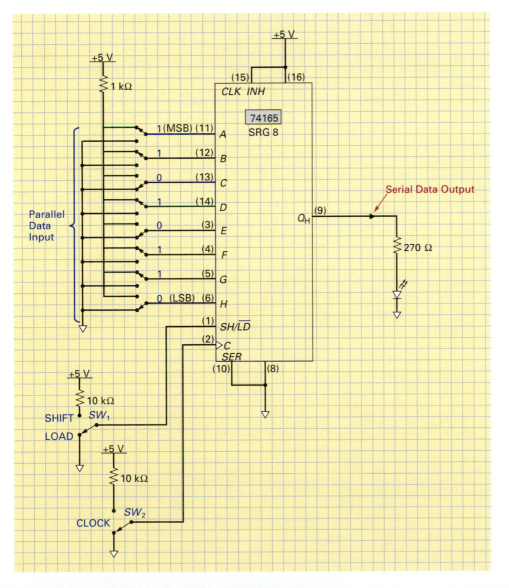

FIGURE 10-8 Application—Parallel-to-Serial Data Converter.

10-2-4 *Bidirectional Universal Shift Register*

A bidirectional shift register can shift stored data either left or right depending on the logic level applied to a control line. The 74194 IC shown in Figure 10-9 is frequently used in applications requiring bidirectional shifting of data. This IC contains forty-six gates and can be made to operate in one of four modes by applying different logic levels to the mode control inputs S_0 and S_1. These modes of operation are listed below.

To parallel-load a 4-bit word, therefore, the S_1 and S_0 mode control inputs must both be HIGH, and the 4-bit word to be loaded must be applied to inputs *A, B, C,* and *D*. Parallel loading is accomplished synchronously (in step or in sync) with an active clock signal, which for the 74194 is a positive clock transition. Once loaded, the stored 4-bit word will appear at the outputs Q_A, Q_B, Q_C, and Q_D. During a parallel-loading operation, serial data flow is inhibited.

S_1	S_0	MODE OF OPERATION
H	H	Parallel (broadside) load
L	H	Shift right (Q_A is shifted toward Q_D)
H	L	Shift left (Q_D is shifted toward Q_A)
L	L	Inhibit clock (do nothing)

To shift data to the right, the mode control input S_1 must be LOW and the mode control input S_0 must be HIGH. In this mode, data within the shift register will be shifted right synchronously from Q_A to Q_D, with new data being inputted from the shift-right serial input (pin 2).

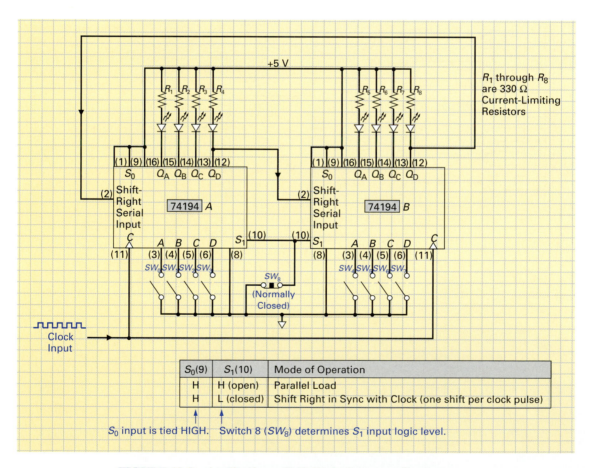

FIGURE 10-9 **Application—Shift-Right Sequence Generator.**

To shift data to the left, the mode control input S_1 must be HIGH and the mode control input S_0 must be LOW. In this mode, data within the shift register will be shifted left synchronously from Q_D to Q_A, with new data being inputted from the shift-left serial input (pin 7).

When both the mode control inputs S_1 and S_0 are LOW, the clocking of the shift register is inhibited. The mode control inputs S_1 and S_0 should only be changed when the clock input is HIGH (inactive).

Application—Shift Register Sequence Generator

Figure 10-9 shows how two 74194 ICs could be connected to form an 8-bit shift-right sequence generator. Switches SW_0 through SW_7 can be set so that they apply any 8-bit pattern or word value (open switch = HIGH, closed switch = LOW). This 8-bit word can be loaded into the shift register by pressing the normally closed push-button (NCPB) SW_8. SW_8 controls the logic level applied to the mode control input S_1 (pin 10), as shown in the table in Figure 10-9. With S_0 connected permanently HIGH, an open SW_8 (switch pressed) will apply an equivalent HIGH input to S_1, which will instruct the two 74194s to parallel-load the 8-bit word, which, when loaded by a positive transition of the clock, will appear at the Q_A, Q_B, Q_C, and Q_D outputs. When SW_8 is closed (switch released), the S_1 mode control input will be pulled LOW, instructing both 74194s to shift right the data stored synchronously with each positive edge of the clock. In this shift-right mode, the Q_D output of 74194 B on the right of Figure 10-9 is connected back to the shift-right serial input of the 74194 A, and therefore the data is continually rotated right.

The LEDs in Figure 10-9 form a common-anode display, which means that a HIGH output will turn OFF an LED, and a LOW output will turn ON an LED.

SELF-TEST EVALUATION POINT FOR SECTION 10-2

Use the following questions to test your understanding of Section 10-2.

1. How many clock pulses would be needed to serially load an 8-bit SISO shift register with an 8-bit word?

2. The _____ shift register could be used for serial-to-parallel conversion, while the _____ shift register could be used for parallel-to-serial conversion. (SISO, SIPO, PISO).

3. How many clock pulses would be needed to parallel-load an 8-bit PISO shift register with an 8-bit word?

4. The _____ shift register can parallel-load, shift right, shift left, input serially, and output serially.

10-3 THREE-STATE OUTPUT REGISTERS

Many registers have three-state or tri-state output gates connected between the flip-flop's Q outputs and the IC's output pins, as illustrated in Figure 10-10(a). The three-state output gate, which was discussed previously in Section 10-1-5, has as its name implies three output states, which are LOW (logic 0), HIGH (logic 1), and float (an open output or HIGH impedance state). With the active-HIGH three-state gates shown in Figure 10-10(a), a HIGH on the enable output control line (E) will connect the Q outputs from the D-type flip-flops directly through to the Y outputs of the IC. On the other hand, when the enable output control line is LOW, the three-state gates are disabled and the Y outputs of the IC will float, since they have effectively been disconnected from the Q outputs. Figure 10-10(b) shows how a small inverted triangle on each output is used to indicate that this 4-bit controlled buffer register (PIPO) has *three-state output control*.

Referring to Figure 10-10(a), let us now see how this PIPO buffer register with three-state output control operates. Other than the enable output control line (E), this register has a load control line (L) and a clock input (C). When the load control line is LOW (inactive), the contents of each of the D-type flip-flops within the register remain unchanged since the data is recirculated from each Q output back to its respective D input for each positive edge of the clock input via the enabled unshaded AND gates. When the load control input is HIGH,

FIGURE 10-10   A Controlled Buffer Register with Three-State Output.

Bus-Organized Digital System

A digital system interconnection technique in which a common bus is time-shared by all the devices connected to the bus.

Bus

A group of conductors used as a path for transmitting digital information from one of several sources to one of several destinations.

however, all of the shaded AND gates are enabled, so the parallel input word on the D_0, D_1, D_2, and D_3 inputs are connected to the flip-flop data inputs and latched into the flip-flops on the next positive edge of the clock.

The three-state gate was originally developed by National Semiconductor in the early 1970s, and its purpose was to reduce the number of wires in a digital system. Three-state gates made possible a **bus-organized digital system.** To explain this concept, Figure 10-11 shows a simple circuit consisting of four input switches, three registers, and four output LEDs. A group of wires used to carry binary data is called a **bus,** and in Figure 10-11 a 4-bit bus is used as a common transmission path between the input device, output device, and the registers. Since the input device and the registers all have three-state outputs, all of these devices (unless enabled) are effectively isolated or disconnected from the bus. If they were not

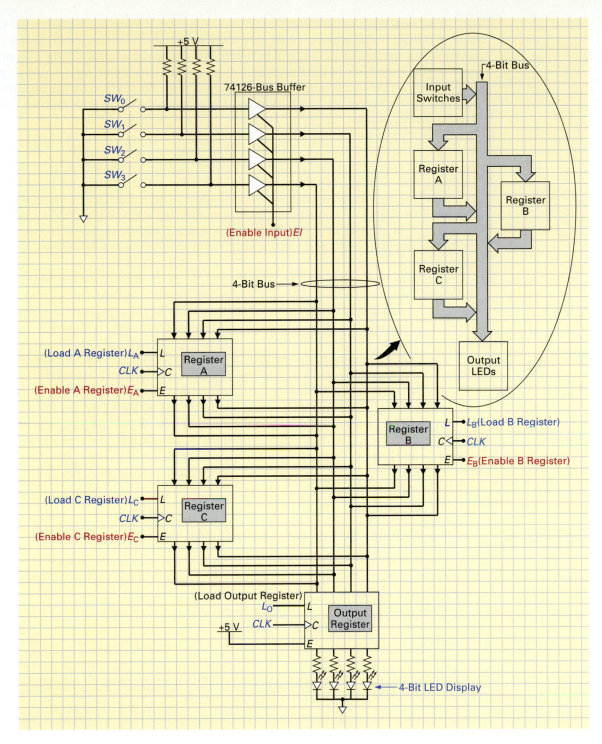

FIGURE 10-11 **A Bus-Organized Digital System.**

disconnected, a *bus conflict* would occur since one register could be driving a bus line HIGH while another register could be trying to drive that same bus line LOW.

To transfer a 4-bit word from register A to register B, we must connect the contents of register A on to the bus, making E_A HIGH, and then load this data into register B by making I_B HIGH. Since the clock (*CLK*) is connected to all registers, the next positive edge of the clock will load the 4-bit word from register A into register B.

If we now wanted to transfer the contents of register B to the output, we would simply make E_B HIGH so that contents of the B register would now be connected to the bus and L_O HIGH so that the data on the bus will be loaded into the output register. On the next positive clock edge, the contents of the B register will be stored in the output register and displayed immediately on the output LEDs, since the output of the output register has no three-state control. The output register does not connect to the common bus, so its output can therefore be permanently enabled.

Before three-state outputs were available, separate groups of wires were needed between all of the devices in a circuit. For example, without three-state control, the circuit in Figure 10-11 would need a set of wires connecting register A to register B, another set connecting A to C, another set connecting B to C, another set connecting the input to A, and so on. As you can easily imagine, even the smallest of digital electronic systems containing only several devices would need a mass of connecting wires. With a bus-organized digital system, only one set of connecting wires is needed, and these can be connected to all devices as long as those devices that transmit data onto the common bus have three-state output control.

In Figure 10-11, only a 4-bit bus was used. In many digital systems 8-bit (8 wires), 16-bit (16 wires), and 32-bit (32 wires) buses are common and are used to transfer 8-bit, 16-bit, or 32-bit parallel words from one device to another. In schematic diagrams, a solid bar with arrows is generally used to represent a bus of any size, as shown in the inset in Figure 10-11.

The 74173 IC shown in Figure 10-12 is a 4-bit (PIPO) buffer register with three-state output control. This IC contains four D-type flip-flops with an active-HIGH clear input (pin 15). The three-state output buffers are controlled by the active-LOW inputs M and N (pins 1 and 2). When both of these output control inputs (M and N) are LOW, the HIGH or LOW logic levels stored in the four flip-flops will be available at the four outputs (pins 3, 4, 5, and 6). The parallel loading of data into the flip-flops is controlled by the active-LOW ANDed inputs $\overline{G}_1$ and $\overline{G}_2$ (pins 9 and 10). When both of these data enable inputs are LOW, the data at the D-inputs (pins 14, 13, 12, and 11) are loaded into their respective flip-flops on the net positive transmission clock input.

Application—A Bus-Organized Digital System

Figure 10-12 shows how four 74173s and one 74126 (quad buffer gates with three-state outputs) could be connected to construct the bus-organized digital system discussed previously in Figure 10-11.

■ **EXAMPLE:**

Referring to the circuit in Figure 10-12, list which control lines need to be made active in order to perform the following operations:

a. loading the 4-bit word from the input bus buffer into register A

b. copying the contents of register B into register C

c. displaying the contents of register C

■ *Solution:*

a. E_I(HIGH), $\overline{L}_A$(LOW), $CLK\uparrow$

b. E_B(HIGH), $\overline{L}_C$(LOW), $CLK\uparrow$

c. E_C(HIGH), $\overline{L}_O$(LOW), $CLK\uparrow$

CHAPTER 10 / SEQUENTIAL LOGIC CIRCUITS

FIGURE 10-12 Application—A 4-Bit Bus-Organized Digital System Containing Four Input Switches and an Input Buffer, Three PIPO Storage Registers, and an Output Register Driving a Four-LED Display.

SELF-TEST EVALUATION POINT FOR SECTION 10-3

Use the following questions to test your understanding of Section 10-3.

1. A three-state output can be either HIGH, LOW, or _____.
2. A group of parallel wires designed to carry binary data is called a _____.
3. What is bus conflict, and how is it avoided?
4. What are the three main inputs to a controlled buffer register with three-state output control?

In this section we will review the register applications discussed so far in this chapter and examine other register uses.

10-4-1   *Memory Registers*

Memory Register
A register capable of storing several binary words.

Read
An operation in which data is acquired from some source.

Write
An operation in which data is transferred to some destination.

Since a register is basically a storage or memory element used to store a binary word, it is often referred to as a **memory register.** A memory unit may contain one memory register capable of storing only one binary word or many memory registers capable of storing many binary words. Figure 10-13 shows how a controlled buffer register PIPO with three-state output control can be configured to function as a memory register. In order to perform as a memory register, a register must be able to achieve two basic operations, which are referred to as **read** (which is an input operation, since you are reading the contents of the register in the same way that we read or input words in a book) and **write** (which is an output operation, since you are writing data into the register in the same way that you write or output words onto a pad). These two operations are controlled by a READ/$\overline{\text{WRITE}}$ ($R/\overline{W}$) control line, which is used to control the buffer register's load and enable control inputs, as shown in Figure 10-13. When the $R/\overline{W}$ control line is LOW, the load control input is enabled while the enable control input is disabled, so new data is stored or written into the memory register from the bus (a write operation). When the ($R/\overline{W}$) control line is HIGH, the enable control input is active while the load control input is inactive, so the 8 bits of stored data within the register are retrieved, or read, from the register onto the bus (read operation).

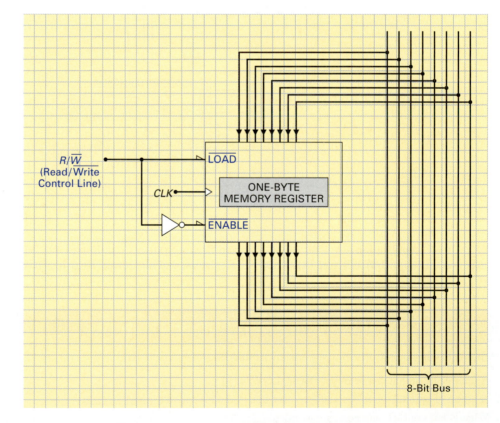

FIGURE 10-13   **Application—A One-Byte (8-Bit) Memory Register.**

10-4-2 *Serial-to-Parallel and Parallel-to-Serial Conversions*

Figure 10-14(a) shows how a SIPO shift register can be used for serial-to-parallel conversion. This shift register shifts in an 8-bit serial input at a rate of 1 bit per clock pulse, and after eight clock pulses all of the byte is stored and all bits are simultaneously available as an 8-bit parallel output word. Figure 10-14(b) shows how a PISO shift register can be used for

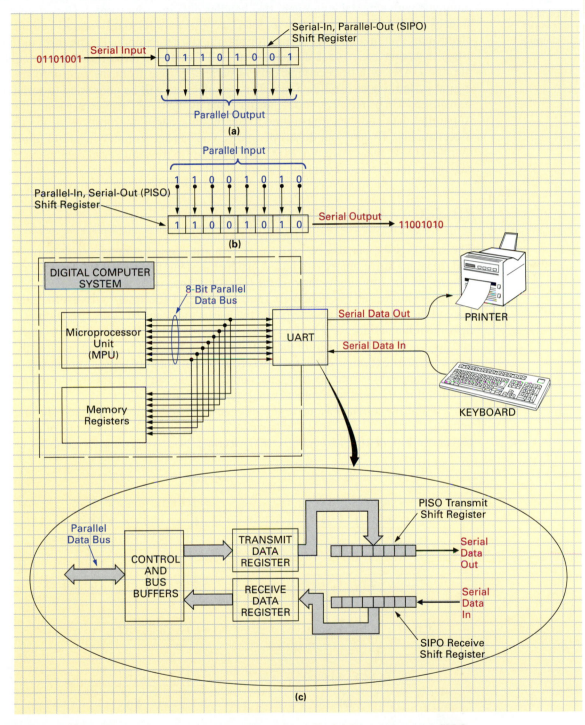

FIGURE 10-14 Application—Serial-to-Parallel and Parallel-to-Serial Conversions. (a) SIPO. (b) PISO. (c) UART.

parallel-to-serial conversion. This shift register simultaneously loads all 8 bits of a parallel input and then shifts the data out serially at a rate of 1 bit per clock pulse, and therefore after eight clock pulses the byte has been outputted.

The ability of a shift register to convert binary data from serial-to-parallel or parallel-to-serial is made use of in an interfacing (data transfer) device known as a **UART,** which is an abbreviation for **universal asynchronous receiver and transmitter.** Data within a digital system is generally transferred from one device to another in parallel format on a bus, as shown in Figure 10-14(c). In many cases these digital electronic systems must communicate with external equipment that handles data in serial format. Under these circumstances, a UART can be used to interface the parallel data format within a digital electronics system to the serial data format of external equipment. The inset in Figure 10-14(c) shows the basic block diagram of a UART's internal circuit. In this particular example, the UART will be receiving data in parallel format from the microprocessor in the digital system via an 8-bit data bus. This parallel data will be converted to serial data and transmitted to a printer. The serial data received by the UART from a keyboard is converted to parallel data and then transmitted to the digital system microprocessor via the parallel 8-bit data bus. To carry out these operations, you can see from the inset in Figure 10-14(c) that the UART makes use of a PISO shift register to transmit serial data and a SIPO shift register to receive serial data in.

10-4-3 *Arithmetic Operations*

The shift register can also be used to perform arithmetic operations such as multiplication and division. Since binary is a base-2 number system, binary numbers can be changed by some power of 2 by simply moving the position of the number. For example, the contents of the upper shift register in Figure 10-15(a) is 00000111 (decimal 7). By shifting the contents once to the left and shifting a zero into the least significant bit (LSB) position, we can multiply the initial number of 7 by 2 to obtain 14 ($7 \times 2 = 14$). If we continue to shift to the left and add zeros, the value within the shift register in Figure 10-15(a) is continually doubled. The multiplying factor is equal to 2^n, where n is the number of times the contents are shifted. In the example in Figure 10-15(a), the contents of the shift register are shifted three times, and therefore the original number is multiplied by $2^3 = 8$ ($7 \times 8 = 56$).

If shifting a binary value to the left causes it to double ($\times 2$), then shifting a binary value to the right should cause it to halve ($\div 2$). As an example, the contents of the upper shift register in Figure 10-15(b) is 00011000 (decimal 24). By shifting the contents once to the

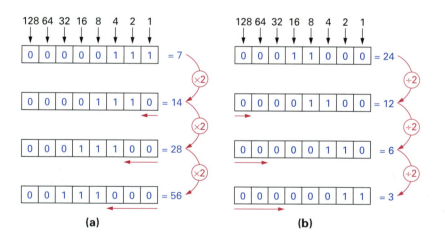

(a)   **(b)**

FIGURE 10-15 **Application—Shifting Binary Numbers. (a) Left to Multiply by 2. (b) Right to Divide by 2.**

right, and shifting a zero into the most significant bit (MSB) position, we can divide the initial number of 24 by 2 to obtain 12 ($24 \div 2 = 12$). If we continue to shift to the right and add zeros, the value within the shift register in Figure 10-15(b) is continually halved. The divide ratio is also equal to 2^n, where n is the number of times the contents are shifted. In the example in Figure 10-15(b), the contents of the shift register are shifted three times, and therefore the original number is divided by $2^3 = 8$ ($24 \div 8 = 3$).

10-4-4   *Shift Register Counters/Sequencers*

One common application of shift registers is as a **sequencer,** in which the parallel outputs of the shift register are used to generate a sequence of equally spaced timing pulses. When connected to operate in this way, shift registers are often referred to as counters, since they generate a specific sequence of logic levels at their parallel outputs. Two of the most widely used shift register *counters/sequencers* are the ring counter and the johnson counter, both of which we will now discuss.

The Ring Counter

Figure 10-16(a) shows how four D-type flip-flops could be connected to form a **ring counter** or **shift register sequencer.** All shift register counters have a feedback connection from the output of the last flip-flop (Q_3) to the input of the first flip-flop (D_0), as shown in Figure 10-16(a). This characteristic circle or "ring" that causes the stored data to be constantly cycled through the shift register accounts for the name *ring counter.*

The circuit shown in Figure 10-16(a) operates as follows. When the active-LOW START control input is brought LOW, Q_0 will be preset HIGH and Q_1, Q_2, and Q_3 will be cleared LOW, as shown in the waveforms in Figure 10-16(b). The stored word (1000) is then shifted right as clock pulses are applied, and since a feedback loop exists between Q_3 and D_0, the single HIGH "start bit" will be continually recirculated. This ring counter action is shown in the waveforms in Figure 10-16(b) and the logic state table in Figure 10-16(c).

Figure 10-17 shows how two 74194s (4-bit bidirectional universal shift registers) could be connected to form an 8-bit ring counter. The ring or feedback connection is made by connecting Q_D of 74194 B back to the shift-right serial input of 74194 A. This ring counter circuit is initiated by pressing SW_1, which will put a HIGH on the mode control S_1. The HIGH applied to the S_1 input will instruct both 74194s to parallel-load their *A, B, C,* and *D* inputs, and since only the *A* input of 74194 A is HIGH, the word 10000000 will be stored. When SW_1 is released, the normally closed switch will connect a LOW to the S_1 mode inputs, instructing them to shift right the stored data. The stored 8-bit value 10000000 will be shifted to the right in sync with the positive edges of the clock input, and the result is shown in the table below.

CLOCK PULSES	74194 A				74194 B			
	Q_A	Q_B	Q_C	Q_D	Q_A	Q_B	Q_C	Q_D
0	1	0	0	0	0	0	0	0
1	0	1	0	0	0	0	0	0
2	0	0	1	0	0	0	0	0
3	0	0	0	1	0	0	0	0
4	0	0	0	0	1	0	0	0
5	0	0	0	0	0	1	0	0
6	0	0	0	0	0	0	1	0
7	0	0	0	0	0	0	0	1

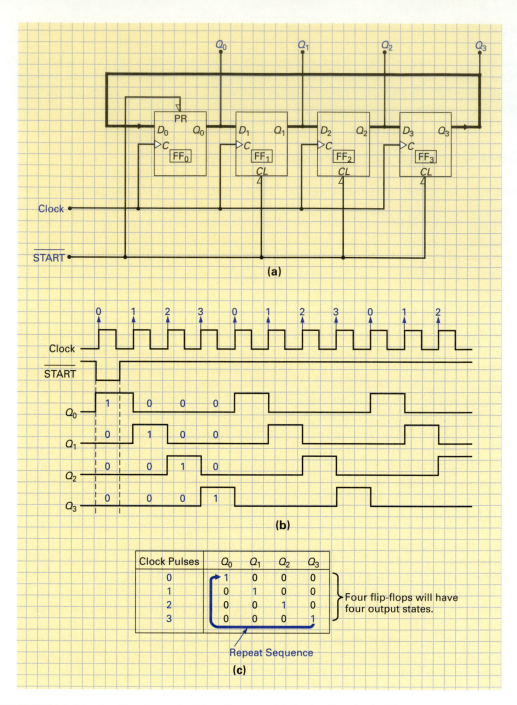

FIGURE 10-16   Application—the Ring Counter. (a) Basic Circuit. (b) Timing Waveforms.
(c) Logic State Table.

In this example circuit, the outputs Q_A, Q_B, Q_C, and Q_D of both 74194s are used to sequence ON and OFF LEDs (common-anode display, HIGH = LED OFF, LOW = LED ON). In a real digital circuit application, the outputs would be connected to logic circuits, and the ON/OFF sequence generated by the ring counter would control when each of these logic circuits is sequenced ON and then OFF. Since the frequency of the clock pulses is fixed, a sequence of equally spaced timing pulses would be generated to control the external logic circuits.

FIGURE 10-17   Application—8-Bit Ring Counter Using Two 74194s.

The Johnson Counter

Figure 10-18(a) shows how four D-type flip-flops could be connected to form a **johnson counter.** As you can see from this illustration, the johnson counter is constructed in almost exactly the same way as the ring counter except that the inverted output of the last flip-flop ($\overline{Q_3}$) is connected to the input of the first flip-flop (D_0). This physical difference in the feedback path of the johnson counter accounts for why this circuit is often referred to as a *twisted-ring counter* instead of being named after its inventor.

Like the ring counter, the johnson counter needs a start circuit to initialize the first output sequence, which, as you can see in the timing waveforms in Figure 10-18(b) and the logic state diagram in Figure 10-18(c), is 1000. Since Q_3 is LOW at the start, $\overline{Q_3}$ will be HIGH, and this HIGH will be fed back to the D_0 input, so HIGH inputs will be clocked into the shift register in a left-to-right motion until the shift register is full of 1s. When Q_3 finally goes HIGH (after the third clock pulse), $\overline{Q_3}$ will go LOW and so will D_0. The shift register will now shift in LOW inputs in a left-to-right motion until it is fully loaded with 0s. When Q_3 finally goes LOW (after the seventh clock pulse), $\overline{Q_3}$ will go HIGH and therefore so will D_0, causing the entire cycle to repeat.

With the ring counter, the number of different output states is governed by the number of flip-flops in the register, and therefore a four-stage ring counter would have four different output states. Since the johnson counter first fills up with 1s (in four clock pulse inputs) and then fills up with 0s (in the next four clock pulses), the number of different output states is equal to twice the number of flip-flops. In the example circuit in Figure 10-18, the four-stage johnson counter will have eight different output states (2×4 flip-flops = 8), as listed in the logic state table in Figure 10-18(c).

Johnson Counter

Also known as a twisted-ring counter because of its inverted feedback path, it is a circuit that first fills with 1s and then fills with 0s and then repeats.

FIGURE 10-18 Application—the Johnson Counter. (a) Basic Circuit. (b) Timing Waveforms. (c) Logic State Table.

Use the following questions to test your understanding of Section 10-4.

1. What is the function of the load and enable inputs to a controlled buffer register with three-state outputs?

2. Define the following terms.
   a. Bus   b. Bus-organized digital system   c. Bus conflict   d. 16-bit bus

3. Give the full name of the abbreviation UART and briefly describe the device's purpose.

4. If a binary number is shifted six positions to the left, the resulting number will be _____ (16, 32, 64, 128) times _____ (larger/smaller).

5. Describe the basic circuit difference between a ring counter and a johnson counter.

10-5   ASYNCHRONOUS COUNTERS

A *counter* is like a register in that it is a sequential logic circuit made up of several flip-flops. A register, however, is designed to store a binary word, whereas the binary word stored in a counter represents the number of clock pulses that have occurred at the clock input. The input clock pulses, therefore, cause the flip-flops of the counter to change state, and by monitoring the outputs of the flip-flops, you can determine how many clock pulses have been applied.

There are basically two types of counters—*asynchronous* and *synchronous*. The key difference between these two counter types is whether or not the operation of the counter is in synchronism with the clock signal. Most of the flip-flops of an asynchronous counter are not connected to the clock signal, and so the operation of this counter is not in sync with the master clock signal. On the other hand, all of the flip-flops of a synchronous counter are connected to the clock signal, and so the operation of this counter is in sync with the master clock signal.

Most of the flip-flops in an **asynchronous counter,** therefore, are not connected to the clock signal, and so the flip-flops do not change state in sync with the master clock signal. To better understand this distinction, let us examine some of the different types of asynchronous counters.

Asynchronous Counter

A counter in which an action starts in response to a signal generated by a previous operation, rather than in response to a master clock signal.

10-5-1   *Asynchronous Binary Up Counters*

Figure 10-19(a) shows how a 4-bit or 4-stage asynchronous binary up counter can be constructed with four *J-K* flip-flops. In this circuit you can see that the *J-K* flip-flops are cascaded, or strung together, in series so that the output of one flip-flop will drive the clock input of the next flip-flop. Since the *J* and *K* inputs of all four flip-flops are tied HIGH, the output of each flip-flop will toggle, or change state, each time a negative edge occurs at the flip-flop's clock input.

The clock input to this circuit and the Q output waveforms for each of the four flip-flops are shown in Figure 10-19(b). The Q_0, Q_1, Q_2, and Q_3 outputs make up a 4-bit word, which we will initially assume is 0000, as shown at the far left of the waveforms and on the first line of the truth table in Figure 10-19(c). The output of FF0 (Q_0) generates the least significant bit (LSB) of the 4-bit output word, while the output of FF3 (Q_3) supplies the most significant bit (MSB) of the 4-bit output word. Since FF0 is driven by the clock input, Q_0 will toggle once for every negative edge of this clock input, as shown in the Q_0 timing waveform. This means that clock input negative edge 1 will cause Q_0 to change from a 0 to a 1, clock input negative edge 2 will cause Q_0 to change from a 1 to a 0, clock input negative edge 3 will cause Q_0 to change from a 0 to a 1, and so on. Since Q_0 is connected to the clock input of FF1, each negative edge of Q_0 will cause the Q_1 output to toggle. Similarly, a negative edge at the Q_1 output will cause the output of FF2 to toggle, and a negative edge at the Q_2 output will cause the output of FF3 to toggle.

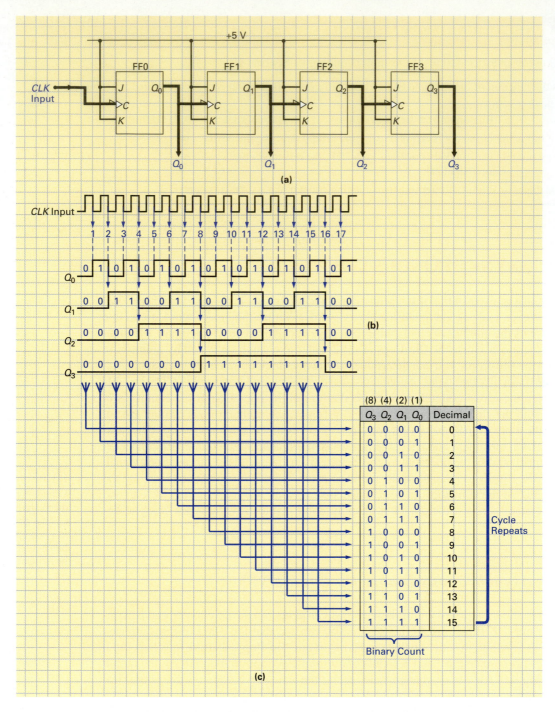

FIGURE 10-19 An Asynchronous Binary Up Counter. (a) Logic Circuit. (b) Waveforms. (c) Truth Table.

The Maximum Count (N) of a Counter

Referring to the truth table in Figure 10-19(c), you may have noticed that the 4-bit binary value appearing at the Q_0, Q_1, Q_2, and Q_3 outputs indicated the number of clock pulses applied to this circuit. For example, before the clock pulse, the output was 0000 (decimal 0). After the first clock pulse, the count was 0001 (decimal 1); after the second clock pulse, the count was 0010 (decimal 2); after the third clock pulse, the count was 0011 (decimal 3); and so on. The **maximum count (N)** of a binary counter is governed by the number of flip-flops in the counter.

Maximum Count

For a counter, the maximum count is $2^n - 1$ or one less than the counter's modulus.

$$N = 2^n - 1$$

N = maximum count before cycle repeats

n = number of flip-flops in the counter circuit

With the counter circuit illustrated in Figure 10-19, the maximum count will be

$$N = 2^n - 1$$
$$= 2^4 - 1$$
$$= 16 - 1$$
$$= 15_{10}(1111_2)$$

This can be verified by referring to the waveforms in Figure 10-19(b) and the last line of the truth table in Figure 10-19(c), which shows the final maximum count of the 4-bit binary output is 1111 (decimal 15). You may have noticed that the maximum binary count that can be output from a binary counter before its cycle repeats is calculated in the same way that we determine the maximum count of a binary word containing a certain number of bits.

EXAMPLE:

Calculate the maximum count of a

a. 1-bit counter
b. 2-bit counter
c. 3-bit counter
d. 4-bit counter
e. 8-bit counter

Solution:

a. A 1-bit counter will have a maximum count of

$$N = 2^n - 1 = 2^1 - 1 = 2 - 1 = 1$$

b. A 2-bit counter will have a maximum count of

$$N = 2^n - 1 = 2^2 - 1 = 4 - 1 = 3$$

c. A 3-bit counter will have a maximum count of

$$N = 2^n - 1 = 2^3 - 1 = 8 - 1 = 7$$

d. A 4-bit counter will have a maximum count of

$$N = 2^n - 1 = 2^4 - 1 = 16 - 1 = 15$$

e. An 8-bit count will have a maximum count of

$$N = 2^n - 1 = 2^8 - 1 = 256 - 1 = 255$$

The Modulus (mod) of a Counter

The **modulus** of a counter describes the number of different output combinations generated by the counter. As an example, the 4-bit counter in Figure 10-19 has a modulus of 16, since the counter generates sixteen different output words or combinations. These sixteen different output combinations (0000 through 1111) are listed in the truth table in Figure 10-19(c). The

Modulus

The maximum number of output combinations generated by a counter, equal to 2^n.

modulus of a counter, therefore, can be calculated in the same way that we determine the number of different combinations generated by a binary word with a certain number of bits.

$$\text{mod} = 2^n$$

$\text{mod} = $ modulus of the counter

$n = $ number of flip-flops in the counter

With the counter circuit illustrated in Figure 10-19, the modulus of the counter will be

$$\text{mod} = 2^n$$
$$= 2^4$$
$$= 16$$

■ EXAMPLE:

Calculate the modulus of a

    a. 1-bit counter

    b. 2-bit counter

    c. 3-bit counter

    d. 4-bit counter

    e. 8-bit counter

■ *Solution:*

    a. A 1-bit counter will have a modulus of $\text{mod} = 2^n = 2^1 = 2$

    b. A 2-bit counter will have a modulus of $\text{mod} = 2^n = 2^2 = 4$

    c. A 3-bit counter will have a modulus of $\text{mod} = 2^n = 2^3 = 8$

    d. A 4-bit counter will have a modulus of $\text{mod} = 2^n = 2^4 = 16$

    e. An 8-bit counter will have a modulus of $\text{mod} = 2^n = 2^8 = 256$

The Frequency Division of a Counter

Frequency Divider

A counter circuit that will divide the input signal frequency.

Referring again to the waveforms in Figure 10-19(b), you can see how a binary counter is also a **frequency divider.** Each of the flip-flops divides its input frequency by 2. For example, two clock pulse cycles (or two negative edges) are needed at the C input of FF0 to produce one cycle at the Q_0 output. Likewise, two complete cycles of the clock are needed at the input of FF1 to produce one cycle at the Q_1 output, and two clock cycles are needed at the input of FF2 to produce one cycle at the Q_2 output. Finally, two clock cycles have to be applied at the input of FF3 in order to produce one cycle at the Q_3 output. Each flip-flop, therefore, acts as a **divide-by-two circuit.** If two flip-flops were connected together, the input clock frequency would be divided by 2, and then divided by 2 again. This overall action would divide the input clock frequency by 4, as shown in Figure 10-19(b), which shows that four negative edges are needed at the clock input in order to produce one complete cycle at the Q_1 output. Therefore, one flip-flop divides by 2, two flip-flops will divide by 4, three flip-flops will divide by 8, four flip-flops will divide by 16, and so on. The frequency division performed by a counter circuit can be calculated with the following formula:

Divide-by-Two Circuit

A counter circuit that will divide the input signal frequency by 2.

$$\textit{Division Factor} = 2^n$$

$n = $ number of flip-flops in the counter circuit

■ **EXAMPLE:**

Calculate the following for a 3-bit asynchronous binary counter:

  a. maximum count

  b. modulus

  c. division factor

■ *Solution:*

a.
$$N = 2^n - 1$$
$$= 2^3 - 1 = 8 - 1 = 7$$

b.
$$\text{mod} = 2^n$$
$$= 2^3 = 8$$

c.
$$\textit{Division Factor} = 2^n$$
$$= 2^3 = 8$$

A 3-bit counter, therefore, has eight different output combinations (a modulus of 8), a maximum count of 7 (111_2), and will divide the input frequency by 8.

The Propagation Delay Time (t_p) of a Counter

The asynchronous counter is also known as a **ripple counter.** This term is used because the clock pulses are applied to the input of only the first flip-flop, and as the count is advanced, the effect "ripples" through to the other flip-flops in the circuit. Since each flip-flop in the counter circuit is triggered by the preceding flip-flop, it can take a certain amount of time for a pulse to ripple through all of the flip-flops and update the outputs to their new value. For example, when the eighth negative clock pulse edge occurs, all four Q outputs need to be changed from 0111 to 1000. If each flip-flop has a propagation delay time (t_p) of 10 ns, it will take 40 ns (4 flip-flops $\times$ 10 ns) to update the counter's count from 0111 to 1000. The counting speed or clock pulse frequency is therefore limited by the propagation delay time of all the flip-flops in the counter circuit. The HIGH input clock frequency limit can be calculated with the following formula:

$$f = \frac{1 \times 10^9}{n \times t_p}$$

    f = upper clock pulse frequency limit
    n = number of flip-flops in the counter circuit
    t_p = propagation delay time of each flip-flop in nanoseconds

Ripple Counter
An asynchronous counter in which the first flip-flop signal change affects the second flip-flop, which in turn affects the third, and so on until the last flip-flop in the series is changed.

■ **EXAMPLE:**

If a 4 MHz clock input were applied to a 4-bit asynchronous (ripple) binary counter, and each flip-flop had a propagation delay time of 32 ns,

  a. would this input frequency be too fast for the counter?

  b. what would be the frequency at the Q_0, Q_1, Q_2, and Q_3 outputs?

■ *Solution:*

a.
$$f = \frac{1 \times 10^9}{n \times t_p}$$
$$= \frac{1 \times 10^9}{4 \times 32} = 7.8 \text{ MHz (upper frequency limit)}$$

A 4 MHz clock input is less than the 7.8 MHz upper frequency limit, and so this input is not too fast.

b. Frequency at Q_0 = 4 MHz ÷ 2 = 2 MHz
   Frequency at Q_1 = 4 MHz ÷ 4 = 1 MHz
   Frequency at Q_2 = 4 MHz ÷ 8 = 500 kHz
   Frequency at Q_3 = 4 MHz ÷ 16 = 250 kHz

10-5-2   *Asynchronous Binary Down Counters*

With the asynchronous binary up counter just described, each input clock pulse would cause the parallel output binary word to be increased by 1, or incremented. By making a small modification to the up counter circuit, it is possible to produce a down counter circuit that will cause the parallel output word to be decreased by 1, or decremented, for each input clock pulse.

Figure 10-20(a) shows how a 4-bit or four-stage asynchronous binary down counter circuit can be constructed with four *J-K* flip-flops. Once again, the *J-K* flip-flops are cascaded, or strung together. In this instance, however, the $\overline{Q}$ outputs of each stage are connected to the clock inputs of the next stage. This small circuit difference causes the count sequence to be the complete opposite of the previously described up counter.

The clock input and Q output waveforms for this 4-bit binary down counter are shown in Figure 10-20(b). Referring to the far left side of the waveforms, you can see that we will begin with all the flip-flops initially RESET, and so the Q_0, Q_1, Q_2, and Q_3 outputs are 0000. If all the Q outputs are LOW, then all of the $\overline{Q}$ outputs must be HIGH, and therefore so are all of the *C* inputs to FF1, FF2, and FF3. Since the *J* and *K* inputs of all four flip-flops are tied HIGH, the output of each flip-flop will toggle each time a negative edge occurs at the circuit's clock (*CLK*) input. When the first clock pulse negative edge is applied to FF0, the Q_0 output will toggle from a binary 0 to a 1. This will cause the $\overline{Q}_0$ output to change from a binary 1 to a 0, and this negative edge or transition will clock the input of FF2, causing it to toggle, sending Q_2 HIGH and $\overline{Q}_2$ LOW. This HIGH-to-LOW transition at $\overline{Q}_2$ will clock the input of FF3, which will toggle FF3, sending its Q_3 output HIGH and its $\overline{Q}_3$ LOW. After the first clock pulse, therefore, the Q_0, Q_1, Q_2, and Q_3 outputs from the down counter will be 1111 (decimal 15), as shown in the waveforms in Figure 10-20(b) and on the first line of the truth table in Figure 10-20(c).

The down counter circuit will then proceed to count down by 1 for every clock pulse input applied. Referring again to the waveforms in Figure 10-20(b), you can see that FF0 will toggle for each negative edge of the *CLK* input, producing a Q_0 output that is half the input frequency (since each flip-flop divides by 2). Referring to the Q_1, Q_2, and Q_3 waveforms, you can see that these outputs seem to toggle in response to each positive edge of the previous flip-flop's Q output. For instance, Q_1 will toggle for every positive edge of the Q_0; Q_2 will toggle for every positive edge of the Q_1; and Q_3 will toggle for every positive edge of the Q_2. This is because a positive edge at any of the Q outputs is a negative edge at the associated $\overline{Q}$ output, and it is the $\overline{Q}$ outputs that clock each following stage.

Referring again to the truth table in Figure 10-20(c), you can see that this mod-16 (a 4-bit counter has sixteen possible output combinations) down counter will count down from 1111 (decimal 15) to 0000 (decimal 0), and then repeat the down-count cycle.

10-5-3   *Asynchronous Binary Up/Down Counters*

Comparing the asynchronous binary up counter in Figure 10-19(a) to the asynchronous down counter in Figure 10-20(a), you can see that the only difference between the two circuits is whether we clock FF1, FF2, and FF3 with the Q or $\overline{Q}$ outputs from the previous flip-flops.

Figure 10-21 shows how three AND-OR arrangements can be controlled by an UP/$\overline{\text{DOWN}}$ control line to create an asynchronous up/down binary counter. If the UP/$\overline{\text{DOWN}}$ control line is made HIGH, all of the shaded AND gates will be enabled,

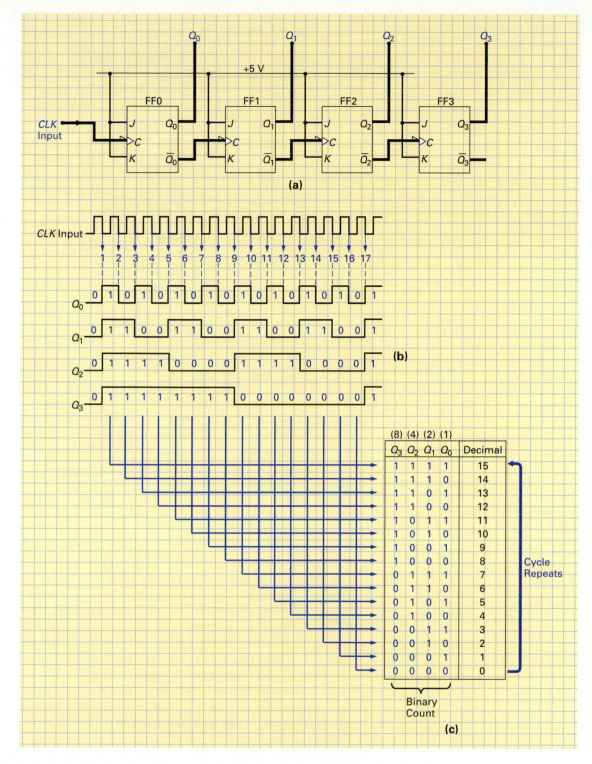

FIGURE 10-20 An Asynchronous Binary Down Counter. (a) Logic Circuit. (b) Waveforms. (c) Truth Table.

connecting the Q outputs through to the flip-flop's C inputs, causing the counter circuit to count up. On the other hand, if the UP/$\overline{\text{DOWN}}$ control line is made LOW, all of the shaded AND gates will be disabled and all of the unshaded AND gates will be enabled, connecting the $\overline{Q}$ outputs through to the flip-flop's C inputs, causing the counter circuit to count down.

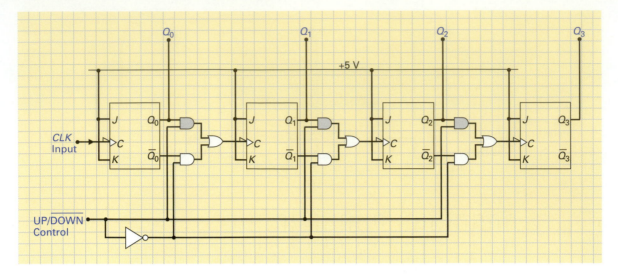

FIGURE 10-21   An Asynchronous Binary Up/Down Counter.

10-5-4   *Asynchronous Decade (mod-10) Counters*

Decade Counter

A counter that has a modulus of 10 and therefore counts from 0 to a maximum of 9.

Figure 10-22(a) shows how the previously discussed asynchronous mod-16 up counter circuit could be modified to construct a mod-10 or **decade counter** circuit. This counter will count from 0000 (decimal 0) to 1001 (decimal 9), and then the cycle will repeat, as shown in the timing waveforms in Figure 10-22(b) and the truth table in Figure 10-22(c). The reason this counter skips the counts 1010 through 1111 (decimal 10 through 15) is due to the action of the NAND gate, which controls the active-LOW clear inputs $(\overline{CLR})$ of all four flip-flops. Since the two inputs to this NAND gate are connected to the Q_1 and Q_3 outputs, when the counter reaches a count of 1010 (decimal 10) both Q_1 and Q_3 will go HIGH, and therefore the output of the NAND gate will go LOW and CLEAR the counter. Referring to the timing waveforms in Figure 10-22(b), you can see that the $(\overline{CLR})$ line is inactive for the counts 0000 through 1001. However, when the tenth clock pulse is applied, both Q_1 and Q_3 are HIGH. This condition of having both Q_1 and Q_3 HIGH is only temporary since the active $\overline{CLR}$ will almost immediately go LOW and therefore CLEAR or RESET all of the flip-flop outputs, producing a 0000 counter output. The truth table in Figure 10-22(c) summarizes the operation of this decade counter by showing how it counts from 0 to 9 and then repeats the cycle.

Figure 10-22(d) illustrates another way to show how a counter cycles through its different output states. In this *state transition diagram,* each circled binary number represents one of the output states, and the solid arrow indicates how one state changes to another as each clock pulse is received. For example, if we start at the count of 0000, after one clock pulse, the counter will go to 0001; after the next clock pulse, the counter will go to 0010; after the next clock pulse, the counter will go to 0011; and so on. The dotted line and shaded circle in this diagram indicate a temporary state that is only momentarily reached. For most purposes, this temporary state is ignored, and the counter is said to go directly from 1001 to 0000, as indicated by the solid arrow.

In summary, a decade counter counts from 0 up to a maximum count of 9, a total of ten different output states (mod-10). Since the counter requires ten input clock pulses before the output of the counter is reset, the Q_3 output frequency will be one-tenth that of the input clock frequency (*CLK* input). A decade counter, therefore, can also function as a divide-by-ten circuit. Remember that the only difference between a counter circuit and frequency divider circuit is how many of the outputs are used. For example, with a binary counter circuit, all of the flip-flops' outputs are used to deliver a parallel output word that represents the count, whereas with a frequency divider circuit, the clock input is applied to the first flip-flop, and the last flip-flop delivers serial output that is some fraction of the input clock frequency.

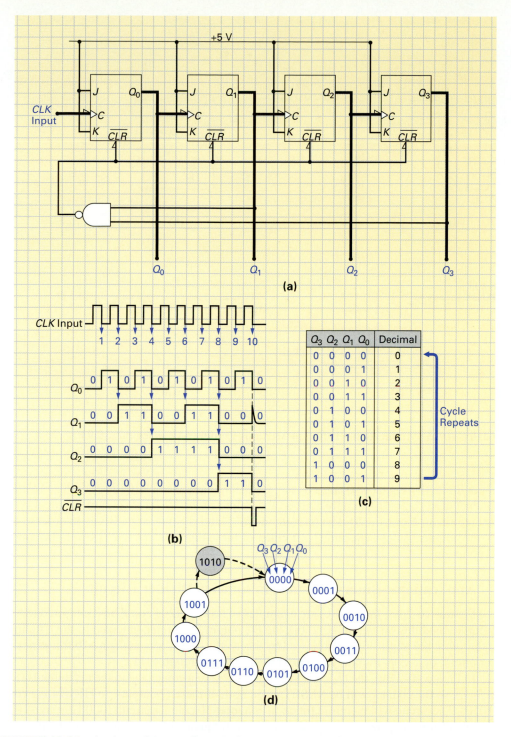

FIGURE 10-22   An Asynchronous Decade Counter. (a) Logic Circuit. (b) Timing Waveforms. (c) Truth Table. (d) State Transition Diagram.

In the counter applications section of this chapter, you will see how the decade counter is frequently chosen in applications that have a decimal display, such as digital clocks, digital voltmeters, and frequency counters. This is because the decade counter allows us to interface the binary information within a digital system to a decimal display that you and I understand.

Figure 10-23 shows the 74293 counter IC. This device contains four master-slave J-K flip-flops that for versatility are connected as a single divide-by-two (÷2) stage and a ripple

FIGURE 10-23 Application—Different Modulus Configurations for the 74293.

counter divide-by-eight ($\div 8$) stage. To achieve the maximum count (0 to 15) or the maximum frequency division ($\div 16$), simply connect the output of the $\div 2$ stage (Q_A) to the input of the $\div 8$ stage (input B or $CLK\ B$). The reset inputs ($R_{0(1)}$ and $R_{0(2)}$) must be HIGH in order to RESET the outputs Q_D, Q_C, Q_B, and Q_A. If either of these inputs is inactive (LOW), the counter will count up.

Application—Binary Counter/Frequency Divider Circuits

Figures 10-23(a) through (c) are examples of how a 74293 could be configured to have a different modulus. In Figure 10-23(a), the Q_A output has been connected to the CK_B input and therefore the $\div 2$ flip-flop is followed by the $\div 8$ flip-flops, producing a $\div 16$ frequency divider circuit or a mod-16 binary counter circuit (0000 through 1111).

In Figure 10-23(b), the Q_B and Q_D outputs have been connected to the reset inputs ($R_{0(1)}$ and $R_{0(2)}$), so the counter will be RESET when both of these Q outputs go HIGH (at a count of 1010). The circuit will therefore function as a mod-10 counter circuit (0000 through 1001) in which all four outputs will be used or as a $\div 10$ frequency divider circuit in which the Q_D output is one-tenth that of the clock input frequency applied to CK_A.

In Figure 10-23(c), only the $\div 8$ ripple counter section of the 74293 is being used to create a mod-6 counter or $\div 6$ frequency divider circuit. Since the Q_C and Q_D outputs have been connected to the reset inputs ($R_{0(1)}$ and $R_{0(2)}$), the counter will RESET the moment the counter reaches a count of 110 (decimal 6). The circuit will therefore function as a mod-6 counter circuit (000 through 101) in which outputs Q_B, Q_C, and Q_D will be used, or as a $\div 6$ frequency divider circuit in which the Q_D output is one-sixth that of the clock input frequency applied to CK_B.

In Figure 10-23(d), only the $\div 2$ section of the 74293 is being used to create a mod-2 counter or $\div 2$ frequency divider.

10-5-5 *Asynchronous Presettable Counters*

Figure 10-24(a) shows how four *J-K* flip-flops and several gates could be connected to construct an **asynchronous presettable counter.** As an example, this circuit is loading a preset data value of 1001 (decimal 9) into the counter directly following the maximum count of 1111 (decimal 15), as shown in the state transition diagram in Figure 10-24(b). Once the value of 1001 has been loaded, the counter will again count up to a maximum of 1111, reload the preset value of 1001 once again, and then repeat the cycle.

The preset value, which in this example is 1001, is applied to the data inputs D_0, D_1, D_2, and D_3. The logic levels applied to these inputs are normally prevented from reaching the *J-K* flip-flop's $\overline{PRE}$ and $\overline{CLR}$ inputs because all of the NAND gates are disabled by a LOW on the LOAD control line. The logic level on the LOAD control line is determined by a NOR, whose four inputs are connected to the counter's four outputs, Q_0, Q_1, Q_2, and Q_3.

To understand the operation of the circuit in Figure 10-24(a), let us begin by assuming that the counter is at its maximum count of 1111. On the next negative edge of the clock, the counter will increment to 0000, and only this output will cause the NOR gate, and therefore the LOAD control line, to go HIGH. With the LOAD control line HIGH, the data inputs and their complements are allowed to pass through the NAND gates to activate either the PRESET or CLEAR inputs of the flip-flops. In this example, the unshaded NAND gates will produce a HIGH output, while the shaded NAND gates will produce a LOW output. Since the flip-flop's $\overline{PRE}$ and $\overline{CLR}$ inputs are both active-LOW, Q_0 will be preset to 1, Q_1 will be cleared to 0, Q_2 will be cleared to 0, and Q_3 will be preset to 1. The 4-bit data value of 1001 is therefore asynchronously loaded into the counter (independent of the clock signal) the moment the counter goes to 0000. This 0000 state is only a temporary state, as shown in the state transition diagram in Figure 10-24(b), since it is almost immediately changed to the loaded state. Successive clock pulses at the input will cause the counter to count from 1001 to 1010, to 1011, to 1100, to 1101, to 1110, and then to the maximum count of 1111. The next clock pulse will reset the counter's count to 0000, enable the LOAD control line, load the data value, and then repeat the cycle.

One of the big advantages with a presettable counter is that it has a *programmable modulus.* For example, in Figure 10-24 we loaded a data value of 1001 (decimal 9) and then counted to 1111 (decimal 15), resulting in a total of seven different output combinations (1001, 1010, 1011, 1100, 1101, 1110, and 1111). As a result, this circuit will function as a mod-7 binary counter or ÷7 frequency divider.

By changing the data value loaded into a presettable counter, we can create a counter of any modulus. The simple formula for calculating the modulus of a presettable counter is

$$M_P = \text{mod} - D$$

$$M_P = \text{modulus of the presettable counter}$$
$$\text{mod} = \text{natural modulus of the counter}$$
$$D = \text{preset data value}$$

In our example circuit in Figure 10-24,

$$M_P = \text{mod} - D$$
$$= 16 - 9 \quad \text{(counter skips states 0 to 8)}$$
$$= 7$$

Asynchronous Presettable Counter

A counter that will count from its loaded preset value up to its maximum value. By changing the preset value, the modulus of the counter can be changed.

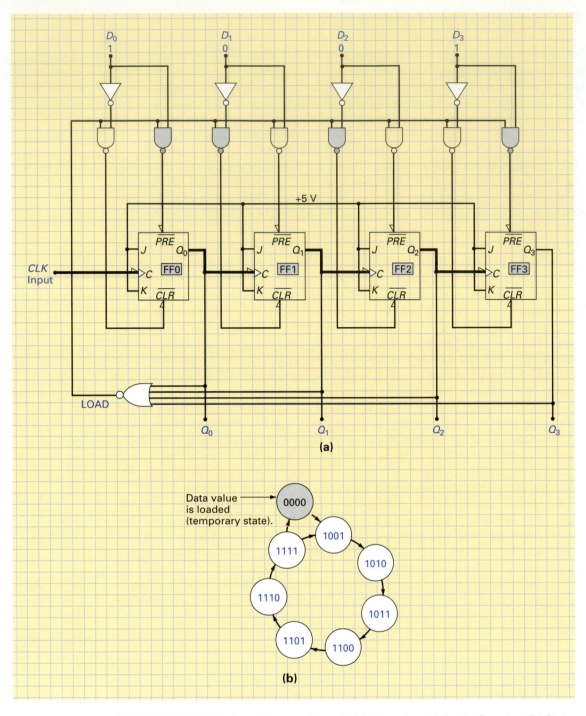

FIGURE 10-24 **An Asynchronous Presettable Counter. (a) Logic Counter. (b) State Transition Diagram for a Preset Value of 1001.**

■ **EXAMPLE:**

Assuming a 6-bit asynchronous presettable counter, calculate

   a. its natural modulus

   b. its preset modulus if 001100_2 is loaded

Solution:

a.
$$\text{mod} = 2^n$$
$$= 2^6 = 64$$

b.
$$M_P = \text{mod} - D$$
$$= 64 - 12 = 52$$

10-6   SYNCHRONOUS COUNTERS

As mentioned previously, asynchronous counters are often referred to as ripple counters. The term *ripple* is used because the flip-flops making up the asynchronous counter are strung together with the output of one flip-flop driving the input of the next. As a result, the flip-flops do not change state simultaneously or in sync with the input clock pulses, since the new count has to ripple through and update all of the flip-flops. This ripple-through action leads to propagation delay times that limit the counter's counting speed. This limitation can be overcome with a **synchronous counter** in which all of the flip-flops within the counter are triggered simultaneously and therefore synchronized to the master timing clock signal.

Synchronous Counter
A counter in which all operations are controlled by the master clock signal.

10-6-1   *Synchronous Binary Up Counters*

Figure 10-25(a) shows how four *J-K* flip-flops and two AND gates can be connected to form a 4-bit synchronous mod-16 up counter. The clock input signal line has been drawn as a heavier line in this illustration to show how all of the flip-flops in a synchronous counter circuit are triggered simultaneously by the common clock input signal. This parallel connection makes the counter synchronous, since all of the flip-flops will now operate in step with the input clock signal.

    The synchronous binary up counter shown in Figure 10-25(a) will operate as follows. The *J* and *K* inputs of FF0 are tied HIGH, and therefore the Q_0 output will continually toggle, like a single-stage asynchronous counter. This fact is verified by looking at the Q_0 column in Figure 10-25(b), which shows how the Q_0 output continually alternates HIGH and LOW. The *J* and *K* inputs of FF1 are controlled by the divide-by-two output of FF0. This means that when Q_0 is LOW, the Q_1 output of FF1 will not change; however, when Q_0 is HIGH, the Q_1 output of FF1 will toggle. This action can be confirmed by referring to the Q_1 column in Figure 10-25(b), which shows that when the Q_0 output goes HIGH, the Q_1 output will toggle or change state on the next active edge of the clock signal. The *J* and *K* inputs of FF2 are controlled by the ANDed outputs of Q_0 and Q_1. This means that only when Q_0 and Q_1 are both HIGH will the output of AND gate *A* be HIGH, and this HIGH will enable FF2 to toggle. Referring to the Q_2 column in Figure 10-25(b), you can see that only when Q_0 and

FIGURE 10-25 A Synchronous Binary Up Counter. (a) Logic Circuit. (b) Truth Table.

Q_1 are HIGH will Q_2 toggle or change state on the next active edge of the clock. The J and K inputs of FF3 are controlled by the ANDed outputs of Q_0, Q_1, and Q_2. This means that only when Q_0, Q_1, and Q_2 are HIGH will the output of AND gate B be HIGH, enabling FF3 to toggle. If you look at the Q_3 column in Figure 10-25(b), you will notice that only when Q_0, Q_1, and Q_2 are all HIGH will Q_3 toggle or change state on the next active edge of the clock.

10-6-2 *Synchronous Counter Advantages*

The key advantage of the asynchronous or ripple counter is its circuit simplicity, as can be seen by comparing the synchronous binary up counter in Figure 10-25(a) to the asynchronous binary up counter in Figure 10-19(a). The key disadvantage of the asynchronous counter is its frequency of operation or counting speed limitation. Since the input is applied only to the first flip-flop, it takes a certain time for all of the outputs to be updated to reflect the new count. This propagation delay time of the counter is equal to the sum of all the flip-flop propagation delay times. This limitation means that we cannot trigger or clock the input of an asynchronous counter before all of the outputs have settled into their new state, and therefore the clock input frequency (or pulses to be counted) has a speed or frequency limitation. Asynchronous counters are sometimes constructed using ECL flip-flops, which have an extremely small propagation delay time, and therefore a high input clock frequency can be applied. These ECL flip-flops, however, are more expensive and have a higher power consumption.

Synchronous counters are a direct solution to the asynchronous counter limitations, since they have small propagation delay times and are less expensive and consume less power than ECL asynchronous counters. The key to the synchronous counter's small propagation delay time is that all of its flip-flops are triggered or clocked simultaneously by a common clock input signal. This means that all of the flip-flops will be changing their outputs at the same time, so the total counter propagation delay time is equal to the delay of only one flip-flop. In reality, we should also take into account the time it takes for a Q output to propagate through a single AND gate and reach the J-K inputs of the following stage. Taking these two factors into consideration, we arrive at the following formula for calculating the delay time for a synchronous counter:

$$t_p = \text{single flip-flop } t_p + \text{single AND gate } t_p$$

EXAMPLE:

Assuming a J-K flip-flop propagation delay time of 30 ns and an AND gate propagation delay time of 10 ns, calculate the counter propagation time for a four-stage asynchronous counter and four-stage synchronous counter.

Solution:

$$
\begin{aligned}
t_{p(asynchronous)} &= \text{single flip-flop } t_p \times \text{number of flip-flops} \\
&= 30 \text{ ns} \times 4 = 120 \text{ ns} \\
t_{p(synchronous)} &= \text{single flip-flop } t_p + \text{single AND gate } t_p \\
&= 30 \text{ ns} + 10 \text{ ns} = 40 \text{ ns}
\end{aligned}
$$

In summary, the two advantages of a synchronous counter are that no transient output states occur due to the ripple through additive propagation delays and that synchronous counters can operate at much higher clock input frequencies.

10-6-3   *Synchronous Presettable Binary Counters*

Like the presettable asynchronous counter, the presettable synchronous counter has a programmable modulus. The 74LS163A, shown in Figure 10-26, is an example of a presettable synchronous up counter IC. A 4-bit binary value can be loaded into this counter to obtain any desired modulus by applying a LOW to the $\overline{\text{LOAD}}$ control input (pin 9). When the $\overline{\text{LOAD}}$ input is made LOW, the Q outputs (Q_A, Q_B, Q_C, and Q_D) will reflect the data inputs (A, B, C, and D) after the next positive edge of the clock input (pin 2). There are two active-HIGH enable inputs (pins 7 and 10), and because of the ANDing of these inputs, both must be HIGH for the counter to count. For example, let us assume that the counter is first loaded with a preset value of 1100 (decimal 12) and then counted up to its maximum count of 1111 (decimal 15). At this time, the ripple carry output (RCO, pin 15) generates an output pulse. This last-stage output is used to enable the input of another counter, if two

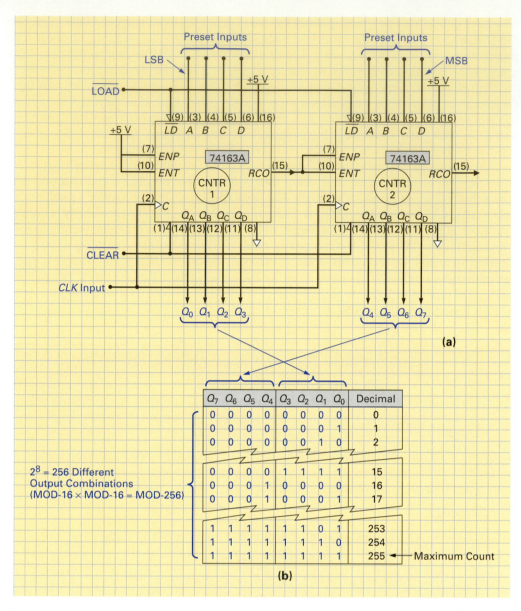

FIGURE 10-26   Application—Cascading Counters to Obtain a Higher Modulus.

counters were needed in a circuit application. For example, when a counter has reached a count of

$$Q_3 \quad Q_2 \quad Q_1 \quad Q_0$$

$$\boxed{1 \mid 1 \mid 1 \mid 1} = 15_{10}$$

a reset-and-carry action occurs from right to left as follows:

Carry Bit $\qquad Q_3 \quad Q_2 \quad Q_1 \quad Q_0$

$$\textcircled{1} \qquad \boxed{0 \mid 0 \mid 0 \mid 0} = 0_{10}$$

If only a 4-bit counter is needed, all of the Q outputs are reset to 0, the cycle repeats, and the carry bit is lost. If, however, two 4-bit counters are needed, the RCO of the LSB counter will be connected to the input of the MSB counter so that the Q_3 output will carry into the Q_4 flip-flop. In this situation the *carry bit* is not lost and the 8-bit counter counts beyond 15 as follows:

Most Significant 4 Bits					Least Significant 4 Bits				
Q_7	Q_6	Q_5	Q_4		Q_3	Q_2	Q_1	Q_0	
0	0	0	0		1	1	1	1	15
0	0	0	1	*RCO*	0	0	0	0	16
0	0	0	1		0	0	0	1	17
0	0	0	1		0	0	1	0	18

Application—Cascaded Counter Circuit

Figure 10-26(a) shows how two mod-16 74163A ICs can be *cascaded,* or strung together, to obtain a mod-256 counter. When connecting synchronous counters in cascade, the *RCO* (so called because it ripples out of the counter's last stage) is connected to the enable inputs of counter 2 (CNTR 2). As a result, when CNTR 1 reaches its maximum count (1111), the *RCO* from CNTR 1 will go HIGH and enable CNTR 2 so that on the next positive edge of the clock CNTR 2 will have a count of 0001 and CNTR 1 will have reset to 0000.

Since the counter in Figure 10-26(a) contains eight outputs (Q_0 through Q_7), it will produce a total of 2^8, or 256, different output combinations, with a maximum count of 255, as shown in Figure 10-26(b). If this circuit were used as a frequency divider, it would produce a single output pulse at the Q_7 output for every 256 input clock pulses (a divide-by-256 circuit).

If some other counter modulus or frequency division were required (anywhere between 2 and 256), the preset inputs could be used to load in any value. For example, if 0001 1000 (decimal 24) were loaded into the 8-bit counter, the counter would have a modulus of

$$M_p = \text{mod} - D$$
$$= 256 - 24 = 232$$

This counter would now count from 24 to 255 (a total of 232 different output combinations), or it would divide the input clock frequency by 232.

10-6-4   *Synchronous Decade (mod-10) Counters*

Like the asynchronous decade counter, a synchronous decade (mod-10) counter generates only the BCD outputs 0000 though 1001 (decimal 0 through 9). In the counter applications section of this chapter, you will see how decade counters are frequently chosen in applications that have a decimal display, such as digital clocks, digital voltmeters, and frequency counters. This is because the decade counter allows us to interface the binary information within a digital system to a decimal display that you and I understand.

The 74LS160A, shown in Figure 10-27, is a good example of a synchronous decade counter IC. The 74LS160A has exactly the same input and output pin assignments as the 74LS163A covered in Figure 10-26; however, this counter has only ten different output combinations.

Application—A 0-to-99 Counter and Display Circuit

Figure 10-27 shows how two 74LS160A synchronous decade counter ICs, two 7447 BCD-to-seven-segment decoder ICs, and two common-anode seven-segment LED displays could be connected to construct a 0-to-99 counter and display circuit. Up until this point, we have always followed the industry schematics standard of having inputs on the left side of the page and outputs on the right side of the page. A difficulty arises, however, when we are showing the various digits of a display, since the units digit is normally always on the right

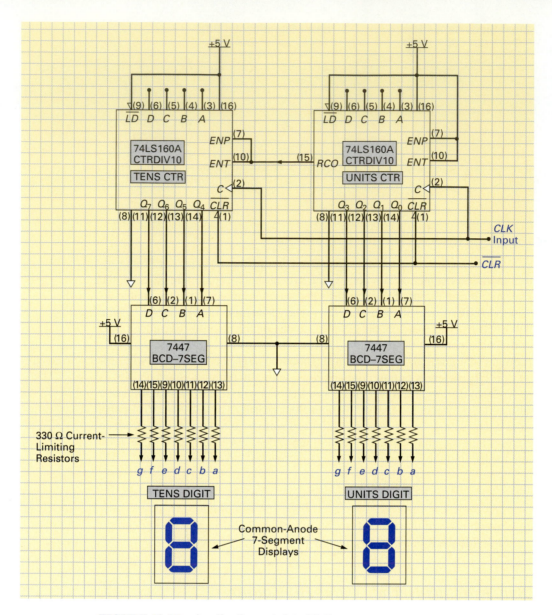

FIGURE 10-27   Application—A 0-to-99 Counter.

side and digits of increasing weight (such as tens, hundreds, and thousands) are to the left of the units digit. In this illustration, we will have the input clock signal applied at the right and propagate to the left so that the digital displays will be in their correct position.

The operation of this circuit is as follows. The units counter counts the input clock pulses, and after every ten clock pulses it activates the *RCO*. This enables the tens counter, which counts up by 1 on the next positive edge of the clock. The two decade counters will produce BCD outputs that range from $0000\ 0000_{BCD}$ (decimal 0) to $1001\ 1001_{BCD}$ (decimal 99). The 4-bit BCD output codes from the decade counters are decoded by the 7447s and displayed as their decimal equivalents on the seven-segment displays.

10-6-5   *Synchronous Up/Down Counters*

Bidirectional Counter

A counter that can be controlled to either count up or count down.

All up/down counters are sometimes referred to as **bidirectional counters** since they can count in one of two directions—up or down. A control line is generally used to determine whether the count is incremented (advanced upward) or decremented (advanced downward).

FIGURE 10-28 Application—Up/Down Flashing Circuit.

For example, if a 2-bit counter's DOWN/$\overline{\text{UP}}$ control line were made LOW, the counter would count up (0,1,2,3,0,1,2, ...), whereas if the DOWN/$\overline{\text{UP}}$ control were made HIGH, the counter would count down (3,2,1,0,3,2,1, ...).

The 74193 IC is described in data books as a presettable mod-16 up/down synchronous counter with asynchronous loading and asynchronous master reset. This IC is shown in Figure 10-28. This IC counter contains four *J-K* flip-flops and a variety of control gates. The counter can be used as a mod-*N* counter, or divider, by applying any preset value to the data inputs (*A*, *B*, *C*, and *D*) and activating the active-LOW load input (pin 11). An asynchronous CLEAR or master reset input (pin 14) is available and will force all outputs LOW when a HIGH level is applied. Synchronous operation is provided by having all flip-flops clocked simultaneously by a LOW-to-HIGH transition (positive edge-triggered). The direction of counting (up or down) is controlled by the two trigger inputs COUNT UP (pin 5) and COUNT DOWN (pin 4). To count up, simply apply the clock input to the COUNT UP input while the COUNT DOWN input is HIGH; whereas to count down, apply the clock input to the COUNT DOWN input while the COUNT UP input is HIGH.

The 74193 up/down counter was also designed to be cascaded, and therefore a CARRY output will pulse HIGH when the counter cycles beyond its maximum count, as follows:

	(8)	(4)	(2)	(1)		(10)	(1)
	1	1	1	0	=	1	4
Carry 1	1	1	1	1	=	1	5
1	0	0	0	0	=	1	6
	1	0	0	0	1	=	1	7

Since the 74193 can also count down, this IC also has a BORROW output, which will pulse HIGH when the counter cycles down to its lowest value, resets to maximum, and then repeats, as follows:

	(16)	(8)	(4)	(2)	(1)		(10)	(1)
	1	0	0	0	1	=	1	7
	1	0	0	0	0	=	1	6
Borrow	1							
	0	1	1	1	1	=	1	5
	0	1	1	1	0	=	1	4

Application—Back-and-Forth Flashing Circuit

Figure 10-28 shows how a 74193 up/down counter, a 74154 four-line to sixteen-line decoder, and a 7400 could be connected to form a back-and-forth flashing circuit. In this circuit, the 74193 is made to count up from 0000 to 1111 and is then switched to count down from 1111 to 0000, and then switched again to repeat the cycle. The NAND gates provide the count-up and count-down control, while the 74154 enables (makes LOW) its 0 through 15 outputs in response to the binary count from the 74193 counter. As a result, the sixteen LEDs in the display are sequenced ON and OFF in a continual back-and-forth action.

SELF-TEST EVALUATION POINT FOR SECTION 10-6

Use the following questions to test your understanding of Section 10-6.

1. Synchronous counters are also referred to as ripple counters. (True/False)
2. What are the main advantages of the asynchronous counter and the synchronous counter?
3. The _____ output from an up/down counter will go active when the counter counts up to its maximum count and then resets, while the _____ output will go active when the counter goes down to 0 and then recycles to its maximum count.
4. By cascading counters we can obtain a higher _____.

10-7 COUNTER APPLICATIONS

The counter is one of the most versatile of all the digital circuits and is used in a wide variety of applications. In most cases, however, it is basically used to count or to divide the input signal frequency. As a counter, it generates a parallel output that represents the number of clock pulses that have occurred. As a frequency divider, it generates a serial output that is a fraction of the input frequency.

In the three circuit applications that follow we will see how counters are used as a counter and as a frequency divider.

10-7-1 A Digital Clock

Counters are probably most frequently used to form a digital clock circuit, like the one seen in Figure 10-29. This digital time-keeping circuit displays the hours, minutes, and seconds of the day.

Starting at the top of this circuit you can see that the power supply unit is converting the 120 V ac input into a +5 V dc output, which is used to power all of the digital logic circuits. Since the 120 V ac input from the power company alternates at an extremely accurate 60 cycles per second (60 Hz), we can make use of this timing signal to generate a one-cycle-

FIGURE 10-29 **Application—A Digital Clock Circuit.**

per-second (1 Hz) master timing, or clock, signal. This is achieved by tapping off the 60 Hz ac from the secondary of the power supply transformer and then feeding this signal into a pulse-shaping circuit that will convert the 60 Hz sine wave input into a sixty-pulses-per-second (60 pps) square wave output. The 60 pps output from the pulse-shaping circuit is then applied to a divide-by-ten counter and then to a divide-by-six counter that in combination will divide the 60 pps input by 60 to produce a one-pulse-per-second output from the divide-by-sixty circuit. This 1 pps clock signal from the divide-by-sixty circuit is the master timing signal for the seconds counter, minutes counter, and hours counter circuits.

The seconds counter circuit contains a mod-10, or ÷10, counter followed by a mod-6, or ÷6, counter. These two counters in combination will count the 1 pps clock input, and after every sixty seconds or sixty pulses will produce a 1 ppm output to the minutes counter circuit. The minutes counter circuit also contains a mod-10, or ÷10, counter followed by a mod-6, or ÷6, counter. These two counters in combination will count the 1 ppm clock input, and after every sixty minutes or sixty pulses will produce a l pph output to the hours counter circuit. The hours counter circuit contains a mod-10, or ÷10, counter followed by a mod-2, or ÷2, counter (which is a single flip-flop). These two counters in combination will count the 1 pph clock input. Once the count advances from 12 (1 0010_{BCD}) to 13 (1 0011_{BCD}), NAND gate A will almost immediately clear and preset load the hours counter to 1 (0 0001_{BCD}), ensuring that the hours count increments from 12:59 to 1:00.

The BCD outputs from the seconds counters, minutes counters, and hours counters are decoded by BCD-to-seven-segment 7447s, and these seven-segment codes drive the common-anode displays.

10-7-2　*A Frequency Counter*

A **frequency counter** is a good example of a digital system that makes use of a clock oscillator, frequency dividers, counters, registers, decoders, and a seven-segment display.

Frequency Counter Block Diagram

Before we discuss the frequency counter circuit diagram, let us first examine the basic block diagram of the frequency counter, shown in Figure 10-30. To start at the far left of the timing diagram, you can see that the counter is cleared by a HIGH pulse just before the count

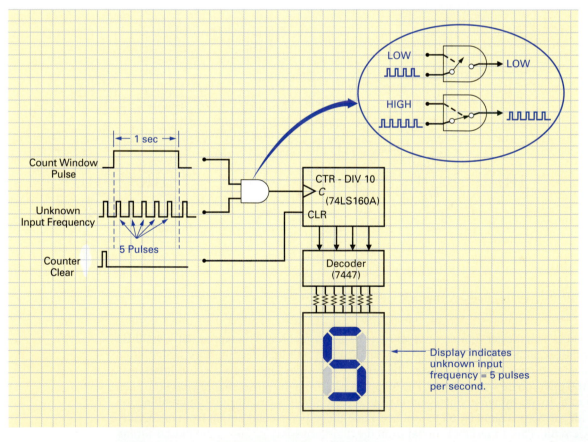

FIGURE 10-30　Simplified Frequency Counter Circuit.

window pulse begins. The count window pulse and the unknown input frequency are both applied to an AND gate, which acts as a controlled switch, as shown in the inset in Figure 10-30. When the count window pulse applies a LOW to the AND gate, it acts as an open switch since any LOW input to an AND gate produces a constant LOW out. On the other hand, when the count window pulse applies a HIGH to the AND gate, it acts as a closed switch connecting the unknown input frequency pulses through to the output. In the example shown in Figure 10-30, five pulses are switched through the AND gate to the clock input of the counter during the one-second count window pulse. These five pulses will trigger the decade counter five times, causing it to count to 5 (0101). This BCD count of 5 from the counter will be decoded by the BCD-to-seven-segment decoder and displayed on the seven-segment LED display. Since five pulses occurred in one second, the unknown input frequency is 5 pps.

The count window pulse is sometimes called the *sample pulse,* because the unknown input frequency is sampled during the time that this signal is active. The simple frequency counter circuit shown in Figure 10-30 is extremely limited, since it can only determine the unknown input frequency of inputs between 1 and 9 pps. To construct a more versatile frequency counter, we will have to include some additional digital ICs in the circuit.

Frequency Counter Circuit Diagram

Figure 10-31 shows a more complete frequency-counter circuit diagram. Although at first glance this appears to be a complex circuit, it can easily be understood by examining one block at a time.

The accuracy of a frequency counter is directly dependent on the accuracy of the count window pulse. To ensure this accuracy, a 100 kHz crystal oscillator is used to generate a very accurate square wave. The output from the crystal oscillator is used to clock a chain of divide-by-ten ICs (74LS160As) in the count window generator circuit block. The 100 kHz input at the right is successively divided by 10 to produce a 1 kHz, 100 Hz, 10 Hz, and 1 Hz output. Any one of these count window pulse frequencies can be selected by SW_1 and applied to a divide-by-two stage (1/2 7476) in the divide-by-two and monostable block. The waveforms shown in the inset in Figure 10-31 illustrate what occurs when SW_1 is placed in the 1 Hz position. The square wave input to the 7476 completes one cycle in one second; however, the positive half cycle of this waveform only lasts for half a second. By including the divide-by-two stage, the 1 Hz input is divided by 2 to produce an output frequency of 0.5 Hz, which means one complete cycle lasts for two seconds, and therefore the positive alternation, or count window pulse, lasts for one second. Including a ÷2 stage in this way means that when SW_1 selects

    a. 1 Hz, the count window pulse out of the ÷2 stage will have a duration of one second.

    b. 10 Hz, the count window pulse out of the ÷2 stage will have a duration of 100 ms (0.1 s).

    c. 100 Hz, the count window pulse out of the ÷2 stage will have a duration of 10 ms (0.01 s).

    d. 1 kHz, the count window pulse out of the ÷2 stage will have a duration of 1 ms (0.001 s).

As we will see later, these smaller count window pulses will allow us to measure higher input frequencies.

The count window pulse out of the ÷2 stage is applied to an AND gate along with the unknown input frequency. In the example timing waveforms shown in the inset in Figure 10-31, you can see that ten pulses are switched through the AND gate during the window of one second. These ten pulses are used to clock a four-stage counter in the counter, register, decoder, and display circuit block. The units counter on the right will count from 0 through 9, and if the count exceeds 9, it will reset to zero and carry a single tens pulse into the tens counter. In this example, this is exactly what will happen, and the final display will show decimal 10, meaning that the unknown input frequency is 10 pps. Similarly, the tens counter will reset-and-carry into the hundreds counter, and the hundreds counter will reset and carry into the thousands counter, as is expected when counting in decimal.

FIGURE 10-31 Application—A Frequency Counter Circuit.

The BCD outputs of the four-stage counter are applied to a set of registers, and the output of these registers drives a four-digit seven-segment display via BCD-to-seven-segment display decoders. Referring to the divide-by-two and monostable block, you can see that the 7476 $\overline{Q}$ output is used to trigger a 74121 one-shot, whose *RC* network has been chosen to generate a 1 µs pulse. Referring again to the timing waveforms in the inset in Figure 10-31, you can see that the positive edge of the 7476 $\overline{Q}$ output triggers the 74121, which generates a positive 1 µs pulse at its *Q* output and a negative 1 µs pulse at its $\overline{Q}$ output. The positive edge of the 74121 Q output clocks the four 74173 4-bit registers, which store the new frequency count. This stored value appears at the Q outputs of the 74173s, where it is decoded by the 7447s and displayed on the seven-segment displays. The 74173s are included to hold the display value constant, except for the few microseconds when the display is being updated to a new value. The negative 1 μs pulse at the 74121 $\overline{Q}$ output is used to clear the four-stage counter once the new frequency count has been loaded into the 74173 registers.

To explain this circuit in a little more detail, let us examine each of the frequency counter's different range selections (controlled by SW_1), which are shown in the table in Figure 10-31.

1 Hz or 1 sec Range. When SW_1 selects this frequency, a count window pulse of one second is generated. Using this window of time and a 4-digit display, we can measure any input frequency from 0001 Hz to 9999 Hz. If this maximum of 9999 Hz is exceeded, the counter will reset all four digits. For example, an input of 9999 Hz will be displayed as 9999, whereas an input of 10000 Hz will be displayed as 0000. To measure an input frequency in excess of 9999 Hz, therefore, we will have to use a smaller window so that fewer pulses are counted.

10 Hz or 100 ms Range. When SW_1 selects this frequency, a count window pulse of 1/10 of a second is generated. This means that if the unknown input frequency is 10,000 Hz, only 1/10 of these pulses will be switched through the AND gate to the counter (1/10 of 10,000 = 1000). This value of 1000 is smaller and can be displayed on a 4-digit display; however, a display of 1000 Hz is not accurate since the input frequency is 10,000 Hz. Since 10,000 Hz = 10 kHz, on this range we will add a decimal point in the mid position and add a prefix of kHz after the value so that the readout is correct, as shown in the table in Figure 10-31. An input of 10,000 Hz, therefore, will be displayed as 10.00 kHz.

100 Hz or 10 ms Range. When SW_1 selects this frequency, a count window pulse of 1/100 of a second is generated. This means that if the unknown input frequency is 100,000 Hz, only 1/100 of these pulses will be switched through the AND gate to the counter (1/100 of 100,000 = 1000). This value of 1000 is smaller and can be displayed on a 4-digit display; however, a display of 1000 Hz is not accurate since the input frequency is 100,000 Hz. Since 100,000 Hz = 100 kHz, on this range we will move the position of the decimal point and keep the prefix of kHz after the value so that the readout is correct, as shown in the table in Figure 10-31. An input of 100,000 Hz, therefore, will be displayed as 100.0 kHz.

1 kHz or 1 ms Range. When SW_1 selects this frequency, a count window pulse of 1/1000 of a second is generated. This means that if the unknown input frequency is 1,000,000 Hz, only 1/1000 of these pulses will be switched through the AND gate to the counter (1/1000 of 1,000,000 = 1000). This value of 1000 is smaller and can be displayed on a 4-digit display; however, a display of 1000 Hz is not accurate since the input frequency is 1,000,000 Hz. Since 1,000,000 Hz = 1 MHz, on this range we will again move the position of the decimal point and add a prefix of MHz after the value so that the readout is correct, as shown in the table in Figure 10-31. An input of 1,000,000 Hz, therefore, will be displayed as 1.000 MHz.

10-7-3 *The Multiplexed Display*

The display circuitry used for the previously described digital clock circuit and the frequency counter circuit has two basic disadvantages: It uses a large number of ICs and it uses a large amount of power. Both of these disadvantages can be solved by using a **multiplexed display circuit** similar to the one shown in Figure 10-32. This circuit needs only one 7447 decoder IC instead of the six that were needed for the digital clock circuit and the four that were needed for the frequency counter circuit. Second, since a multiplexed display has only one digit ON at a time, the total circuit current, and therefore power consumption, is dramatically less. For example, each LED in each seven-segment display draws approximately 20 mA, and therefore each seven-segment display can draw a maximum of 140 mA. If all six digits

Multiplexed Display Circuit

A display circuit in which the drive circuitry is time-shared.

FIGURE 10-32 Application—A 6-Digit Multiplexed Display Circuit.

in the digital clock circuit are at their maximum, a total current of 6×140 mA = 840 mA (0.84 A) will be needed. With a multiplexed display, only one seven-segment display is enabled at a time, and therefore the maximum display current at any one time is only 140 mA. If you combine this with the fact that each 7447 draws 260 mA, and only one 7447 (instead of four or six) is needed, you can see that the current savings are dramatic.

The multiplexed display circuit shown in Figure 10-32 operates in the following way. To *multiplex* is to simultaneously transmit two or more signals over a single channel. With the multiplexed display circuit in Figure 10-32, the 74293 mod-8 counter controls the switching operation by controlling both the 74138 1-of-8 decoder and the four 74151 1-of-8 multiplexers. Referring to the table below the circuit in Figure 10–32, you can see that when the 74293 reaches a count of 001 (decimal 1), the 74138 will make LOW its active-LOW 1 output, which will turn ON Q_1 and switch the +5 V supply to only the first seven-segment display. The 001 output from the 74293 is also applied to the four 74151 multiplexers, which will respond by switching all of their 1 inputs through to their Y outputs. This will connect the A_0, A_1, A_2, and A_3 4-bit word from the left-most 74175 register through to the 7447, where it will be decoded from BCD to a corresponding seven-segment code. This seven-segment code is applied to all six seven-segment displays; however, since only one of the displays is receiving a +5 V supply, the data will only be displayed on the left-most seven-segment display.

Referring again to the table in Figure 10-32, you can see that as the count of the 74293 advances, the displays are enabled in turn by the 74138, and the corresponding input words from the registers are switched through by the 74151 multiplexers. If the cycle is repeated at a fast enough rate, your eyes will not see the ON/OFF switching of each display, only a continuous display. **Time division multiplexing (TDM),** or the timesharing of several signals over one set of wires or a single channel, is a method frequently used in many electronic circuits and systems.

Time Division Multiplexing (TDM)

A timesharing technique in which several signals can be transmitted within a single channel.

SELF-TEST EVALUATION POINT FOR SECTION 10-7

Use the following questions to test your understanding of Section 10-7.

1. When a counter is used as a _____ it provides a parallel output that represents the number of input clock pulses applied. As a _____, a counter provides a serial output that is a fraction of the input frequency.

2. Do the counters in the seconds counter, minutes counter, and hours counter circuit in Figure 10-29 function as counter or frequency dividers?

3. Do the mod-10 counters in the count window generator circuit in Figure 10-31 function as counters or frequency dividers?

4. Do the mod-10 counters in the counter, register, decoder, and display circuit in Figure 10-31 function as counter or frequency dividers?

10-8   TROUBLESHOOTING COUNTER CIRCUITS

To review, the procedure for fixing a failure can be broken down into three steps—diagnose, isolate, and repair. Since this chapter has been devoted to counters, let us first examine the operation of a typical counter circuit and then apply our three-step troubleshooting process to this circuit.

10-8-1   *A Counter Circuit*

As an example, we will apply our three-step troubleshooting procedure to the frequency counter circuit previously discussed and shown in Figure 10-31.

Step 1: Diagnose

As before, it is extremely important that you first understand the operation of a circuit and how all of the devices within it are supposed to work so that you are able to determine whether or not a circuit malfunction really exists. If you were preparing to troubleshoot the circuit in Figure 10-31, your first step should be to read through the circuit description and review the operation of each integrated circuit until you feel completely confident with the correct operation of the entire circuit. Once you are fully familiar with the operation of the circuit, you will easily be able to diagnose the problem as either an *operator error* or a *circuit malfunction*.

Distinguishing an operator error from an actual circuit malfunction is an important first step, and a wrong diagnosis can waste much time and effort. For example, the following could be interpreted as circuit malfunctions, when in fact they are simply operator errors.

Example 1.
Symptom: SW_1 has selected the 1 ms range, input frequency = 100 Hz, display shows 0000.

Diagnosis: Operator Error—A 1 ms window will not sample any of the input since each cycle of the input lasts for 10 ms (0.01 s). To determine low input frequencies, the operator will need to use a wider window.

Example 2.
Symptom: SW_1 has selected the 100 ms range, input frequency = 12.3 kHz, display shows 1230.

Diagnosis: Operator Error—Since no circuitry is included to turn ON the decimal point in the correct position and display kHz, the display has to be interpreted based on the range selected. For example, 1230 on the 100 ms range = 12.30 kHz.

Once you have determined that the problem is not an operator error, but is in fact a circuit malfunction, proceed to step 2 and isolate the circuit failure.

Step 2: Isolate

No matter what circuit or system failure has occurred, you should always follow a logical and sequential troubleshooting procedure. Let us review some of the isolating techniques and then apply them to our example circuit in Figure 10-31.

a. Use a cause-and-effect troubleshooting process, which means study the effects you are getting from the faulty circuit and then logically reason out what could be the cause.

b. Check first for obvious errors before leaping into a detailed testing procedure. Is the power OFF or not connected to the circuit? Are there wiring errors? Are all of the ICs correctly oriented?

c. Using a logic probe or voltmeter, test that power and ground are connected to the circuit and are present at all points requiring power and ground. If the whole circuit, or a large section of the circuit, is not operating, the problem is normally power. Using a multimeter, check that all of the dc voltages for the circuit are present at all IC pins that should have power or a HIGH input and are within tolerance. Second, check that 0 V, or ground, is connected to all IC ground pins and to all inputs that should be tied LOW.

d. Use your senses to check for broken wires, loose connectors, overheating or smoking components, pins not making contact, and so on.

e. Test the clock signals using the logic probe or the oscilloscope. Although clock signals can be easily checked with a logic probe, the oscilloscope is best for checking the signal's amplitude, frequency, pulse width, and so on. The oscilloscope will also display timing problems and noise spikes (glitches), which could be false-triggering a circuit into operation at the wrong time. With the frequency counter circuit in Figure 10-31, you should first test the 100 kHz output from the crystal oscillator and then test in sequence the 10 kHz, 1 kHz, 100 Hz, 10 Hz, and 1 Hz outputs from the count window generator circuit. Problems in frequency divider chains are generally easily detected since an incorrect frequency output normally indicates that the previous stage has malfunctioned.

f. Perform a static test on the circuit. With our circuit example in Figure 10-31, you should disconnect the inputs to the counter, register, decoder, and display circuit and then use a logic pulser to clock the counter and a logic probe to test that the valid count is appearing at the outputs of each stage of the counter. For example, after nineteen clock pulse inputs from the logic pulser, the logic probe should indicate the following logic levels at the counter outputs.

$$0000 \quad 0000 \quad 1000 \quad 1001_{BCD}$$
$$0 \qquad 0 \qquad 1 \qquad 9$$

By pulsing the register clock input, this value should appear at the output of the 74173s, be decoded by the 7447s, and be displayed. By pulsing the counter clear input, you should be able to clear all of the outputs of the counter. The divide-by-two and monostable circuit can also be tested in this way by disconnecting the inputs and using the logic pulser as a signal source and the logic probe as a signal detector.

g. With a dynamic test, the circuit is tested while it is operating at its normal clock frequency (which in Figure 10-31 is 100 kHz), and therefore all of the inputs and outputs are continually changing. Although a logic probe can detect a pulse waveform, the oscilloscope will display more signal detail, and since it can display more than one signal at a time, it is ideal for making signal comparisons and looking for timing errors. If the clock section is suspected, you can check the rest of the circuit by disconnecting the clock input to the circuit and then using a function generator to generate a 100 kHz clock.

In some instances, you will discover that some circuits will operate normally when undergoing a static test and yet fail a dynamic test. This effect usually points to a timing problem involving the clock signals and/or the propagation delay times of the ICs used within the circuit.

h. Noise due to electromagnetic interference (EMI) can false-trigger an IC, such as the counters in Figure 10-31. This problem can be overcome by not leaving any of the IC's inputs unconnected and, therefore, floating. Connect unneeded active-HIGH inputs to ground and active-LOW inputs to $+V_{CC}$.

i. Apply the half-split method of troubleshooting first to a circuit and then to a section of a circuit to help speed up the isolation process. With our circuit example in Figure 10-31, a good midpoint check would be to test the different frequency outputs from the count window generator circuit.

Also remember that a load problem can make a source appear at fault. If an output is incorrect, it would be safer to disconnect the source from the rest of the circuit and then recheck the output. If the output is still incorrect, the problem is definitely within the source; however, if the output is correct, the load is probably shorting the line either HIGH or LOW.

j. Substitution can be used to help speed up your troubleshooting process. Once the problem is localized to an area containing only a few ICs, substitute suspect ICs with known-working ICs (one at a time) to see if the problem can be quickly remedied.

k. Many electronic system manufacturers provide troubleshooting trees in the system technical manual. These charts are a graphical means to show the sequence of tests to be performed on a suspected circuit or system. Figure 10-33 shows a simple troubleshooting tree for our frequency counter circuit.

Step 3: Repair

Once the fault has been found, the final step is to repair the circuit and then perform a final test to see that the circuit and the system are now fully operational.

10-8-2   *Sample Problems*

Once you have constructed a circuit like the circuit in Figure 10-31, introduce a few errors to see what effects or symptoms they produce. Then logically reason out why a particular error or cause has a particular effect on the circuit. Never short any two points together unless you have carefully thought out the consequences, but generally it is safe to open a path and see the results. Here are some problems for our example circuit in Figure 10-31.

    a. Remove the 100 kHz crystal.

    b. Disconnect one of the *RCO* outputs in the count window generator circuit.

    c. Disconnect the clock input to one of the 74173s.

    d. Disconnect the $\overline{Q}$ output of the divide-by-two circuit.

FIGURE 10-33   A Troubleshooting Tree for the Frequency Counter Circuit.

    e. Disconnect the *Q* output of the 74121.

    f. Disconnect the *ENT* and *ENP* inputs of the first 74LS160A counter from the +5 V and connect these lines to ground.

    g. Disconnect one of the inputs to a 7447.

SELF-TEST EVALUATION POINT FOR SECTION 10-8

Use the following questions to test your understanding of Section 10-8.

1. What are the three basic troubleshooting steps?

2. Which of the 74LS160As in Figure 10-31 function as frequency dividers and which function as counters?

3. To test the count window generator circuit in Figure 10-31, the clock oscillator input is disconnected and a logic pulser is used to inject a 500 kHz input. Does a problem exist if outputs of 5 kHz, 500 Hz, 50 Hz, and 5 Hz are measured with the oscilloscope at each of the SW_1 positions?

4. What symptom would you get from the circuit in Figure 10-31 if you were to apply a permanent HIGH to the unknown frequency input?

REVIEW QUESTIONS

Multiple-Choice Questions

1. The _____ of a counter describes the number of different output combinations generated by the counter.
  a. division factor   **c.** maximum count
  b. modulus   **d.** propagation delay time

2. The _____ of a counter describes the time it will take for all of the outputs to be updated.

  a. division factor   **c.** maximum count
  b. modulus   **d.** propagation delay time

3. The _____ of a counter describes how much lower the final flip-flop's output frequency will be compared to the input clock frequency.
  a. division factor   **c.** maximum count
  b. modulus   **d.** propagation delay time

4. A 4-bit SISO shift register will take _____ clock pulse(s) to store a 4-bit word.
   a. one   **c.** three
   b. two   **d.** four

5. A 4-bit PISO shift register will take _____ clock pulse(s) to store a 4-bit word.
   a. one   **c.** three
   b. two   **d.** four

6. Which of the following shift registers could be used for serial-to-parallel data conversion?
   a. SISO   **d.** PIPO
   b. PISO   **e.** All of the above
   c. SIPO

7. Which of the following shift registers could be used for parallel-to-serial data conversion?
   a. SISO   **d.** PIPO
   b. PISO   **e.** All of the above
   c. SIPO

8. Which of the following is considered a serial-in, serial-out shift register operation?
   a. Shift-right
   b. Shift-left
   c. Rotate-right
   d. Rotate-left
   e. All of the above

9. When an operation is in step with the clock signal, it is called a(an) _____ operation.
   a. asynchronous
   b. UART
   c. sequenced
   d. synchronous

10. When an operation is not in step with the clock signal, it is called a(an) _____ operation.
   a. asynchronous
   b. UART
   c. sequenced
   d. synchronous

11. A buffer register with three-state output control will generally have three control line inputs called _____, _____ and _____.
   a. V_{CC}, clock, S_1
   b. clock, enable, load
   c. SRI, load, S_0
   d. shift-left, shift-right, inhibit

12. New data is _____ into a memory register, while stored data is _____ out of a memory register.
   a. read, written
   b. written, read

13. A universal asynchronous receiver/transmitter (UART) has a _____ and a _____ internally for interfacing between a bus and a single line.
   a. serial-to-parallel shift register, parallel-to-serial shift register
   b. memory register, arithmetic shift register
   c. ring counter, johnson counter
   d. buffer register, PIPO register

14. What will happen to the value of a binary number stored in a shift register if it is shifted to the left by three positions and zeros are shifted in through the LSB position?
   a. Value will be divided by 8.
   b. Value will be multiplied by 6.
   c. Value will be multiplied by 8.
   d. Value will be divided by 4.

15. With a ring counter, the _____ output of the last flip-flop is connected back to the input of the first flip-flop, whereas with a johnson counter the _____ output of the last flip-flop is connected back to the input of the first flip-flop.
   a. Q, CLEAR
   b. $\overline{Q}$, Q
   c. PRESET, $\overline{Q}$
   d. Q, $\overline{Q}$

16. The _____ of a counter describes the highest binary value that will appear at the parallel outputs.
   a. division factor
   b. modulus
   c. maximum count
   d. propagation delay time

17. A mod-10 or decade counter will divide the input clock frequency by a factor of _____.
   a. 16   **c.** 10
   b. 8   **d.** 100

18. A four-stage binary counter will have a natural modulus of _____.
   a. 16   **c.** 10
   b. 8   **d.** 100

19. A(an) _____ counter will start counting from a loaded value up to its natural modulus.
   a. synchronous
   b. asynchronous
   c. presettable
   d. cascaded

20. The maximum counting speed of a counter is limited by the counter's _____.
   a. division factor
   b. modulus
   c. maximum count
   d. propagation delay time

21. What is the primary advantage of a synchronous counter?
   a. High counting speed
   b. Long propagation delay
   c. Circuit simplicity
   d. Ripple action

22. What is the primary advantage of an asynchronous counter?
   a. High counting speed
   b. Long propagation delay
   c. Circuit simplicity
   d. Ripple action

23. Cascaded up counters generate a _____ output to the next significant counter.
   a. carry output   **c.** RCO
   b. borrow output   **d.** both a. and c. are true

24. Cascaded down counters generate a _____ output to the next significant counter.
    a. carry output
    b. borrow output
    c. *RCO*
    d. both a. and c. are true

25. A state transition diagram shows a counter's _____.

    a. output states   **c.** propagation delay
    b. division factor   **d.** modulus

Practice Problems

To practice your circuit recognition and operation, refer to Figure 10-34 and answer the following four questions.

26. Describe the 74175 IC in one sentence.

27. Which type of flip-flop is being used to form these two register circuits?

28. The circuit configuration in Figure 10-34(a) is a _____.

    a. PIPO   **c.** SIPO
    b. SISO   **d.** PISO

29. The circuit configuration in Figure 10-34(b) is a _____.

    a. PIPO   **c.** SIPO
    b. SISO   **d.** PISO

30. Calculate the frequency division factor of the following counters:
    a. mod-12 counter
    b. mod-4096 counter
    c. a counter with 8,192 different output states
    d. a counter with a maximum count of 15

31. Calculate the maximum count of a
    a. 4-bit counter.
    b. 6-bit counter.
    c. 24-bit counter.

32. Calculate the modulus of a
    a. 4-bit counter.
    b. 6-bit counter.
    c. 24-bit counter.

33. Assuming a flip-flop propagation delay of 25 ns, calculate the total asynchronous counter propagation delay time and the upper clock pulse frequency limit for the following:
    a. 4-bit counter
    b. 6-bit counter
    c. 24-bit counter

To practice your circuit recognition and operation skills, refer to Figure 10-35 and answer the following seven questions.

34. Is this circuit a synchronous or an asynchronous counter?

35. What is this counter's natural modulus?

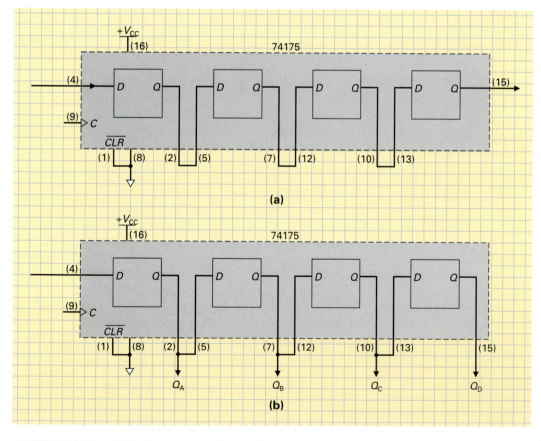

(a)

(b)

FIGURE 10-34   **74175 Register Configurations.**

FIGURE 10-35   Application—A 5-Stage Up Counter.

FIGURE 10-36   Application—Count Up to a Selected Value and Stop Circuit.

36. What is this counter's maximum count?

37. What is this counter's frequency division between the clock input and Q_4?

38. What would be the Q_4 output frequency if the clock input were 2 kHz?

39. What would be this counter's upper clock pulse frequency limit if each flip-flop had a propagation delay of 22 ns?

40. Assuming an 8-bit asynchronous presettable counter, calculate
   a. its natural modulus.
   b. its preset modulus if 0011 0100 is loaded.

To practice your circuit recognition and operation skills, refer to Figure 10-36 and answer the following question.

41. Briefly describe the operation of the
   a. 74193 IC in this circuit including the NAND gate's function at the clock input and the function of SW_1.
   b. 74154 IC in this circuit.
   c. If the alligator clip is connected to the 74154's pin 14 output and SW_1 is pressed, how will this circuit operate?

Web Site Questions

Go to the web site http://www.prenhall.com/cook, select the textbook *Practical Digital Electronics,* select this chapter, and then follow the instructions when answering the multiple choice practice problems.

Arithmetic Operations and Circuits

The Wizard of Menlo Park

Thomas Alva Edison was born to Samuel and Nancy Edison on February 11, 1847. As a young boy he had a keen and inquisitive mind, yet he did not do well at school, so his mother, a former schoolteacher, withdrew him from school and tutored him at home. In later life he said that his mother taught him to read well and instilled a love for books that lasted the rest of his life. In fact, the inventor's personal library of more than 10,000 volumes is preserved at the Edison Laboratory National Monument in West Orange, New Jersey.

At the age of twenty-nine, after several successful inventions, Edison put into effect what is probably his greatest idea—the first industrial research laboratory. Choosing Menlo Park in New Jersey, which was then a small rural village, Edison had a small building converted into a laboratory for his fifteen-member staff and a house built for his wife and two small daughters. When asked to explain the point of this lab, Edison boldly stated that it would produce "a minor invention every ten days and a big thing every six months or so." At the time, most of the scientific community viewed his prediction as preposterous; however, in the next ten years Edison would be granted 420 patents, including those for the electric light bulb, the motion picture, the phonograph, the universal electric motor, the fluorescent lamp, and the medical fluoroscope.

Over 1,000 patents were granted to Edison during his lifetime; and his achievements at what he called his "invention factory" earned him the nickname "the wizard of Menlo Park." When asked about his genius he said, "Genius is two percent inspiration and ninety-eight percent perspiration."

Introduction

At the heart of every digital electronic computer system is a circuit called a microprocessor unit (MPU). To perform its function, the MPU contains several internal circuits, and one of these circuits is called the arithmetic-logic unit, or ALU. This unit is the number-crunching circuit of every digital system and, as its name implies, it can be controlled to perform either arithmetic operations (such as addition and subtraction) or logic operations (such as AND and OR).

In order to understand the operation of an ALU, we must begin by discussing how binary numbers can represent any value, whether it is large, small, positive, or negative. We must also discuss how these numbers can be manipulated to perform arithmetic operations such as addition, subtraction, multiplication, and division. All of these topics will be covered in the first section of this chapter.

In the second section of this chapter, we will discuss how logic circuits can be connected to form arithmetic circuits, and how these circuits can perform arithmetic operations.

In the third section of this chapter, we will combine all of the topics covered in this chapter and see how an ALU can perform either arithmetic operations or logic operations.

In the fourth section of this chapter, we will practice circuit troubleshooting techniques by applying our three-step troubleshooting procedure to a typical arithmetic circuit.

Before we discuss the operation of arithmetic circuits, we must first understand the arithmetic operations these circuits perform. In the first section we will see how two binary numbers are added together, how one binary number can be subtracted from another, how one binary number is multiplied by another, and how one binary number is divided into another—in short, the principles of binary arithmetic.

11-1-1 *Binary Arithmetic*

The addition, subtraction, multiplication, and division of binary numbers is performed in exactly the same way as the addition, subtraction, multiplication, and division of decimal numbers. The only difference is that the decimal number system has ten digits, whereas the binary number system has two.

As we discuss each of these arithmetic operations, we will use decimal arithmetic as a guide to binary arithmetic so that we can compare the known to the unknown.

Binary Addition

Figure 11-1(a) illustrates how the decimal values 29,164 and 63,729 are added together. First the units 4 and 9 are added, producing a sum of 13, or 3 carry 1. The carry is then added to the 6 and 2 in the tens column, producing a sum of 9 with no carry. This process is then repeated for each column, moving right to left, until the total sum of the addend, augend, and their carries has been obtained.

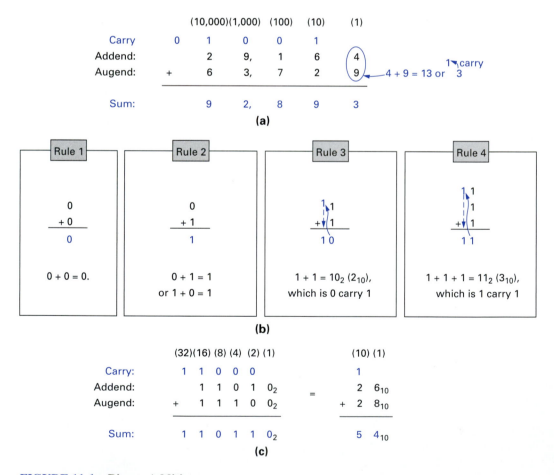

FIGURE 11-1 Binary Addition.

When adding two decimal values, many combinations are available since decimal has ten different digits. With binary, however, only two digits are available and therefore only four basic combinations are possible. These four rules of binary addition are shown in Figure 11-1(b). To help understand these rules, let us apply them to the example in Figure 11-1(c) in which we will add 11010 (decimal 26) to 11100 (decimal 28). Like decimal addition, we will begin at the right in the ones column and proceed to the left. In the ones column $0 + 0 = 0$ with a carry of 0 (Rule 1). In the twos column $1 + 0$ and the 0 carry from the ones column $= 1$ with no carry (Rule 2). In the fours column we have $0 + 1$ plus a carry of 0 from the twos column. This results in a fours column total of 1, with a carry of 0 (Rule 2). In the eights column we have $1 + 1$ plus a 0 carry from the fours column, producing a total of 10_2 ($1 + 1 =$ decimal 2, which is 10 in binary). In the eights column this 10 (decimal 2) is written down as a total of 0 with a carry of 1 (Rule 3). In the sixteens column, we have $1 + 1$ plus a carry of 1 from the eights column, producing a total of 11_2 ($1 + 1 + 1 =$ decimal 3, which is 11 in binary). In the sixteens column, this 11 (decimal 3) is written down as a total of 1 with a carry of 1 (Rule 4). Since there is only a carry of 1 in the thirty-twos column, the total in this column will be 1.

Converting the binary addend and augend to their decimal equivalents as seen in Figure 11-1(c), you can see that 11010 (decimal 26) plus 11100 (decimal 28) will yield a sum or total of 110110 (decimal 54).

■ **EXAMPLE:**

Find the sum of the following binary numbers.

a. $1011 + 1101 = ?$ b. $1000110 + 1100111 = ?$

■ *Solution:*

	Carry:		1	1	1	1			
	Augend			1	0	1	1_2	(11_{10})
a.	Addend:	+		1	1	0	1_2	(13_{10})
	Sum:		1	1	0	0	0_2	(24_{10})

	Carry:		1	0	0	0	1	1	0		
	Augend			1	0	0	0	1	1	0_2	(70_{10})
b.	Addend:	+		1	1	0	0	1	1	1_2	(103_{10})
	Sum:		1	0	1	0	1	1	0	1_2	(173_{10})

Binary Subtraction

Figure 11-2(a) reviews the decimal subtraction procedure by subtracting 4,615 from 7,003. Starting in the units column, you can see that since we cannot take 5 away from 3, we will have to go to a higher-order minuend unit and borrow. Since the minuend contains no tens or hundreds, we will have to go to the thousands column. From this point a chain of borrowing occurs as a thousand is borrowed and placed in the hundreds column (leaving 6 thousands), one of the hundreds is borrowed and placed in the tens column (leaving 9 hundreds), and one of the tens is borrowed and placed in the units column (leaving 9 tens). After borrowing 10, the minuend units digit has a value of 13, and if we now perform the subtraction, the result or difference of $13 - 5 = 8$. Since all of the other minuend digits are now greater than their respective subtrahend digits, there will be no need for any further borrowing to obtain the difference.

When subtracting one decimal value from another, many combinations are available since decimal has ten different digits. With binary, however, only two digits are available and therefore only four basic combinations are possible. These four rules of binary subtraction are shown in Figure 11-2(b). To help understand these rules, let us apply them to the example in Figure 11-2(c), in which we will subtract 11100 (decimal 28) from 1010110 (decimal 86). Like

FIGURE 11-2 Binary Subtraction.

decimal subtraction, we will begin at the right in the ones column and proceed to the left. In the ones column $0 - 0 = 0$ (Rule 1). In the twos column $1 - 0 = 1$ (Rule 2), and in the fours column $1 - 1 = 0$ (Rule 3). In the eights column we cannot subtract 1 from nothing or 0, so therefore we must borrow 1 from the sixteens column, making the minuend in the eights column 10 (decimal 2). The difference can now be calculated in the eights column since $10 - 1 = 1$ (decimal $2 - 1 = 1$, Rule 4). Due to the previous borrow, the minuend in the sixteens column is now 0 and therefore we need to borrow a 1 from the thirty-twos column. Since the thirty-twos column is also 0, we will need to borrow from the sixty-fours column. Borrowing 1 from the sixty-fours column will leave a minuend of 0 and make the minuend in the thirty-twos column equal to 10 (decimal 2). Borrowing 1 from the 10 (borrowing decimal 1 from 2) in the thirty-twos column will leave a minuend of 1 and make the minuend in the sixteens column equal to 10 (decimal 2). We can now subtract the subtrahend of 1 from the minuend of 10 (decimal 2) in the sixteens column to obtain a difference of 1. Due to the previous borrow, the minuend in the thirty-twos column is equal to 1, and $1 - 0 = 1$ (Rule 2). Finally, since the minuend and subtrahend are both 0 in the sixty-fours column, the subtraction is complete.

Converting the binary minuend and subtrahend to their decimal equivalents, as seen in Figure 11-2(c), you can see that 1010110 (decimal 86) minus 11100 (decimal 28) will result in a difference of 111010 (decimal 58).

EXAMPLE:

Find the difference between the following binary numbers.

a. $100111 - 11101 = ?$

b. $101110111 - 1011100 = ?$

Solution:

Minuend:		$^0\cancel{1}$	$^1\cancel{0}$	^{10}	1	1	1_2	(39_{10})
a. Subtrahend:	$-$		1	1	1	0	1_2	(29_{10})
Difference:		0	0	1	0	1	0_2	(10_{10})

Minuend:		1	0	1	$^0\cancel{1}$	$^{10}\cancel{1}$	^{10}	1	1	1_2	(375_{10})
b. Subtrahend:	$-$			1	0	1	1	1	0	0_2	(92_{10})
Difference:		1	0	0	0	1	1	0	1	1_2	(283_{10})

Binary Multiplication

Figure 11-3(a) reviews the decimal multiplication procedure by multiplying 463 by 23. To perform this operation, you would begin by multiplying each digit of the multiplicand by the units digit of the multiplier to obtain the first partial product. Second, you would multiply each digit of the multiplicand by the tens digit of the multiplier to obtain the second partial product. Finally, all of the partial products are added to obtain the final product.

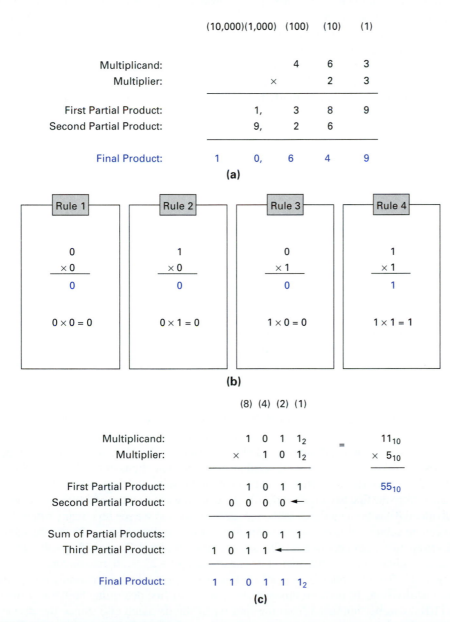

FIGURE 11-3   Binary Multiplication.

When multiplying one decimal value by another, many combinations are available since decimal has ten different digits. With binary multiplication, however, only two digits are available and therefore only four basic combinations are possible. These four rules of binary multiplication are shown in Figure 11-3(b). To help understand these rules, let us apply them to the example in Figure 11-3(c), in which we will multiply 1011 (decimal 11) by 101 (decimal 5). Like decimal multiplication, we begin by multiplying each bit of the multiplicand by the ones column multiplier bit to obtain the first partial product. Second, we multiply each bit of the multiplicand by the twos column multiplier bit to obtain the second partial product. Remember that the LSB of the partial product must always be directly under its respective multiplier bit, and therefore the LSB of the second partial product should be under the twos column multiplier bit. In the next step, we add together the first and second partial products to obtain a sum of the partial products. With decimal multiplication we usually calculate all of the partial products, and then add them together to obtain the final product. This procedure can also be used with binary multiplication; however, it is better to add only two partial products at a time since the many carries in binary become hard to keep track of and easily lead to errors. In the next step, we multiply each bit of the multiplicand by the fours column multiplier bit to obtain the third partial product. Remember that the LSB of this third partial product must be placed directly under its respective multiplier bit. The final step is to add the third partial product to the previous partial sum to obtain the final product or result.

If you study the partial products obtained in this example, you will notice that they were either exactly equal to the multiplicand (when the multiplier was 1) or all 0s (when the multiplier was 0). You can use this shortcut in future examples; however, be sure to always place the LSB of the partial products directly below their respective multiplier bits.

■ EXAMPLE:

Multiply 1101_2 by 1011_2.

■ *Solution:*

Multiplicand:					1	1	0	1_2		13_{10}
Multiplier:				×	1	0	1	1_2		× 11_{10}
First Partial Product:					1	1	0	1		143_{10}
Second Partial Product:			1	1	0	1	←			
Sum of Partial Product:		1	0	0	1	1	1			
Third Partial Product:		0	0	0	0	←	←			
Sum of Partial Product:		1	0	0	1	1	1			
Fourth Partial Product:	1	1	0	1	←	←	←			
Final Product:	1	0	0	0	1	1	1	1_2		

Binary Division

Figure 11-4(a) reviews the decimal division procedure by dividing 830 by 23. This long-division procedure begins by determining how many times the divisor (23) can be subtracted from the first digit of the dividend (8). Since the dividend is smaller, the quotient is 0. Next, we see how many times the divisor (23) can be subtracted from the first two digits of the dividend (83). Since 23 can be subtracted three times from 83, a quotient of 3 results, and the product of 3 × 23, or 69, is subtracted from the first two digits of the dividend, giving a remainder of 14. Then the next digit of the dividend (0) is brought down to the remainder (14), and we see how many times the divisor (23) can be subtracted from the new remainder (140). In this example, 23 can be subtracted six times from 140, a quotient of 6 results, and the product of 6 × 23, or 138, is subtracted from the remainder, giving a new remainder of 2. Therefore, 830 ÷ 23 = 36, remainder 2.

Figure 11-4(b) illustrates an example of binary division, which is generally much simpler than decimal division. To perform this operation, we must first determine how many times the divisor (101_2) can be subtracted from the first bit of the dividend (1_2). Since the dividend is smaller than the divisor, a quotient of 0_2 is placed above the first digit of the dividend. Next, we

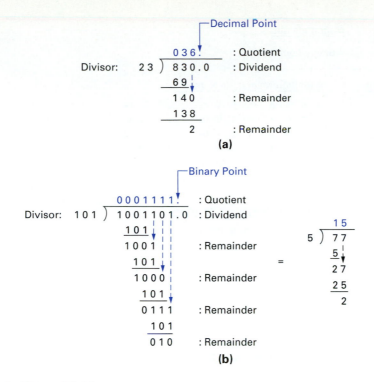

FIGURE 11-4   **Binary Division.**

see how many times the divisor (101_2) can be subtracted from the first two bits of the dividend (10_2), and this again results in a quotient of 0_2. Similarly, the divisor (101_2) cannot be subtracted from the first three bits of the dividend, so the quotient is once again 0_2. Finally, we find a dividend (1001_2) that is greater than the divisor (101_2), and therefore a 1_2 is placed in the quotient, and the divisor is subtracted from the first four bits of the dividend, resulting in a remainder of 100_2. Then the next bit of the dividend (1_2) is brought down to the remainder (100_2), and we see how many times the divisor (101_2) can be subtracted from the new remainder (1001_2). Since 101_2 is smaller than 1001_2, a quotient of 1_2 results. The divisor is subtracted from the remainder, resulting in a new remainder of 100_2. Then the next bit of the dividend (0_2) is brought down to the remainder (100_2), and we see how many times the divisor (101_2) can be subtracted from the new remainder (1000_2). Since 101_2 is smaller than 1000_2, a quotient of 1 results. The divisor is subtracted from the remainder, resulting in a new remainder of 11_2. Then the next bit of the dividend (1_2) is brought down to the remainder (11_2), and we see how many times the divisor (101_2) can be subtracted from the new remainder (111_2). Since 101_2 is smaller than 111_2, a quotient of 1 results. The divisor is subtracted from the remainder, resulting in a final remainder of 10. Therefore, $1001101_2 \div 101_2 = 1111_2$, remainder 10 ($77 \div 5 = 15$, remainder 2).

■   **EXAMPLE:**

Divide 1110101_2 by 110_2.

■   *Solution:*

```
                0  0  1  0  0  1  1₂    :Quotient          019₁₀
Divisor:    110)1  1  1  0  1  0  1₂    :Dividend        6₁₀)117₁₀
                1  1  0  ↓  ↓  ↓  ↓
                0  0  1  0  1  0  ↓     :Remainder
                         1  1  0  ↓                         6
                         1  0  0  1     :Remainder          57
                         1  1  0                            54
                         0  0  1  1₂    :Remainder          3₁₀
```

FIGURE 11-5   The Sign-Magnitude Number System.

11-1-2   *Representing Positive and Negative Numbers*

Many digital systems, such as calculators and computers, are used to perform mathematical functions on a wide range of numbers. Some of these numbers are positive and negative signed numbers; therefore, a binary code is needed to represent these numbers.

The Sign and Magnitude Number System

The **sign and magnitude number system** is a binary code system used to represent positive and negative numbers. A sign magnitude number contains a *sign bit* (0 for positive, 1 for negative) followed by the *magnitude bits.*

As an example, Figure 11-5(a) shows some 4-bit sign-magnitude positive numbers. All of these sign-magnitude binary numbers—0001, 0010, 0011, and 0100—have an MSB, or sign bit, of 0, and therefore they are all positive numbers. The remaining three bits in these 4-bit words indicate the magnitude of the number based on the standard 421 binary column weight ($001_2 = 1_{10}$, $010_2 = 2_{10}$, $011_2 = 3_{10}$, $100_2 = 4_{10}$).

As another example, Figure 11-5(b) shows some 4-bit sign-magnitude negative numbers. All of these sign-magnitude binary numbers—1100, 1101, 1110, and 1111—have an MSB, or sign bit, of 1, and therefore they are all negative numbers. The remaining three bits in these 4-bit words indicate the magnitude of the number based on the standard 421 binary column weight ($100_2 = 4_{10}$, $101_2 = 5_{10}$, $110_2 = 6_{10}$, $111_2 = 7_{10}$).

If you need to represent larger decimal numbers, you simply use more bits, as shown in Figure 11-5(c) and (d). The principle still remains the same in that the MSB indicates the sign of the number, and the remaining bits indicate the magnitude of the number.

The sign-magnitude number system is an ideal example of how binary numbers could be coded to represent positive and negative numbers. This code system, however, requires complex digital hardware for addition and subtraction and is therefore seldom used.

■ EXAMPLE:

What are the decimal equivalents of the following sign-magnitude number system codes?

a. 0111   b. 1010   c. 0001 1111   d. 1000 0001

Solution:

a. +7 b. −2 c. +31 d. −1

EXAMPLE:

What are the 8-bit sign-magnitude number system codes for the following decimal values?

a. +8 b. −65 c. +127 d. −127

■ *Solution:*

a. 0000 1000 b. 1100 0001 c. 0111 1111 d. 1111 1111

The One's Complement (1's Complement) Number System

The **one's complement number system** became popular in early digital systems. It also used an MSB sign bit that was either a 0 (indicating a positive number) or a 1 (indicating a negative number).

Figure 11-6(a) shows some examples of 8-bit one's complement positive numbers. As you can see, one's complement positive numbers are no different from sign-magnitude positive numbers. The only difference between the one's complement number system and the sign-magnitude number system is the way in which negative numbers are represented. To understand the code used to represent negative numbers, we must first understand the meaning of the term *one's complement.* To one's complement a binary number means to invert or change all of the 1s in the binary word to 0s, and all of the 0s in the binary word to 1s. Performing the one's complement operation on a positive number will change the number from a one's complement positive number to a one's complement negative number. For example, by one's complementing the number 00000101, or +5, we will get 11111010, which is −5. In Figure 11-6(b), you will see the negative one's complement codes of the positive one's complement codes shown in Figure 11-6(a).

The next question you may have is, how can we determine the decimal value of a negative one's complement code? The answer is to do the complete opposite, or complement, to that of the positive one's complement code. This means count the 1s in a positive one's complement code, and count the 0s in a negative one's complement code. For example, with a positive one's complement code (MSB = 0), the value is determined by simply adding together all of the column weights that contain a 1 (00000110 has a 1 in the fours column and

One's Complement Number System

A binary code used to represent positive and negative numbers in which negative values are determined by inverting all of the binary digits of an equivalent positive value.

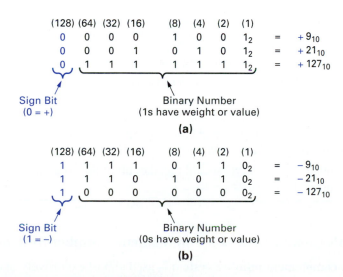

(a)

Sign Bit (0 = +)   Binary Number (1s have weight or value)

(b)

Sign Bit (1 = −)   Binary Number (0s have weight or value)

FIGURE 11-6 The One's Complement Number System.

twos column, so this number is $4 + 2$, or $+6$). On the other hand, for a negative one's complement code (MSB $= 1$), the value is determined by doing the opposite and adding together all of the column weights that contain a 0 (11110101 has a 0 in the eights column and twos column, so this number is $8 + 2$, or -10).

EXAMPLE:

What are the decimal equivalents of the following one's complement number system codes?

    a. 0010 0001   b. 1101 1101

Solution:

    a. MSB $= 0$ (number is positive, so add together all of the column weights that contain a 1). The number 0010 0001 has a 1 in the thirty-twos column and a one in the units column, so 0010 0001 $= +33$.

    b. MSB $= 1$ (number is negative, so add together all of the column weights that contain a 0). The number 1101 1101 has a 0 in the thirty-twos column and a 0 in the twos column, so 1101 1101 $= -34$.

Since the term *one's complement* can be used to describe both an operation (change all of the 1s to 0s, and all of the 0s to 1s) and a positive and negative number system, some confusion can occur. To avoid this, remember that if you are asked to "one's complement a number," it means to change all of the 1s in the binary word to 0s and all of the 0s to 1s. If, on the other hand, you are asked to "find the decimal equivalent of a one's complement number," you should decode the sign bit and binary value to determine the decimal equivalent value.

EXAMPLE:

One's complement the following numbers:

    a. 0101 1111   b. 1110 0101

Solution:

To one's complement a number means to change all of the 1s to 0s and all of the 0s to 1s.
    a. 1010 0000   b. 0001 1010

EXAMPLE:

Find the decimal equivalent of the following one's complement numbers:

    a. 0101 1111   b. 1110 0101

Solution:

    a. MSB $= 0$ (number is positive, so add together all of the column weights that contain a 1). The number 0101 1111 $= +95$.

    b. MSB $= 1$ (number is negative, so add together all of the column weights that contain a 0). The number 1110 0101 $= -26$.

Two's Complement Number System

A binary code used to represent positive and negative numbers in which negative values are determined by inverting all of the binary digits of an equivalent positive value and then adding 1.

The Two's Complement (2's Complement) Number System

The **two's complement number system** is used almost exclusively in digital systems to represent positive and negative numbers.

(128)	(64)	(32)	(16)		(8)	(4)	(2)	(1)			
0	0	0	0		1	1	1	1_2	=	$+15_{10}$
0	1	1	1		1	1	1	1_2,	=	$+127_{10}$

Sign Bit
(0 = +)

Binary Number
(1s have weight or value)

(a)

(128)	(64)	(32)	(16)		(8)	(4)	(2)	(1)			
1	1	1	1		0	1	0	1_2	=	-11_{10}

Sign Bit
(1 = −)

Binary Number
(0s have weight or value + 1)

(b)

FIGURE 11-7 The Two's Complement Number System.

Positive two's complement numbers are no different from positive sign-magnitude and one's complement numbers in that an MSB sign bit of 0 indicates a positive number and the remaining bits of the word indicate the value of the number. Figure 11-7(a) shows two examples of positive two's complement numbers.

Negative two's complement numbers are determined by two's complementing a positive two's complement number code. To two's complement a number, first one's complement it (change all the 1s to 0s and 0s to 1s) and then add 1. For example, if we were to two's complement the number 0000 0101 (+5), we would obtain the two's complement number code for −5. The procedure to follow to two's complement 0000 0101 (+5), would be as follows:

0000 0101	Original Number (+5)
1111 1010	One's Complement (inverted)
+ 1	Plus 1
1111 1011	Two's Complement Code for −5

The new two's complement code obtained after this two's complement operation is the negative equivalent of the original number. In our example, therefore:

Original Number: 0000 0101 = +5
After Two's Complement Operation: 1111 1011 = −5

Any negative equivalent two's complement code therefore can be obtained by simply two's complementing the same positive two's complement number. For example, to find the negative two's complement code for −125, we would perform the following operation:

Two's	0111 1101	Original Number (+125)
Complement →	1 000 0010	One's Complement (inverted)
Operation	+ 1	Plus 1
	1000 0011	Two's Complement Code for −125

■ **EXAMPLE:**

Find the negative two's complement codes for

a. −26   b. −67

	0001 1010	Two's Complement Code for +26
a.	1110 0101	One's Complement (inverted)
	+ 1	Plus 1
	1110 0110	Two's Complement Code for −26
	0100 0011	Two's Complement Code for +67
b.	1 011 1100	One's Complement (inverted)
	+ 1	Plus 1
	1011 1101	Two's Complement Code for −67

Performing the two's complement operation on a number twice will simply take the number back to its original value and polarity. For example, let's take the two's complement code for +6 (0000 0110) and two's complement it twice.

First Two's	0000 0110	Original Number (+6)
Complement →	1111 1001	One's Complement (inverted)
Operation	+ 1	Plus 1
	1111 1010	Two's Complement Code for −6

Second Two's	0000 0101	One's Complement (inverted)
Complement →	+ 1	Plus 1
Operation	0000 0110	Back to Original Number (+6)

The decimal equivalent of a positive two's complement number (MSB = 0) is easy to determine, since the binary value following the sign bit applies the same as any binary number, as shown in Figure 11-7(a). On the other hand, the decimal equivalent of a negative two's complement number (MSB = 1) is not determined in the same way. To determine the value of a negative two's complement number, refer to the remaining bits following the sign bit, add the weight of all columns containing a 0, and then add 1 to the result. For example, the negative two's complement number in Figure 11-7(b) has 0s in the eights and twos columns, and therefore its value will equal the sum of 8 and 2, plus the additional 1 (due to two's complement). As a result,

$$\overset{(8)(4)(2)(1)}{1\ 1\ 1\ 1\ 0\ 1\ 0\ 1} = (8 + 2) + 1 = -11$$

■ **EXAMPLE:**

Determine the decimal equivalent of the following negative two's complement numbers:

a. 1110 0001   b. 1011 1011

c. 1000 0000   d. 1111 1111

■ **Solution:**

a. $\overset{(\pm)(64)(32)(16)(8)(4)(2)(1)}{1\ 1\ 1\ 0\ 0\ 0\ 0\ 1}_2 = (16 + 8 + 4 + 2) + 1 = -31_{10}$

b. $\overset{(\pm)\ (64)\ (32)\ (16)\ (8)\ (4)\ (2)\ (1)}{1\ 0\ 1\ 1\ 1\ 0\ 1\ 1}_2 = (64 + 4) + 1 = -69_{10}$

c. $\overset{(\pm)\ (64)\ (32)\ (16)\ (8)\ (4)\ (2)\ (1)}{1\ 0\ 0\ 0\ 0\ 0\ 0\ 0}_2 = (64 + 32 + 16 + 8 + 4 + 2 + 1) + 1 = -128_{10}$

d. $\overset{(\pm)\ (64)\ (32)\ (16)\ (8)\ (4)\ (2)\ (1)}{1\ 1\ 1\ 1\ 1\ 1\ 1\ 1}_2 = (0) + 1 = -1_{10}$

4-Bit Word	Standard Binary Value	2's Complement Value	
0 0 0 0	0	0	
0 0 0 1	1	+1	
0 0 1 0	2	+2	
0 0 1 1	3	+3	8 Positive
0 1 0 0	4	+4	Numbers
0 1 0 1	5	+5	
0 1 1 0	6	+6	
0 1 1 1	7	+7	
1 0 0 0	8	−8	
1 0 0 1	9	−7	
1 0 1 0	10	−6	
1 0 1 1	11	−5	8 Negative
1 1 0 0	12	−4	Numbers
1 1 0 1	13	−3	
1 1 1 0	14	−2	
1 1 1 1	15	−1	

$2^4 =$ 16 Words

FIGURE 11-8 **4-Bit Two's Complement Codes.**

As with the sign-magnitude and one's complement number systems, if you need to represent larger decimal numbers with a two's complement code, you will have to use a word containing more bits. For example, a total of 16 different combinations is available when we use a 4-bit word ($2^4 = 16$), as shown in Figure 11-8. With standard binary, these 16 different combinations give us a count from 0 to 15, as shown in the center column in Figure 11-8. As far as two's complement is concerned, we split these 16 different codes in half and ended up with 8 positive two's complement codes and 8 negative two's complement codes, as shown in the right column in Figure 11-8. Since decimal zero uses one of the positive two's complement codes, the 16 different combinations give us a range of -8_{10} to $+7_{10}$ (the maximum positive value, $+7$, is always one less than the maximum negative value, -8, since decimal zero uses one of the positive codes).

Using an 8-bit word will give us 256 different codes ($2^8 = 256$), as shown in Figure 11-9. Dividing this total in half will give us 128 positive two's complement codes and 128 negative two's complement codes. Since decimal zero uses one of the positive two's complement codes, the 256 different combinations gives us a range of -128_{10} to $+127_{10}$.

■ **EXAMPLE:**

What would be the two's complement range for a 16-bit word?

■ *Solution:*

$2^{16} = 65,536$; half of $65,536 = 32,768$; therefore, the range will be $-32,768$ to $+32,767$.

11-1-3 *Two's Complement Arithmetic*

To perform its function, a computer's MPU contains several internal circuits, and one of these circuits is called the **arithmetic-logic unit,** or **ALU.** This unit is the number-crunching circuit of every digital system and, as its name implies, it can be controlled to perform either arithmetic operations (such as addition or subtraction) or logic operations (such as AND and OR). The arithmetic-logic unit has two parallel inputs, one parallel output, and a set of function-select control lines, as shown in Figure 11-10.

Arithmetic-Logic Unit (ALU)

The section of a computer that performs all arithmetic and logic operations.

8-Bit Word	Standard Binary Value	2's Complement Value
0 0 0 0 0 0 0 0	0	0
0 0 0 0 0 0 0 1	1	+1
0 0 0 0 0 0 1 0	2	+2
0 0 0 0 0 0 1 1	3	+3
0 0 0 0 0 1 0 0	4	+4
0 0 0 0 0 1 0 1	5	+5
0 0 0 0 0 1 1 0	6	+6
0 0 0 0 0 1 1 1	7	+7
≈	≈	≈
0 1 1 1 1 0 1 1	123	+123
0 1 1 1 1 1 0 0	124	+124
0 1 1 1 1 1 0 1	125	+125
0 1 1 1 1 1 1 0	126	+126
0 1 1 1 1 1 1 1	127	+127
1 0 0 0 0 0 0 0	128	−128
1 0 0 0 0 0 0 1	129	−127
1 0 0 0 0 0 1 0	130	−126
1 0 0 0 0 0 1 1	131	−125
1 0 0 0 0 1 0 0	132	−124
1 0 0 0 0 1 0 1	133	−123
≈	≈	≈
1 1 1 1 1 0 1 1	251	−5
1 1 1 1 1 1 0 0	252	−4
1 1 1 1 1 1 0 1	253	−3
1 1 1 1 1 1 1 0	254	−2
1 1 1 1 1 1 1 1	255	−1

$2^8 =$ 256 Words

128 Positive Numbers

128 Negative Numbers

FIGURE 11-9   **8-Bit Two's Complement Codes.**

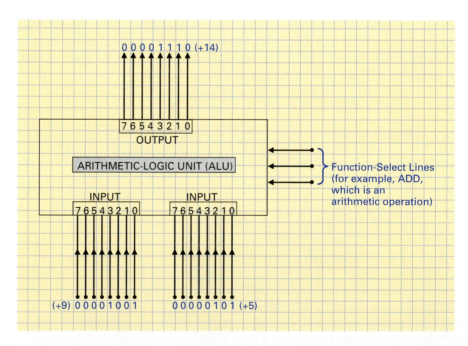

FIGURE 11-10   **Block Diagram of the Arithmetic-Logic Unit (ALU).**

310

As far as arithmetic operations are concerned, the ALU has to be able to perform the four basic operations, which, as you know, are addition, subtraction, multiplication, and division. One would expect, therefore, that an ALU would contain a separate circuit for each of these operations. This is not so, since all four operations can be performed using a binary-adder circuit. To explain this point, because the binary-adder circuit's function is to perform addition, and since multiplication is simply repeated addition, we can also use the binary-adder circuit to perform any multiplication operations. For example, 3×4 means that you are to take the number 4 and add it three times ($4 + 4 + 4 = 12$, therefore $3 \times 4 = 12$). But what about subtraction and division? The answer to this problem is the two's complement number system. As you will see in this section, by representing all values as two's complement numbers, we can perform a subtraction operation by slightly modifying the second number and then adding the two numbers together. This means that our binary-adder circuit can also be used to perform subtraction operations, and since division is simply repeated subtraction, we can also use the subtraction operation of the binary-adder circuit to perform any division operation. For example, $12 \div 4$ means that you are to take the number 12 and see how many times you can subtract 4 from it ($12 - 4 - 4 - 4 = 0$; therefore there are three 4s in 12, $12 \div 4 = 3$).

By using two's complement numbers, therefore, we can use a binary-adder circuit within an ALU to add, subtract, multiply, and divide. Having one circuit able to perform all four mathematical operations means that the circuit will be small in size, fast to operate, consume very little power, and be cheap to manufacture. In this section we will see how we can perform all four mathematical operations by always adding two's complement numbers.

Adding Positive Numbers

To begin with, let us prove that the addition of two positive two's complement numbers will render the correct result.

EXAMPLE:

Add $+9$ (0000 1001) and $+5$ (0000 0101).

Solution:

$$
\begin{array}{r}
0000\ 1001 \\
+\ 0000\ 0101 \\
\hline
0000\ 1110
\end{array}
$$

MSB of 0 = Positive number $\rightarrow$

When both of the numbers to be added are positive two's complement numbers (both have an MSB of 0), it makes sense that the sum will have an MSB of 0, indicating a positive result. This is always true, unless the range of the two's complement word is exceeded. For example, the maximum positive number that can be represented by an 8-bit word is $+127$ (0111 1111). If the sum of the two positive two's complement numbers to be added exceeds the upper positive range limit, there will be a *two's complement overflow* into the sign bit. For instance:

$$
\begin{array}{ll}
(\pm)(64)(32)(16)\ (8)\ (4)\ (2)\ \ (1) & \\
0\ \ 1\ \ 1\ \ 0\ \ 0\ \ 1\ \ 0\ \ 0 & (+100) \\
+\ 0\ \ 0\ \ 0\ \ 1\ \ 1\ \ 1\ \ 1\ \ 0 & +\ (+30) \\
\hline
1\ \ 0\ \ 0\ \ 0\ \ 0\ \ 0\ \ 1\ \ 0 & (-126)
\end{array}
$$

MSB of 1 = negative number $\rightarrow$

If two positive numbers are applied to an arithmetic circuit, and it is told to add the two, the result should also be a positive number. Most arithmetic circuits are able to detect this problem if it occurs by simply monitoring the MSB of the two input words and the MSB of the output word, as shown in Figure 11-11. If the two input sign bits are the same but the output sign bit is different, then a two's complement overflow has occurred.

FIGURE 11-11 Two's Complement Overflow Detection.

Adding Positive and Negative Numbers

By using the two's complement number system, we can also add numbers with unlike signs and obtain the correct result. As an example, let us add +9 (0000 1001) and −5 (1111 1011).

$$
\begin{array}{rl}
& 0000\ 1001 \quad (+9) \\
+ & 1111\ 1011 \quad +(-5) \\
\hline
\text{Ignore the final carry} \rightarrow 1 \quad & 0000\ 0100 \quad (+4)
\end{array}
$$

A final carry will always be generated whenever the sum is a positive number; however, if this carry is ignored because it is beyond the 8-bit two's complement word, the answer will be correct.

■ **EXAMPLE:**

Add +10 (0000 1010) and −12 (1111 0100).

■ *Solution:*

$$
\begin{array}{rl}
& 0000\ 1010 \quad (+10) \\
+ & 1111\ 0100 \quad + (-12) \\
\hline
& 1111\ 1110 \quad (-2)
\end{array}
$$

Adding Negative Numbers

As a final addition test for the two's complement number system, let us add two negative numbers to see if we can obtain the correct negative sum.

■ **EXAMPLE:**

Add −5 (1111 1011) and −4 (1111 1100).

$$\begin{array}{rr} 1111\ 1011 & (-5) \\ +\ 1111\ 1100 & +\ (-4) \\ \hline \text{Ignore the final carry} \rightarrow 1 \quad 1111\ 0111 & (-9) \end{array}$$

The sum in this example is the correct two's complement code for -9.

Two's complement overflow can also occur in this condition if the sum of the two negative numbers exceeds the negative range of the word. For example, with an 8-bit two's complement word, the largest negative number is -128, and if we add -100 (1001 1100) and -30 (1110 0010), we will get an error, as follows:

$$\begin{array}{rr} 1001\ 1100 & (-100) \\ +\ 1110\ 0010 & +\ (-30) \\ \hline \text{Ignore the final carry} \rightarrow 1 \quad 0111\ 1110 & (+126) \end{array}$$

Once again, the two's complement overflow detect circuit, shown in Figure 11-11, will indicate such an error since two negative numbers were applied to the input and a positive number appeared at the output.

In all of the addition examples mentioned so far (adding positive numbers, adding positive and negative numbers, and adding negative numbers) you can see that an ALU added the bits of the input bytes without any regard for whether they were signed or unsigned numbers. This brings up a very important point, which is often misunderstood. The ALU adds the bits of the input words using the four basic rules of binary addition, and it does not know if these input and output bit patterns are coded two's complement words or standard unsigned binary words. It is therefore up to us to maintain a consistency, which means that if we are dealing with coded two's complement words at the input of an ALU, we must interpret the output of an ALU as a two-coded two's complement word. On the other hand, if we are dealing with standard unsigned binary words at the input of an ALU, we must interpret the output of an ALU as a standard unsigned binary word.

Subtraction Through Addition

The key advantage of the two's complement number system is that it enables us to perform a subtraction operation using the adder circuit within the ALU. To explain how this is done, let us take an example and assume that we want to subtract 3 (0000 0011) from 8 (0000 1000), as follows:

$$\begin{array}{llll} \text{Minuend:} & 0000\ 1000 & (+8) \\ \text{Subtrahend:} & -\ 0000\ 0011 & -\ (+3) \\ \hline \text{Difference:} & 0000\ 0101 & (+5) \end{array}$$

If you subtract the subtrahend of $+3$ from the minuend of $+8$, you will obtain the correct result of $+5$ (0000 0101). However, we must find the difference in this problem by using addition and not subtraction. The solution to this is achieved by *two's complementing the subtrahend and then adding the result to the minuend.* The following shows how this is done:

$$\begin{array}{llll} \text{Minuend:} & 0000\ 1000 & (+8) \\ \text{Subtrahend:} & -\ \boxed{0000\ 0011} & -\ (+3) \end{array}$$

$$\begin{array}{ll} 0000\ 0011 & \text{2's Complement Code for } +3 \\ 1111\ 1100 & \text{1's Complement (inverted)} \\ +\qquad 1 & \text{Plus 1} \\ \hline 1111\ 1101 & \text{2's Complement Code for } -3 \end{array}$$

$$\begin{array}{llll} \text{Minuend:} & & 0000\ 1000 & (+8) \\ \text{Two's Complement Subtrahend:} & +\ & 1111\ 1101 & +\ (-3) \\ \hline \text{Difference (ignore carry):} & 1\ & 0000\ 0101 & +5 \end{array}$$

FIGURE 11-12 Flow Chart for Subtraction through Addition.

As you can see from this operation, adding the -3 to $+8$ achieves exactly the same result as subtracting $+3$ from $+8$. We can therefore perform a subtraction by two's complementing the subtrahend of a subtraction problem and then adding the result to the minuend.

Digital electronic calculators and computer systems perform the previously described steps automatically. That is, they first instruct the ALU to one's complement the subtrahend input (function-select control inputs = invert, which is a logic operation), then add 1 to the result (function-select control inputs = + 1, which is an arithmetic operation), and finally add the result of the two previous operations (which is the two's complemented subtrahend) to the minuend (function select = ADD, arithmetic operation). Although this step-by-step procedure or **algorithm** seems complicated, it is a lot simpler than having to include a separate digital electronic circuit to perform all subtraction operations. Having one circuit able to perform both addition and subtraction means that the circuit will be small in size, fast to operate, consume very little power, and be cheap to manufacture.

In Figure 11-12, you can see a **flow chart** summarizing the subtraction through addition procedure. Flow charts provide a graphic way of describing a step-by-step procedure (algorithm) or a program of instructions.

Algorithm

A set of rules for the solution of a problem.

Flow Chart

A graphic representation of a step-by-step procedure or program of instructions.

Multiplication Through Repeated Addition

Most digital electronic calculators and computers do not include a logic circuit that can multiply, since the adder circuit can also be used to perform this arithmetic operation. For example, to multiply 6×3, you simply add 6 three times.

Multiplicand:	6		6
Multiplier:	$\times\ 3$	$=$	$+\ 6$
Product:	18		$+\ 6$
			18

As in subtraction through addition, an algorithm must be followed when a digital system needs to perform a multiplication operation. Figure 11-13 shows the flow chart for multipli-

Example

Multiplicand: 6
Multiplier: $\times 3$
Product: 18

FIGURE 11-13   **Flow Chart for Multiplication through Repeated Addition.**

cation through repeated addition. Let us apply this procedure to our previous example of 6 $\times$ 3 and see how the multiplicand 6 is added three times to obtain a final product of 18. The steps would be as follows:

Start: Multiplicand = 6,   Multiplier = 3,   Product = 0.
After First Pass: Multiplicand = 6,   Multiplier = 2,   Product = 6.
After Second Pass: Multiplicand = 6,   Multiplier = 1,   Product = 12.
After Third Pass: Multiplicand = 6,   Multiplier = 0,   Product = 18.
Stop

In this example we cycled through the flow chart three times (until the multiplier was 0) and then finally obtained a product of 18.

Although this algorithm seems complicated, it is a lot simpler than having to include a separate digital circuit for multiplication. Having one circuit able to perform addition, subtraction, and multiplication means that the circuit will be small in size, fast to operate, consume very little power, and be cheap to manufacture.

Division Through Repeated Subtraction

Up to this point, we have seen how we can use the two's complement number system and the adder circuit within an ALU for addition, subtraction, and multiplication. Just as multiplication can be achieved through repeated addition, division can be achieved through repeated subtraction. For example, to divide 31 by 9 you simply see how many times 9 can be subtracted from 31, as follows:

$$
\begin{array}{r}
3 \\
9\overline{)31} \\
27 \\
\hline
4
\end{array}
\begin{array}{l}
\text{:Quotient} \\
\text{:Dividend} \\
\\
\text{:Remainder}
\end{array}
\quad = \quad
\begin{array}{r}
31 \\
-9 \\
\hline
22 \\
-9 \\
\hline
13 \\
-9 \\
\hline
4
\end{array}
\begin{array}{l}
\text{:First Subtraction} \\
\\
\text{:Second Subtraction} \\
\\
\text{:Third Subtraction}
\end{array}
$$

Since we were able to subtract 9 from 31 three times with a remainder of 4, then $31 \div 9 = 3$, remainder 4.

As in multiplication, an algorithm must be followed when a digital system needs to perform a division. Figure 11-14 shows the flow chart for division through repeated subtraction. Let us apply this procedure to our previous example of $31 \div 3$, and see how the quotient is incremented to 3, leaving a remainder of 4. The steps would be as follows:

Start:	Dividend = 31,	Divisor = 9,	Quotient = 0,	Remainder = 31.
After First Pass:	Dividend = 22,	Divisor = 9,	Quotient = 1,	Remainder = 22.
After Second Pass:	Dividend = 13,	Divisor = 9,	Quotient = 2,	Remainder = 13.
After Third Pass:	Dividend = 4,	Divisor = 9,	Quotient = 3,	Remainder = 4.

Stop

When this program of instructions cycles or loops back for the third pass, the divisor of 9 is subtracted from the remaining dividend of 4, resulting in -5. When the question, Is the dividend negative? is asked, the answer will be yes, so the branch on the right is followed. This branch of the flow chart will add the divisor of 9 to the dividend of -5 to restore it to the correct remainder of 4 before the program stops. To summarize, these are the steps followed by this program:

$$
\begin{array}{rl}
(+31) & \\
-(+\ 9) & \\
\hline
(+22) & :\text{Quotient} = 1 \\
-(+\ 9) & \\
\hline
(+13) & :\text{Quotient} = 2 \\
-(+\ 9) & \\
\hline
(+\ 4) & :\text{Quotient} = 3 \\
-(+\ 9) & \\
\hline
(-\ 5) & \\
+(+\ 9) & :\text{Add Divisor to Dividend} \\
\hline
(+\ 4) & :\text{Remainder} = 4
\end{array}
$$

FIGURE 11-14 **Flow Chart for Division through Repeated Subtraction.**

FIGURE 11-15   The Floating-Point Number System.

When a digital system is instructed to perform a division, it will use the ALU for all arithmetic and logic operations. Since division is achieved through repeated subtraction, and subtraction is achieved through addition, the ALU's adder circuit can also be used for division.

Although this algorithm seems complicated, it is a lot simpler than having to include a separate digital circuit for division. As previously mentioned, having one circuit able to perform addition, subtraction, multiplication, and division means that the circuit will be small in size, fast to operate, consume very little power, and be cheap to manufacture.

11-1-4   *Representing Large and Small Numbers*

While two's complementing enables us to represent positive and negative numbers, the range is limited to the number of bits in the word. For example, the range of an 8-bit two's complement word is from -128 to $+127$, while the range of a 16-bit two's complement word is from $-32,768$ to $+32,767$. To extend this range, we can keep increasing the number of bits; however, these large words become difficult to work with. Another problem arises when we wish to represent fractional numbers such as 1.57.

These problems are overcome by using the **floating-point number system,** which uses scientific notation to represent a wider range of values. Figure 11-15 explains the two-byte (16-bit) floating-point number system. The first byte is called the **mantissa,** and it is always an unsigned binary number that is interpreted as a value between 0 and 1. The second byte is called the **exponent,** and it is always a two's complement number and therefore represents a range from 10^{-128} to 10^{+127}. As an example, let us see how the floating-point number system can be used to represent a large-value number such as 27,000,000,000,000 and a small-value number such as 0.0000000015.

Floating-Point Number System

A number system in which a value is represented by a mantissa and an exponent.

Mantissa

The fractional part of a logarithm.

Exponent

A small number placed to the right and above a value to indicate the number of times that symbol is a factor.

EXAMPLE:

Give the two-byte floating-point binary values for 27,000,000,000,000.

Solution:

The value 27,000,000,000,000 would be written in scientific notation as 0.27×10^{14} and when coded as a floating-point number, would appear as follows:

$$27,000,000,000,000$$
$$= \quad\quad\quad 0.27 \times 10^{14}$$

0001 1011		0000 1110

Mantissa	Exponent (two's
(unsigned binary)	complement) =
= 27, which is	$+14$, which is
interpreted as 0.27.	interpreted as 10^{14}.

In this example, the mantissa byte (standard binary number) is equal to 27 (0001 1011) and is therefore interpreted as 0.27, and the exponent byte (two's complement number) is equal to +14 (0000 1110) and is therefore interpreted as 10^{14}.

■ **EXAMPLE:**

Give the two-byte floating-point binary values for 0.0000000015.

■ *Solution:*

The value 0.0000000015 would be written in scientific notation as 0.15×10^{-8} and when coded as a floating-point number, would appear as follows:

$$0.0000000015$$
$$= 0.15 \times 10^{-8}$$

0000 1111	1111 1000
Mantissa	Exponent (two's
(unsigned binary)	complement) =
= 15, which is	−8, which is
interpreted as 0.15.	interpreted as 10^{-8}.

In this example, the mantissa byte (standard binary number) is equal to 15 (0000 1111) and is therefore interpreted as 0.15, and the exponent byte (two's complement number) is equal to −8 (1111 1000) and is therefore interpreted as 10^{-8}.

As with all binary number coding methods, we must know the number representation system being used if the bit pattern is to be accurately deciphered. For instance, the 16-bit word in Figure 11-15 would be very different if it were first interpreted as an unsigned binary number and then compared to a two's complement number and to a floating-point number.

Use the following questions to test your understanding of Section 11-1.

1. Perform the following binary arithmetic operations:
   a. 1011 + 11101 = ?   b. 1010 − 1011 = ?   c. 101 × 10 = ?   d. 10111 ÷ 10 = ?

2. Two's complement the following binary numbers:
   a. 1011   b. 0110 1011   c. 0011   d. 1000 0111

3. What is the decimal equivalent of the following two's complement numbers?
   a. 1011   b. 0110   c. 1111 1011   d. 0000 0110

4. The two's complement number system makes it possible for us to perform addition, subtraction, multiplication, and division all with a/an _____ circuit.

11-2   ARITHMETIC CIRCUITS

In the previous section, we discovered that there is no need to have a separate circuit to add, another circuit to subtract, another circuit to multiply, and yet another circuit to divide. If we use two's complement numbers, these four mathematical operations can be performed with a simple binary-adder circuit. In this section, we will examine the different types of binary-adder circuits and a typical binary-adder IC.

A	B	Sum	Carry	
0	0	0	0	0 + 0 = 0 with no carry.
0	1	1	0	0 + 1 = 1 with no carry.
1	0	1	0	1 + 0 = 1 with no carry.
1	1	0	1	$1 + 1 = 2_{10}$ or 10_2, which is 0 carry 1.

(a)

(b)

FIGURE 11-16 The Half Adder (HA). (a) Truth Table. (b) Logic Circuit.

11-2-1 *Half-Adder Circuit*

Figure 11-16(a) reviews the four rules of binary addition. Studying the sum column of the truth table, you may recognize that the output generated for each input combination is exactly the same as that generated by an XOR gate (odd number of ls at the input gives a 1 at the output). Referring now to the carry column in the truth table, you may recognize that in this case the output generated for each input combination is exactly the same as that generated by an AND gate (any 0 in gives a 0 out). Figure 11-16(b) shows how these two gates could be connected to sense and add an A and B input and generate both a sum and carry output that follows the truth table in Figure 11-16(a).

The circuit in Figure 11-16(b) is called a **half-adder circuit,** and if we ever need to add only two input bits, this circuit would be ideal. In most binary addition operations, however, we will need to add three input bits at a time. To explain this, let us add 11_2 (3_{10}) and 01_2 (1_{10}).

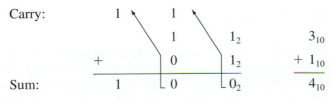

Half-Adder Circuit

A digital circuit that can sum a 2-bit binary input and generate a SUM and CARRY output.

CI	A	B	Sum	Carry	
0	0	0	0	0	Carry Input (CI) of $0 + 0 + 0 = 0$ carry 0.
0	0	1	1	0	$0 + 0 + 1 = 1$ carry 0.
0	1	0	1	0	$0 + 1 + 0 = 1$ carry 0.
0	1	1	0	1	$0 + 1 + 1 = 10_2$ or 2_{10}, which is 0 carry 1.
1	0	0	1	0	$1 + 0 + 0 = 1$ carry 0.
1	0	1	0	1	$1 + 0 + 1 = 10_2$ or 2_{10}, which is 0 carry 1.
1	1	0	0	1	$1 + 1 + 0 = 10_2$ or 2_{10}, which is 0 carry 1.
1	1	1	1	1	$1 + 1 + 1 = 11_2$ or 3_{10}, which is 1 carry 1.

Carry Input (CI) from Previous Column

(a)

(b)

FIGURE 11-17 The Full Adder (FA). (a) Truth Table. (b) Logic Circuit.

In the first column (units column) we only need to add two bits, $1 + 1$, and therefore a half-adder circuit could be used to perform this addition. In the second column, however, we need to add three bits, $1 + 0$ plus the 1 carry from the first column. A different logic circuit that can add a 3-bit input (*A, B* and carry) and then generate a sum and carry output is needed.

Full-Adder Circuit

A digital circuit that can sum a 3-bit binary input and generate a SUM and CARRY output.

11-2-2 *Full-Adder Circuit*

A full-adder circuit is able to add a 3-bit input and generate a sum and carry output. Figure 11-17(a) summarizes the addition operation that must be performed by a full-adder circuit. The full-adder circuit shown in Figure 11-17(b) has the three inputs *A, B,* and carry input (CI) and the two outputs sum and carry output (CO). The full-adder circuit contains two half adders, which in combination will generate the correct sum and carry output logic levels for any of the input combinations listed in Figure 11-17(a).

If we ever need to add only three input bits, the full-adder circuit would be ideal. In most binary addition operations, however, we will need to add several 3-bit columns. To explain this, let us add 0111_2 (7_{10}) and 0110_2 (6_{10}).

In the first column (units) we only need to add 2 bits; therefore, a half-adder circuit could be used to find the sum of this 2-bit column. In the remaining columns, however, we will always have a carry input from the previous column, and therefore a parallel-adder circuit will be needed to sum these 3-bit columns.

11-2-3   *Parallel-Adder Circuit*

Figure 11-18(a) shows a 4-bit **parallel-adder circuit.** The half adder (HA) on the right side will add A_1 and B_1 and produce at its output S_1 (sum 1 output) and C_1 (carry 1 output). All of the remaining adders will have to be full adders (FA) since these columns will always need

Parallel-Adder Circuit
A digital circuit that can simultaneously add the corresponding parts of two binary input words.

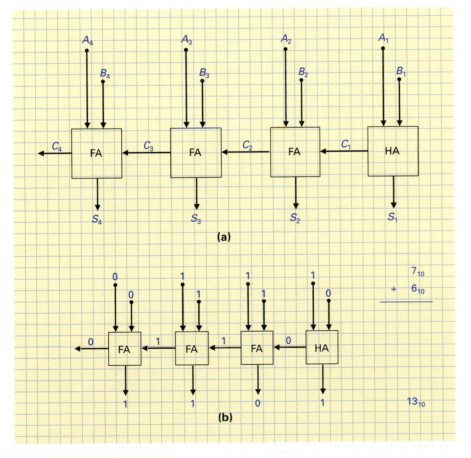

FIGURE 11-18   The Parallel Adder.

FIGURE 11-19 **Application—An 8-Bit Parallel-Adder Circuit.**

to add three input bits: the A input, B input, and the carry input from the previous column. This 4-bit parallel-adder circuit will therefore perform the following addition:

Carry:	C_4	C_3	C_2	C_1
Addend:		A_4	A_3	A_2	A_1
Augend:	+	B_4	B_3	B_2	B_1
Sum:	C_4	S_4	S_3	S_2	S_1

Figure 11-18(b) shows how this 4-bit adder circuit will add an A_4, A_3, A_2, and A_1 input of 0111_2 (7_{10}) and a B_4, B_3, B_2, and B_1 input of 0110_2 (6_{10}). The result will be as follows:

Carry:	0 1 1 0
Addend:	0 1 1 1
Augend:	+ 0 1 1 0
Sum:	0 1 1 0 1

By adding more full adders to the left side of a parallel-adder circuit, you can construct a parallel-adder circuit of any size. For example, an 8-bit parallel-adder circuit would contain one half adder (in the units position) and seven full adders, while a 16-bit parallel-adder circuit would need one half adder and fifteen full adders.

Figure 11-19 shows the 74LS83, which is an example of a commonly used 4-bit parallel-adder IC. The inputs to the 7483 IC are A_4, A_3, A_2, and A_1; B_4, B_3, B_2, and B_1; and an LSB C_0 (pin 13), which is the carry input. The outputs are an MSB carry (C_4) and the sum bits, which are symbolized with the Greek capital letter *sigma* (Σ) and are labeled $\Sigma_4, \Sigma_3, \Sigma_2$, and Σ_1. The sigma symbol is also used to denote an adder IC.

The 7483 IC is called a *fast carry* or *look-ahead carry* IC because of the time delay that occurs as the carry ripples from the LSB column to the second column, from the second to the third, and from the third to the fourth. For example, consider the following addition:

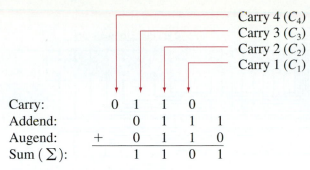

	Carry 4 (C_4)
	Carry 3 (C_3)
	Carry 2 (C_2)
	Carry 1 (C_1)

Carry:	0	1	1	0	
Addend:		0	1	1	1
Augend:	+	0	1	1	0
Sum (Σ):		1	1	0	1

If each full adder within the 7483 had a propagation delay of 30 ns, it would take 30 ns for the first FA to generate C_1, another 30 ns for the second FA to generate C_2, another 30 ns for the third FA to generate C_3, and another 30 ns for the fourth FA to generate C_4. This means that the sum will not be available for 120 ns (4 $\times$ 30 ns) after the 4-bit A input and B input have been applied. This *carry propagation delay,* or *carry ripple delay,* becomes a real problem with 16-bit and 32-bit adder circuits. To overcome this problem, IC logic designers include several logic gates that look at the lower-order bits of the two inputs and generate higher-order carry bits. Most commercially available adder ICs feature full internal look-ahead across all 4-bits, generating the carry in typically 10 ns.

Application—An 8-Bit Parallel Circuit

Figure 11-19 shows how two 74LS83 4-bit adders could be cascaded to form an 8-bit parallel-adder circuit. The carry input of the first 74LS83 is connected LOW, while the carry output of the first 74LS83 is connected to the carry input of the second 74LS83. This circuit will add the two 8-bit A and B inputs and produce an 8-bit sum output. In the following section we will be using this circuit to form a complete two's complement adder/subtractor circuit.

SELF-TEST EVALUATION POINT FOR SECTION 11-2

Use the following questions to test your understanding of Section 11-2.

1. A half adder can be used to add _____ bits, while a full adder can add _____ bits.
2. Which of the following adder circuits could be used to add two binary words?
   a. Half adder   b. Full adder   c. Parallel adder   d. Diamond head adder
3. How would you describe the 7483 IC?
4. The sum outputs of an adder circuit are normally symbolized with the Greek capital letter _____.

11-3   ARITHMETIC CIRCUIT APPLICATIONS

In this section we will be examining some of the applications for arithmetic circuits.

11-3-1   *Basic Two's Complement Adder/Subtractor Circuit*

Figure 11-20 shows how two 7483 ICs could be connected to form an 8-bit two's complement adder/subtractor. In this circuit, there is also an 8-bit input switch circuit, an 8-bit output LED display circuit, and two 8-bit registers (A and B). This circuit operates as follows.

A set of eight input switches and two 74126 three-state buffers (controlled by the active-HIGH input enable, E_I) are used to connect any 8-bit word (set by the switches) onto

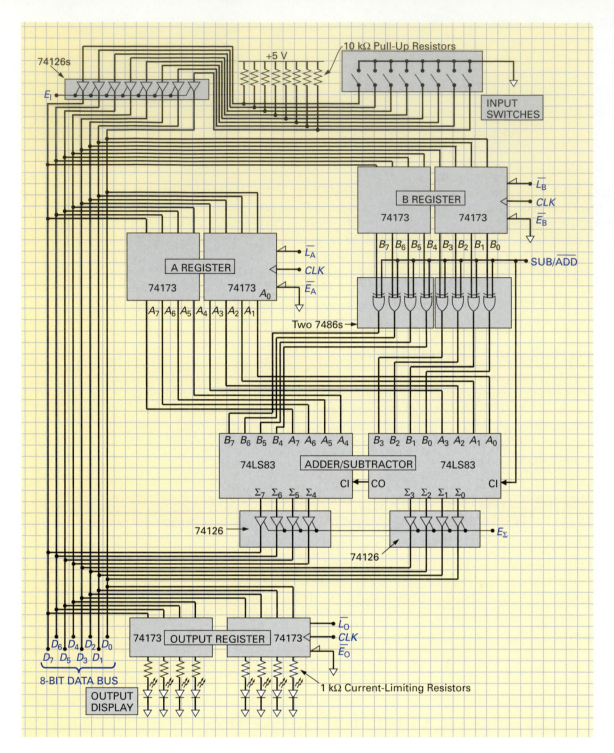

FIGURE 11-20 **Application—An 8-Bit Bus-Organized Two's Complement Adder/Subtractor Circuit.**

the 8-bit data bus (D_0 through D_7). The 8-bit input data can be loaded into the A register (by pulsing LOW, $\overline{L}_A$), the B register (by pulsing LOW, $\overline{L}_B$), or the output register (by pulsing LOW, $\overline{L}_O$).

Referring to the two 74LS83s (which form an 8-bit adder circuit), you can see that the A register supplies the 8-bit A input (A_7 through A_0), while the B register supplies the

8-bit B input (B_7 through B_0). The 8-bit sum output from the two 74LS83s (Σ_7 through Σ_0) can be switched onto the 8-bit data bus by pulsing the sum enable control line (E_Σ) HIGH.

The two 7486s connected to the B input of the adder/subtractor contain eight XOR gates which are included to function as controlled inverters under the control of the SUB/$\overline{ADD}$ input. When the SUB/$\overline{ADD}$ control line is LOW, inputs B_7 through B_0 will pass through the XOR gates without inversion, and therefore the sum output of the 74LS83s will be

$$\Sigma_7 \text{ through } \Sigma_0 = A_7 \text{ through } A_0 + B_7 \text{ through } B_0$$

The LSB carry input (CI) is also connected to the SUB/$\overline{ADD}$ control line, and since this input is LOW in this instance, it will have no effect on the adder circuit's result.

On the other hand, when the SUB/$\overline{ADD}$ control line is HIGH, the XOR gates will act as inverters and invert the output from the B register. The B input to the 74LS83s will therefore be the one's complement of the original word stored in the B register. Since the SUB/$\overline{ADD}$ control line is HIGH in this instance, the LSB carry input will also be HIGH, and therefore a 1 will be added to the LSB of the adder. As a result, when the SUB/$\overline{ADD}$ control line is HIGH, the B input word is first one's complemented and then a 1 is added to the LSB, so the overall effect is that the B input is two's complemented. Since the B input word is two's complemented and added to the A input word, the result will be a two's complement subtraction through addition operation or

$$\Sigma_7 \text{ through } \Sigma_0 = A_7 \text{ through } A_0 - B_7 \text{ through } B_0$$

By using the input switches, the operator is able to load an 8-bit two's complement word into the A register (E_1 = HIGH, $\overline{L}_A$ = LOW) and then load a different 8-bit two's complement word into the B register (E_1 = HIGH, $\overline{L}_B$ = LOW). Then, by controlling the logic level on the SUB/$\overline{ADD}$ control line (HIGH = subtract, LOW = add) you can control whether the sum or the difference appears at the output of the 74LS83s. To display the output of the 74LS83s on the 8-bit LED display, simply enable the 74126 three-state output buffers (E_Σ = HIGH) and then load the result from the data bus into the output register ($\overline{L}_O$ = LOW).

11-3-2 *An Arithmetic-Logic Unit (ALU) IC*

The 74181 IC is a good example of an arithmetic-logic unit (ALU) that can perform sixteen different arithmetic operations and sixteen different logic operations on two 4-bit input words.

Application—Single-Chip Arithmetic-Logic Unit (ALU Circuit)

Figure 11-21 shows some of the details for this IC. Referring to the Function Select Table, you can see that the mode control input M, pin 8, determines whether the ALU performs a logic operation (M = HIGH) or arithmetic operation (M = LOW). The four select inputs (S_3, S_2, S_1, and S_0) are used to select which one of the sixteen different logic functions or arithmetic operations are performed on the A input (A_3, A_2, A_1, or A_0) and B input (B_3, B_2, B_1, or B_0) words. For example, if M = HIGH (logic function) and S_3, S_2, S_1, and S_0 = HHHL ($F = A + B$, $F = A$ OR B), then the output word F (F_3, 210) will be the result of ORing input word A with input word B. As another example, if M = LOW (arithmetic operation), S_3, S_2, S_1, and S_0 = HLLH ($F = A$ plus B) when the active-LOW carry input is HIGH ($\overline{C}_n$ = HIGH, no carry), then the output word F will be the sum of input word A plus input word B. Figure 11-21 shows how a 74181 4-bit ALU IC could be connected so that its sixteen different logic functions and sixteen different arithmetic operations could be tested.

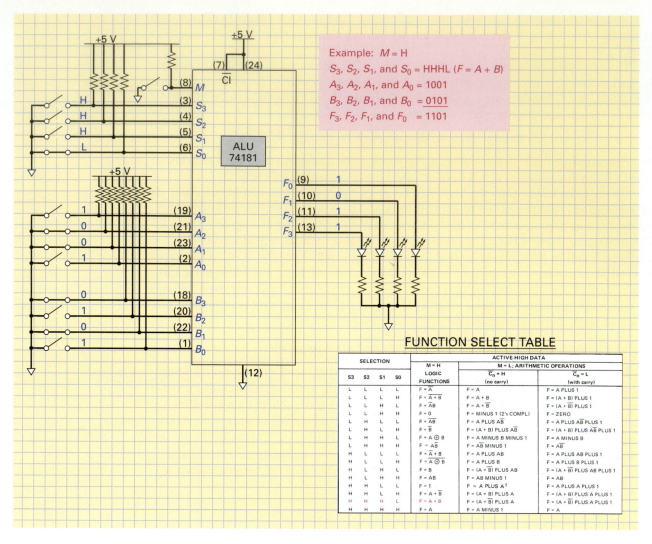

SELECTION				ACTIVE-HIGH DATA		
				M = H	M = L; ARITHMETIC OPERATIONS	
S3	S2	S1	S0	LOGIC FUNCTIONS	$\overline{C}_n$ = H (no carry)	$\overline{C}_n$ = L (with carry)
L	L	L	L	$F = \overline{A}$	F = A	F = A PLUS 1
L	L	L	H	$F = \overline{A+B}$	F = A + B	F = (A + B) PLUS 1
L	L	H	L	$F = \overline{A}B$	F = A + $\overline{B}$	F = (A + $\overline{B}$) PLUS 1
L	L	H	H	F = 0	F = MINUS 1 (2's COMPL)	F = ZERO
L	H	L	L	$F = \overline{AB}$	F = A PLUS A$\overline{B}$	F = A PLUS A$\overline{B}$ PLUS 1
L	H	L	H	$F = \overline{B}$	F = (A + B) PLUS A$\overline{B}$	F = (A + B) PLUS A$\overline{B}$ PLUS 1
L	H	H	L	$F = A \oplus B$	F = A MINUS B MINUS 1	F = A MINUS B
L	H	H	H	$F = A\overline{B}$	F = A$\overline{B}$ MINUS 1	F = A$\overline{B}$
H	L	L	L	$F = \overline{A}+B$	F = A PLUS AB	F = A PLUS AB PLUS 1
H	L	L	H	$F = \overline{A \oplus B}$	F = A PLUS B	F = A PLUS B PLUS 1
H	L	H	L	F = B	F = (A + $\overline{B}$) PLUS AB	F = (A + $\overline{B}$) PLUS AB PLUS 1
H	L	H	H	F = AB	F = AB MINUS 1	F = AB
H	H	L	L	F = 1	F = A PLUS A†	F = A PLUS A PLUS 1
H	H	L	H	$F = A+\overline{B}$	F = (A + B) PLUS A	F = (A + B) PLUS A PLUS 1
H	H	H	L	F = A + B	F = (A + $\overline{B}$) PLUS A	F = (A + $\overline{B}$) PLUS A PLUS 1
H	H	H	H	F = A	F = A MINUS 1	F = A

FIGURE 11-21   **Single-Chip 74181 ALU Test Circuit.**

SELF-TEST EVALUATION POINT FOR SECTION 11-3

Use the following questions to test your understanding of Section 11-3.

1. With the two's complement adder/subtractor circuit in Figure 11-20, which ICs are responsible for one's complementing during a subtraction operation?

2. With the two's complement adder/subtractor circuit in Figure 11-20, how is the additional 1 added so that we can two's complement the *B* input word?

3. What operation would a 74181 IC perform if its control inputs were
   a. M = H, S = LHLH?
   b. M = L, S = LHHL, $\overline{C}_n$ = L?

4. What would appear at the output of a 74181 if A = 0011, B = 0110, M = L, S = HLLH?

Multiple-Choice Questions

1. The sum of 101_2 and 011_2 is
   a. 0111_2.
   b. 8_{10}.
   c. 1001_2.
   d. 6_{10}.

2. Which digital logic unit is used to perform arithmetic operations and logic functions?
   a. MPU
   b. ALU
   c. CPU
   d. ICU

3. With binary addition, $1 + 1 = ?$
   a. 10_2
   b. 0 carry 1
   c. 2_{10}
   d. all of the above

4. With binary subtraction, $0 - 1 = ?$
   a. 10_2
   b. 1 borrow 0
   c. 1 borrow 1
   d. all of the above

5. What would be the result of $1011101_2 \times 01_2$?
   a. 1011101_2
   b. 1110101_2
   c. 000111_2
   d. 1011100_2

6. The four rules of binary multiplication are $0 \times 0 =$ _____, $1 \times 0 =$ _____, $0 \times 1 =$ _____, and $1 \times 1 =$ _____.
   a. 0,1,0,1
   b. 0,0,1,1
   c. 0,0,0,0
   d. 0,0,0,1

7. What would be the result of $1001_2 \div 0011_2$?
   a. 3_{10}
   b. 3, remainder zero
   c. 0011_2
   d. All of the above

8. The one's complement of $0101\ 0001_2$ would be
   a. $0101\ 0010_2$.
   b. $1010\ 1110_2$.
   c. $1010\ 1111_2$.
   d. all of the above.

9. What would be the decimal equivalent of 1001_2 if it were a two's complement number?
   a. $+9$
   b. -9
   c. $+7$
   d. -7

10. Which number system makes it possible for us to add, subtract, multiply, and divide binary numbers all with an adder circuit?
    a. Sign-magnitude
    b. One's complement
    c. Two's complement
    d. Floating-point

11. Which binary code system enables us to represent extremely large numbers or small fractions with fewer bits?
    a. Sign-magnitude
    b. One's complement
    c. Two's complement
    d. Floating-point

12. With a half-adder circuit, the sum output of the 2-bit input is generated by a/an _____ gate, while the carry output of the 2-bit input is generated by a/an _____ gate.
    a. XOR, AND
    b. XNOR, NAND
    c. NOR, NAND
    d. AND, XOR

13. Which circuit is used to add a 2-bit input and a carry from the previous column?
    a. Half-adder
    b. Full-adder
    c. Parallel-adder
    d. All of the above

14. The 7483 IC is a _____ -bit adder circuit.
    a. 2
    b. 6
    c. 4
    d. 3

15. The 74181 IC could be described as a/an _____.
    a. 4-bit FA
    b. 8-bit FA
    c. 4-bit PA
    d. 8-bit PA

Practice Problems

16. Add the following 8-bit numbers:
    a. 0110 1000 and 0001 1111
    b. 0111 1111 and 0100 0011
    c. 0001 1010 and 0011 1010

17. Subtract the following 8-bit numbers:
    a. 0001 1010 from 1010 1010
    b. 0001 1111 from 0011 0000
    c. 0010 0000 from 1000 0000

18. Multiply the following binary numbers:
    a. 0101 1011 by 011
    b. 0001 1111 by 1011
    c. 1110 by 010

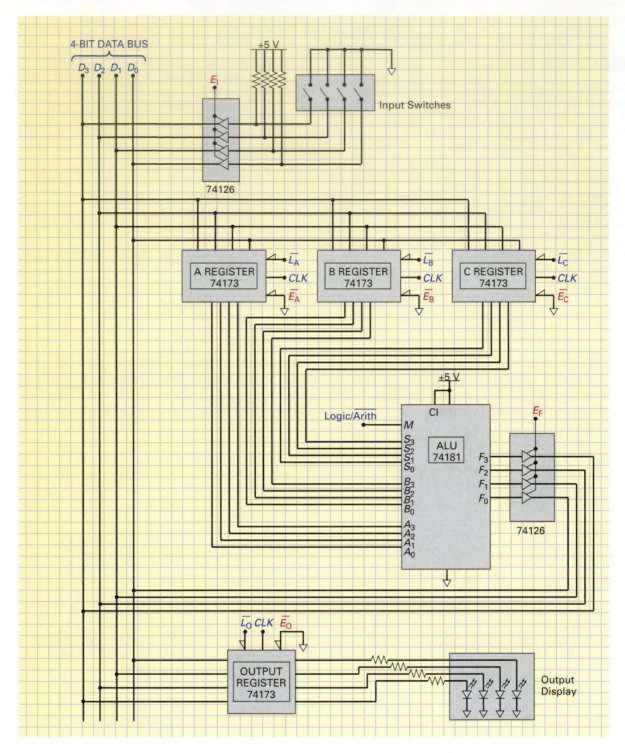

FIGURE 11-22 **Application—A 4-Bit Arithmetic-Logic Circuit.**

19. Divide the following binary numbers:
 a. 0110 into 0111 0110
 b. 101 into 1010 1010
 c. 1001 into 1011 1110

20. Convert each of the following decimal numbers to an 8-bit sign-magnitude number:
 a. +34
 b. −83
 c. +83

21. One's complement the following numbers:
 a. 0110 1001
 b. 1010 1010
 c. 1111 0000

22. Convert all of the following two's complement numbers to their decimal equivalents:
 a. 0001 0000
 b. 1011 1110
 c. 1111 0101

23. Convert each of the following decimal numbers to an 8-bit two's complement number:
    a. +119
    b. −119
    c. −34
24. Two's complement the following numbers:
    a. 0101 1010
    b. 1000 0001
    c. 1111 1101
25. What would be the two's complement range of a 6-bit binary number?

To practice your circuit recognition and operation skills, refer to Figure 11-22 and answer the following five questions.

26. Give a one-line definition of each of the ICs used in this circuit.
27. What is the purpose of the C register in this circuit?
28. Describe the sequence of events that would have to be followed if we wanted to
    a. input a word into the A register.
    b. input a word into the B register.
    c. display the output of the ALU on the LED display.
    d. connect a 4-bit word from the input switches directly to the output display.

29. Describe the sequence of events that would have to be followed if we wanted to
    a. input and add two 4-bit words and display the result.
    b. input and OR two 4-bit words and display the result.
    c. input and AND two 4-bit words and display the result.
30. Describe the sequence of events that would have to be followed if we wanted to input and add two 4-bit words and then store the result in the A register.

Web Site Questions

Go to the web site http://www.prenhall.com/cook, select the textbook *Practical Digital Electronics,* select this chapter, and then follow the instructions when answering the multiple choice practice problems.

Semiconductor Memory

Flying Too High

The term "software"—as opposed to "hardware" or physical components of the computer—was first used in the 1960s to describe the program of instructions that tells a computer what to do. At first, the binary words representing instructions and data were laboriously entered by setting up countless switches and dials. In modern computers, binary coded instructions and data are stored as magnetic signals on disks and then input into the computer's internal memory where the program will tell the computer what to do.

Computer software today is organized into several basic types. There are "translator" programs written to convert high-level computer languages, which are like human speech, into machine language, the ones and zeros understood by the machine or computer. There are also "utility" programs designed to sort and merge data files. Two of the biggest groups of software are "operating systems" and "application programs." An operating system coordinates the operation of the computer system controlling the transfer of data between the computer and its peripheral devices, such as the keyboard, monitor, printer, and disk drive. Application programs are the software that change the personality of a computer from a word processor generating a contract, to an accounting system controlling payroll, to an industrial controller directing a robotics assembly line.

In this vignette, let us examine some of the software-industry folklore relating to operating systems. Most of the 8-bit computers of the late 1970s used an operating system developed by Gary Kildall in 1974 called CP/M, standing for "Control Program for Microcomputers." Just before IBM officially entered into the personal computer market in August of 1981, two IBM executives went to a prearranged meeting with Kildall, who was at the time flying his private plane. After being contacted by radio, Kildall declined to come down and discuss an operating system for their proposed personal computer or "PC." After rescheduling the meeting, the executives were surprised to discover that he had flown off to the Caribbean for a vacation. When he came back, IBM had made a deal with Bill Gates of Microsoft for an operating system. Called MS-DOS, which stood for "Microsoft—Disk Operating System" (DOS means an operating system that is loaded into the computer from a disk), it became an instant success largely due to its blessing from IBM. Hundreds of computer manufacturers started producing computers that were compatible with the IBM PC and therefore used Gates's MS-DOS. Even in an expanding market, Kildall's CP/M could not hold on to any business and is today extinct.

Bill Gates's Microsoft is the largest computer software company in the world, and history is once again repeating itself. Not liking their dependency on Microsoft, IBM has developed their own operating system, which is now in its second version called "OS/2." Another battle begins.

12

Introduction

In Chapter 4 we saw how several flip-flops could be connected to form a register circuit, and how a register could be used to store a binary word. When a register is used to store binary data, it is called a memory register, and in digital systems, memory storage is a very important function.

In a digital electronic computer system, the microprocessor unit (MPU) decodes and executes binary instruction codes and directs the flow of binary data codes in order to perform its function. The instruction codes and the data codes within a digital computer system are stored in either a *primary memory* or a *secondary memory.* These two basic memory categories are shown in Figure 12-1.

The computer's primary memory is made up of two semiconductor memory IC types called ROM (read-only memory) and RAM (random-access memory). The MPU has the binary data that it needs to quickly access data stored in these high-speed ROM and RAM ICs. The ROM and RAM ICs, however, have a high cost per bit of storage, which means that their high speed of operation comes at a high cost. Secondary memory devices, such as floppy disks, hard disks, and optical disks, will store data at a cheaper cost per bit; therefore these devices are used to store the large volume of data that is not presently being used by the MPU.

In this chapter, we will be discussing in detail the semiconductor ROM and RAM primary memory devices.

FIGURE 12-1   Memory Devices.

12-1   SEMICONDUCTOR READ-ONLY MEMORIES (ROMs)

A **read-only memory,** or **ROM,** is an electronic circuit that permanently stores binary data. In this section we will examine the operation, characteristics, applications, and testing of ROM ICs, the different types of which were shown in Figure 12-1.

12-1-1   *A Basic Diode ROM*

Figure 12-2(a) shows how a simple ROM circuit could be constructed using an 8-position rotary switch, a set of diodes, and four pull-up resistors. Each of the eight rows (rows 0 through 7) is equivalent to a 4-bit register, and when selected, the 4-bit binary word stored will be present at the 4-bit output (D_3, D_2, D_1, and D_0).

When the rotary switch is placed in position 0, all three diodes in row 0 will be turned ON, producing a LOW output at D_3, D_2, and D_1. The 4-bit word stored at memory location 0 is therefore 0001.

When the rotary switch is placed in position 1, a different pattern of diodes will be turned ON, producing a LOW output at D_3, D_1, and D_0. The 4-bit output word stored at memory location 1 is therefore 0100.

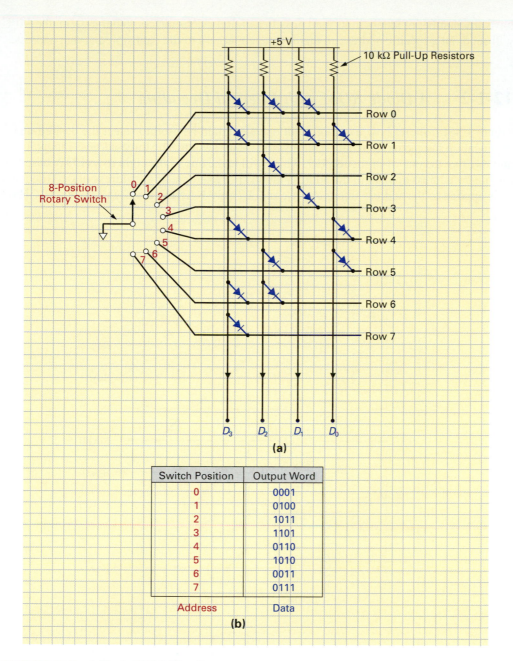

+5 V

10 kΩ Pull-Up Resistors

Row 0

Row 1

Row 2

8-Position
Rotary Switch

Row 3

Row 4

Row 5

Row 6

Row 7

D_3 D_2 D_1 D_0

(a)

Switch Position	Output Word
0	0001
1	0100
2	1011
3	1101
4	0110
5	1010
6	0011
7	0111
Address	Data

(b)

FIGURE 12-2   A Basic ROM Circuit.

Figure 12-2(b) shows all of the 4-bit words stored in each of the eight memory locations. By adding or removing diodes from the matrix, you can change the 4-bit word stored in any of the eight memory locations.

Referring to the bottom of the table in Figure 12-2(b), you can see that the switch position is called the **address,** while the contents of each memory location are called the **data.** Each and every memory location is assigned its own unique address, so we can access the data in any one of the desired memory locations. This method is similar to the postal system, which gains access to us through our unique address. The address of a memory location and the data stored in that memory location are completely unrelated. For example, at address 2 in Figure 12-2, we have a stored data value of 1011 (decimal 11). Similarly, the postal system address and data at that address are also unrelated, in that address 10752 A Street can

Address

A binary code that designates a particular location in either a storage, input, or output device.

Data

A general term used for meaningful information.

have any number of letters in its mail box. In a digital electronic computer system, the MPU will send out a specific address so that it can gain access to the data stored at that address.

12-1-2   *A Diode ROM with Internal Decoding*

Figure 12-3(a) shows how a one-of-eight decoder could be used to select the desired memory location instead of using an eight-position rotary switch. The three address inputs A_2, A_1 and A_0 drive the one-of-eight decoder, which will make one of the eight rows LOW based on the binary address applied to the address inputs. For example, if an address of 010 (decimal 2) is applied to the address inputs A_2, A_1, and A_0, only NAND gate C will have a HIGH on

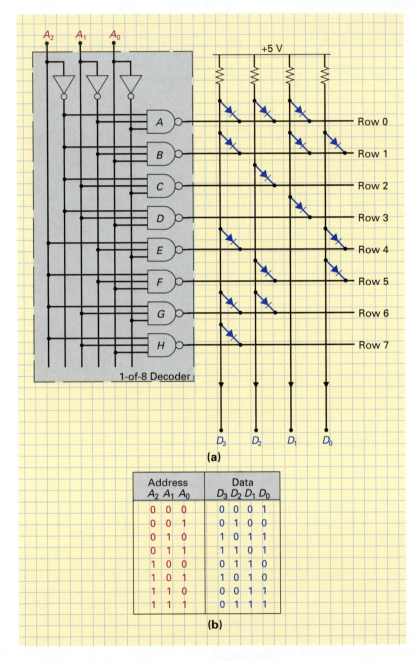

(a)

Address			Data			
A_2	A_1	A_0	D_3	D_2	D_1	D_0
0	0	0	0	0	0	1
0	0	1	0	1	0	0
0	1	0	1	0	1	1
0	1	1	1	1	0	1
1	0	0	0	1	1	0
1	0	1	1	0	1	0
1	1	0	0	0	1	1
1	1	1	0	1	1	1

(b)

FIGURE 12-3   A Basic ROM Circuit with Internal Decoding.

all three of its inputs. This will generate a LOW out on row 2, producing a data output of 1011 at D_3, D_2, D_1, and D_0, respectively. The data stored at each address is listed in the table in Figure 12-3(b).

12-1-3   *Semiconductor ROM Characteristics*

In this section we will examine characteristics that are common to all semiconductor ROMs. These include number of address lines, memory size and organization, read-only operation, volatility, access time, and input/output timing.

Number of Address Lines

Although the ROM circuit in Figure 12-3(a) is more complex than the ROM circuit in Figure 12-2(a), the decoder enables the circuit to be electronically accessed using three address lines instead of manually accessed by eight switched input lines. When an on-chip decoder is included with the ROM memory circuit, the number of address lines (n) applied to the IC will indicate how many memory locations are available within the IC (2^n). For example, the three address lines applied to the ROM circuit in Figure 12-3(a) could address one of eight memory locations ($2^n = 2^3 = 8$ memory locations).

■ **EXAMPLE:**

Calculate the number of memory locations that can be accessed by a ROM memory IC, if it has

   a. four address lines applied

   b. eight address lines applied

   c. sixteen address lines applied

■ *Solution:*

   a. Four address lines can address $2^4 = 16$ memory locations.

   b. Eight address lines can address $2^8 = 256$ memory locations.

   c. Sixteen address lines can address $2^{16} = 65,536$ memory locations.

As you can see from these examples, there is a direct relationship between the number of address lines applied to a memory circuit and the number of memory locations available within that memory circuit. The left and center columns in Table 12-1 show the correlation between the two. The right column in Table 12-1 shows the abbreviation that is used to represent these large values. The label "k" indicates 1000 in decimal; however, here we are dealing with binary (base-2), where K indicates 1,024. A 2K memory, therefore, has 2,048 memory locations ($2 \times 1,024$), and a 1M (one meg) memory has 1,048,576 memory locations ($1,024 \times 1,024$).

Memory Size and Memory Configuration

The term **memory size** describes the number of bits of data stored in the memory. For example, the basic ROM in Figure 12-3 is a 32-bit memory since it can store thirty-two binary digits of data. The term **memory configuration** describes how the memory circuit is organized into groups of bits or words. For example, the thirty-two storage bits in the ROM circuit in Figure 12-3 are organized into eight 4-bit words (8 by 4, 8 × 4).

Memory Size
A term used to describe the total number of binary digits stored in a memory.

Memory Configuration
A term used to describe the organization of the storage bits within a memory.

TABLE 12-1   The Relationship Between Address Bits and Memory Locations

NUMBER OF ADDRESS BITS (n)	EXACT NUMBER OF MEMORY LOCATIONS (2^n)	ABBREVIATED NUMBER OF MEMORY LOCATIONS
8	256	256
9	512	512
10	1,024	1K
11	2,048	2K
12	4,096	4K
13	8,192	8K
14	16,384	16K
15	32,768	32K
16	65,536	64K
17	131,072	128K
18	262,144	256K
19	524,288	512K
20	1,048,576	1M
21	2,097,152	2M
22	3,194,304	4M
23	8,388,608	8M
24	16,777,216	16M
25	33,554,432	32M
26	67,108,864	64M

Storage Density

A relative term used to compare the memory storage ability of one device to another.

Stored

A term used to describe when data is placed in a memory location or output device.

Memory Write Operation

To store information into a location in memory or an output device.

Retrieved

A term used to describe when data is extracted from a memory location or input device.

Memory Read Operation

To extract information from a location in memory or an input device.

Programming

A process of preparing a sequence of operating instructions for a computer system and then loading these instructions in the computer's memory.

The term **storage density** is used to compare one memory to another. For example, if memory A IC can store thirty-two bits and memory B IC can store sixty-four bits, then memory B is said to contain more storage bits per chip.

EXAMPLE:

The 8355 ROM IC shown in Figure 12-4 has eleven address inputs and eight data outputs. What is the size and configuration of this read-only memory?

Solution:

The eleven address line inputs to this IC indicate that there are 2^{11}, or 2,048, different memory locations. The eight data lines to this IC indicate that an 8-bit word is stored at each of the memory locations. Therefore, the 8355 ROM IC has a memory size of 16,384 bits (2,048 $\times$ 8) and is configured as 2,048 by 8.

The Read-Only Operation of a ROM

When a binary word is **stored** in a memory location, the process is called a **memory-write operation.** The term *write* is used, since writing a word into memory is similar to writing a word onto a page.

When a binary word is **retrieved** from a memory location, the process is called a **memory-read operation.** The term *read* is used, since reading a word out of memory is similar to reading a word from a page.

With a ROM, the binary information contained in each memory location is stored or written into the ROM only once. This process of entering data into the ROM is called **programming** or *burning in* the ROM. Once a ROM IC has been programmed, it is installed in a circuit within a digital system, and from that point on the MPU only reads or *fetches* data out of the ROM. Once it has been installed in a circuit, therefore, the ROM will only ever operate as a read-only memory.

FIGURE 12-4 A Read-Only Memory IC.

Bipolar and MOS ROMs

Figure 12-5(a) shows the basic internal structure of a ROM IC. This **architecture** is essentially the same, irrespective of whether the device is a bipolar ROM or MOS ROM.

As far as operation, the *chip enable* $(\overline{CE})$ or *chip select* $(\overline{CS})$ is normally an active-LOW control input that either enables $(\overline{CE} = \text{LOW})$ or disables $(\overline{CE} = \text{HIGH})$ the ROM. It achieves this by disabling the address decoder's AND gates and the three-state output buffers.

When enabled, the one-of-eight decoder selects one of the eight rows depending on the address applied to the three address inputs A_2, A_1, and A_0, $(2^3 = 8)$. The decoder will give a HIGH output at only one of its AND gate outputs, and this HIGH will be applied to four blocks in the ROM matrix. Each of these blocks contains a transistor and a *programmable link,* as shown in the inset in Figure 12-5(a). It is the presence or absence of the programmable link that determines whether a 1 or 0 is stored in each of these blocks. If this were a bipolar ROM, the programmable link would connect the row line from the decoder to the base of an NPN bipolar transistor. If this were a MOS ROM, the programmable link would connect the row line from the decoder to the gate of a field-effect transistor. In both cases, the presence of a link will allow the HIGH on the row line from the decoder to turn ON the transistor and connect a HIGH onto the column line and to the data output line below. On the other hand, the absence of a link will keep the transistor OFF and therefore maintain the column line and the data output line below as LOW.

The logic symbol for this 32-bit ROM, which is organized as an 8 by 4 (eight 4-bit words), is shown in Figure 12-5(b).

Most semiconductor ROM ICs use MOSFETs because their relatively small storage cell size makes it possible for a greater number of storage bits per IC. MOS ROMs also cost less per bit and consume less power than bipolar ROMs. Bipolar TTL and ECL ROM ICs are available; however, their small storage capacity, high cost per bit, and high power consumption make them undesirable except for high-speed applications.

Access Time and Input/Output ROM Timing

The **access time** of a memory is the time interval between the memory receiving a new address input and the data stored at that address being available at the data outputs. To explain this in more detail, Figure 12-5(c) shows the typical timing waveforms for a ROM read operation. The upper waveform represents the address inputs, the center waveform represents the active-LOW chip-enable $(\overline{CE})$ input, and the lower waveform represents the data outputs. Since the address and data waveforms represent more than one line, they are shown as a block on the timing waveform since some of the lines are HIGH and some are LOW. Whenever these blocks criss-cross, as can be seen with the address waveform at time t_1, a change in logic levels has occurred on the lines.

Access Time

The time interval between a device being addressed and the data at that address being available at the device's output.

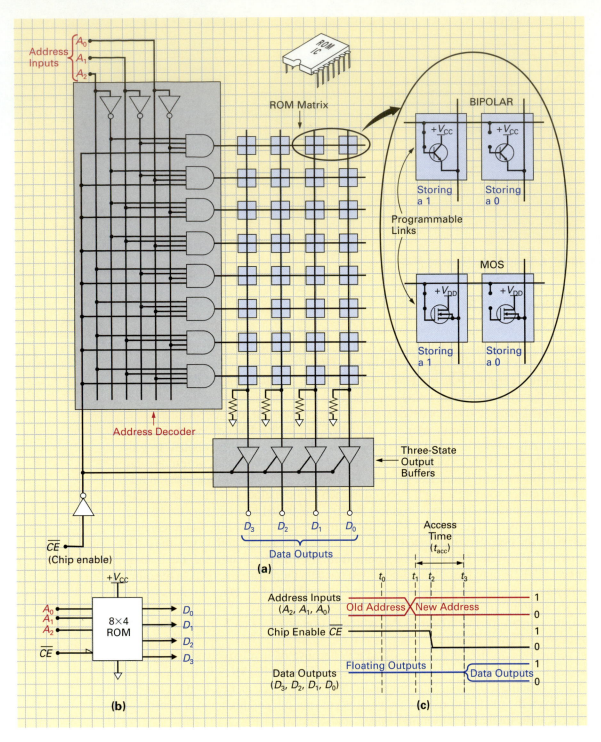

FIGURE 12-5 Typical ROM IC. (a) Basic Architecture. (b) Logic Symbol. (c) Timing Waveform.

Referring to the timing points shown in Figure 12-5(c), you can see that at time t_0, the chip-enable input is HIGH and therefore the ROM IC is disabled. At time t_1, a new address is placed on the 3-bit address bus (A_2, A_1, and A_0), and the ROM's internal address decoder will begin to decode the new address input to determine which memory location is to be enabled. At time t_2, the $\overline{CE}$ input is made active, enabling the final stage of the address decoder and the output three-state buffers. At time t_3, the data outputs (D_3, D_2, D_1, and D_0) change from a floating output state to a logic level that reflects the data stored in the addressed memory location.

Since the access time of a memory is the time interval between the memory receiving a new address input and the data being available at the data output, the access time (t_{acc}) is the time interval between t_1 and t_3 in Figure 12-5(c). Typically, bipolar ROMs have access times in the 10 ns to 50 ns range, while MOS ROMs will have access times from 35 ns to 500 ns.

12-1-4   *ROM Types*

There are three basic types of ROM—mask programmable ROMs (MROMs), programmable ROMs (PROMs), and erasable programmable ROMs (EPROMs, EEPROMs, and flash memories). Let us now examine each of these types and see how they differ.

Mask Programmable ROMs (MROMs)

Mask programmable ROMs, or **MROMs,** are permanently programmed by the manufacturer by simply adding or leaving out diodes or transistors, as previously shown. A customer sends the manufacturer a truth table stating what data should be stored at each address, and then the manufacturer generates a photographic negative called a *mask* that is used to produce the interconnections on the ROM matrix. These masks are normally expensive to develop and therefore this type of ROM only becomes economical if a large number of ROMs are required.

The disadvantage of these mask programmed ROMs is that they cannot be reprogrammed with any design modifications, since programming is permanent.

As an example, the TMS47256 is available from Texas Instruments as an NMOS version (TMS47256) with an access time of 200 ns and a standby power dissipation of 82.5 mW, or as a CMOS version (TMS47C256) with an access time of 150 ns and a standby power dissipation of 2.8 mW.

Programmable ROMS (PROMS)

Although the MROM is extremely expensive, the cost per IC can become minimal when a large volume of ROMs is needed (thousands). On the other hand, if only a small number of ROMs is needed, one option is to use a fusible-link **programmable ROM, or PROM.** Unlike the MROM whose data is etched in stone, so to speak, by the manufacturer, the PROM comes with a clean slate and can be programmed by the user. Like the MROM, however, once the PROM has been programmed, the etched data cannot be erased.

Figure 12-6(a) shows how the architecture of a PROM is similar to that of the MROM shown in Figure 12-5(a). The difference with the PROM, however, is that each of its programmable links is fusible. All of these fusible links are initially intact when the PROM arrives from the manufacturer, and it is up to the user to either blow a fuse link (to store a 0) or leave it intact (to store a 1).

The programming or burning-in process is performed by a *PROM programmer unit,* similar to the one shown in Figure 12-6(b). By placing the PROM into the IC socket on the unit and entering the desired truth table using the keyboard, the PROM programmer unit will then automatically step through each address and blow or leave the fuse intact in each memory location. Once the data has been written into every memory location, the programmer can read back the stored data to verify that it matches the desired truth table.

Both bipolar (low-density, high-speed, high power dissipation) and MOS (high-density, low-speed, low power dissipation) PROMs are available. For example, the 74186 is a 512-bit bipolar PROM, which is organized as sixty-four words of 8 bits each, has an access time of 50 ns. As another example, the TMS27PC256 is a 256K-bit CMOS PROM which is organized as 32K words of eight bits each and has an access time of 120 ns.

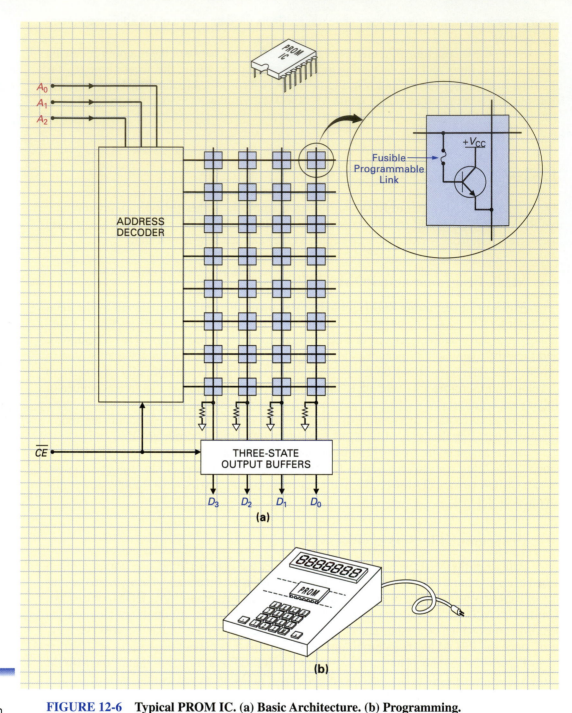

FIGURE 12-6 Typical PROM IC. (a) Basic Architecture. (b) Programming.

Erasable Programmable ROMs (EPROMs, EEPROMs, and Flash)

The ultraviolet-light **erasable PROM (EPROM),** the **electrically erasable PROM (EEPROM),** and *flash memories* are MOS circuits that enable the user to store data in the same way as the standard PROM, using the PROM programmer unit. The important difference with these devices is that if a modification to the stored data needs to be made at a later time, the contents of the IC can be erased and then reprogrammed with the new data.

The EPROM. With the ultraviolet erasable PROM, or EPROM, the data is erased by shining a high-intensity ultraviolet light through a quartz window on the top of the IC for about twenty minutes. Data in this IC type is stored as a charge or no charge, and the ultraviolet light causes these charges to leak off. The disadvantage with the EPROM is that the light erases the entire contents of the IC, and therefore if you only wish to change one byte or one bit, you will still have to erase everything and then reprogram the IC with the entire truth table.

The EEPROM. The electrically erasable PROM, or EEPROM, uses an electrical pulse to program data (like the EPROM); however, unlike the EPROM, it also makes use of an electrical pulse for erasing the stored data. This means that an ultraviolet light source is not needed for this type of PROM, since programming and erasing can be performed without removing the IC package from the circuit.

When programming an EEPROM, a 21 V pulse of typically 10 ms is needed to store a byte (an EPROM requires a 21 V pulse of typically 50 ms). When many thousands of bytes are involved, this makes the EEPROM programming and verifying process a lot less time-consuming. The key advantage of the EEPROM, however, is its ability to erase only a single byte of stored data, or to erase the entire device with a single 21 V/10 ms pulse. This means that a single byte modification can be performed without having to erase the entire device and then completely reprogram, as was the case with the EPROM. The erase time of 10 ms for the EEPROM is also a lot faster than that of the ultraviolet EPROM, which requires twenty minutes of UV exposure to erase its contents.

The EPROM is normally cheaper than the EEPROM, and both are generally used to develop a product. Once the design is finalized, however, a mask ROM, or MPROM, will be ordered in bulk for mass production.

The flash memory. To compare, EPROMs have fast access times, a high storage density, and a low cost per bit of storage. The EPROM's disadvantage, however, is that it has to be removed from its circuit to be erased, which takes twenty minutes. On the other hand, EEPROMs have fast access times and can be erased and reprogrammed without being removed from the circuit. The EEPROM's disadvantage, however, is that it has a low storage density and a much higher cost per bit of storage.

A **flash memory** has the advantages of both the EPROM and the EEPROM by combining the in-circuit erasability of the EEPROM with the high storage density and low cost of the EPROM. It is called a flash memory because of its high-speed erase and write data times. This new technology is rapidly improving, and at present the cost of a flash memory is much less than an EEPROM but not as low as an EPROM. As far as write time is concerned, a typical flash memory will have a write time of 10 μs per byte, compared to 100 μs per byte for a fast EPROM, and 5 ms for an EEPROM.

Flash Memory

A reprogrammable ROM device that has extremely fast erase-data and write-data speeds.

12-1-5 *ROM Applications*

As you study all of the ROM applications discussed in this section, you will notice that all the ROM is really doing is converting a coded input word at the address inputs into a desired output word at the data outputs.

Code Conversion

Figure 12-7 shows how a ROM could be used as a code converter. The input code is applied to the address input pins of the ROM. This code will be decoded by the ROM's internal decoder, which will select one of its internal memory locations and then apply the data content or desired output code to the data output pins.

Some standard ROM code converters are available. For example, there is a ROM code converter to convert the 7-bit alphanumeric ASCII (American Standard Code for Information Interchange) code into another code such as the 7-bit alphanumeric BCDIC (Binary Coded Decimal Interchange Code) code, which is used by some computers instead of ASCII. There is also a ROM code converter available that will achieve the opposite by converting BCDIC codes to ASCII codes.

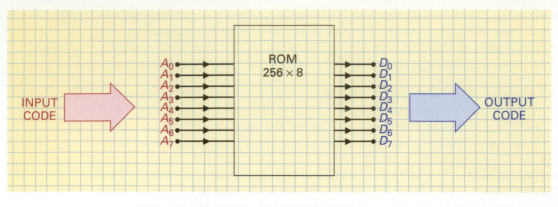

FIGURE 12-7   **The ROM as a Code Converter.**

Arithmetic Operations

As we discovered in the previous chapter, a simple adder circuit and the two's complement number system can be used to perform addition, subtraction, multiplication, and division. More complex mathematical functions, such as trigonometric and logarithmic operations, would require a more complex circuit and algorithm. In many of these instances, a ROM is used to implement the arithmetic operation. To use a simple example, Figure 12-8 shows how a ROM could be used to multiply two 2-bit words. In this circuit the 2-bit multiplicand and 2-bit multiplier are applied to the ROM's 4-bit address input. Since the correct product for every possible input combination is stored in the ROM, the correct result will always be present at the data outputs.

Look-Up Tables

A look-up table is not that different from a code converter. For example, the MM4232 ROM will convert a 9-bit binary input representing an angle between 0° and 90° to an 8-bit binary output that is equivalent to the sine of the input angle.

Logic Circuit Replacement

A ROM can also be used to replace a logic circuit. For example, an eight-input/four-output logic circuit could be replaced by an 8-bit address input/4-bit data output ROM, which could be programmed to give the desired output for any of the 8-bit input combinations.

A_3 A_2	A_1 A_0	D_3 D_2 D_1 D_0	
0 0	×	0 0	=	0 0 0 0	$0 \times 0 = 0$
0 0	×	0 1	=	0 0 0 0	$0 \times 1 = 0$
0 0	×	1 0	=	0 0 0 0	$0 \times 2 = 0$
0 0	×	1 1	=	0 0 0 0	$0 \times 3 = 0$
0 1	×	0 0	=	0 0 0 0	$1 \times 0 = 0$
0 1	×	0 1	=	0 0 0 1	$1 \times 1 = 1$
0 1	×	1 0	=	0 0 1 0	$1 \times 2 = 2$
0 1	×	1 1	=	0 0 1 1	$1 \times 3 = 3$
1 0	×	0 0	=	0 0 0 0	$2 \times 0 = 0$
1 0	×	0 1	=	0 0 1 0	$2 \times 1 = 2$
1 0	×	1 0	=	0 1 0 0	$2 \times 2 = 4$
1 0	×	1 1	=	0 1 1 0	$2 \times 3 = 6$
1 1	×	0 0	=	0 0 0 0	$3 \times 0 = 0$
1 1	×	0 1	=	0 0 1 1	$3 \times 1 = 3$
1 1	×	1 0	=	0 1 1 0	$3 \times 2 = 6$
1 1	×	1 1	=	1 0 0 1	$3 \times 3 = 9$

FIGURE 12-8   **The ROM as a Multiplier.**

Character Generator

Another typical application for ROMs is as a character generator. For example, the Motorola MC6571 ROM is programmed with all of the 7 by 9 dot matrix codes used to display a number, letter, or symbol on a video display monitor screen.

Keyboard Encoder

Keyboard encoders are also not that different from code converters. For example, the MM5740 ROM is used within some computer keyboards to convert the keyboard switch matrix code generated when a key is pressed to a corresponding ASCII code that can be recognized by the computer.

Nonvolatile Computer Memory

The term **nonvolatile** describes a device that will retain its information in the absence of power. ROMs are mainly used in computer systems to store instruction codes and data codes that must not be lost when the system power is turned off. Several ROMs are often used to store information, such as system start-up instructions, keyboard encoding codes, character generator codes, and look-up tables.

Nonvolatile

A memory storage device that will retain its stored data in the absence of power.

12-1-6 *ROM Testing*

A **ROM listing** sheet lists the data values that should be stored in every memory location of a ROM. A ROM can be tested, therefore, by simply accessing or addressing each memory location, and then checking that the stored data in that memory location matches what is listed in the ROM listing.

Figure 12-9 shows a simple ROM test circuit. In this circuit, a push-button switch is used to clock a counter, causing it to advance its count. The counter's output is used to supply a

ROM Listing

A table showing the binary codes stored in every memory location of a ROM.

FIGURE 12-9 **Simple ROM Test Circuit.**

sequential address to the ROM being tested. The data output of the ROM is displayed on a seven-segment hexadecimal display. This circuit, therefore, will allow us to step through each address and check the stored data value against the value shown in the ROM listing. This check will verify the operation and contents of the ROM.

The test circuit in Figure 12-9 would be ideal for testing a small 16 by 8 ROM; however, most digital system ROMs will be much larger, having thousands of memory locations, making a complete test with this circuit long and laborious. In these cases, a ROM tester can be used to check the memory's contents. This test instrument automatically reads the contents of the ROM being tested and compares it to the stored data within a known-good reference ROM to check for any false operation or storage errors.

Most ROMs within digital electronic computer systems are tested by a built-in **self-diagnostic program.** This test program will generally use the **check-sum method** to verify the ROM's contents. With the check-sum method, the computer reads out and adds all of the data stored in every one of the ROM's memory locations to obtain a sum (ignoring carries). The resultant sum is compared with a sum value that was stored in a specific address in the ROM when it was programmed. If a difference exists between the present check-sum and the previously stored check-sum, then there is definitely a fault with the programmed data.

SELF-TEST EVALUATION POINT FOR SECTION 12-1

Use the following questions to test your understanding of Section 12-1.

1. Is a ROM a volatile or nonvolatile device?
2. What is a memory's address input, and how is the number of address input lines related to the number of memory locations?
3. Define the following terms:
   a. storage density   **b.** memory configuration   **c.** burning in   **d.** access time
4. Give the full names for the following abbreviations:
   a. MROM   **b.** PROM   **c.** EPROM   **d.** EEPROM
5. How is a ROM used as a code converter?
6. What is a ROM listing, and how can it be used to test a ROM?

12-2   SEMICONDUCTOR READ/WRITE MEMORIES (RWMs)

As its name implies, the previously discussed read-only memory, or ROM, can only read out or retrieve data from its memory locations once it is installed within a digital system. Data, such as ASCII coded text, numerical codes, or microprocessor instruction codes, are permanently stored in the ROM's memory locations before the device is installed in a digital circuit.

Read/write memories (RWMs) can perform either a read or a write operation after they are installed in a digital circuit. Unlike the ROM, which contains permanent storage locations, a RWM acts as a temporary storage location for binary data. When data is stored in a RWM, the data is said to be written into the memory, whereas when data is retrieved from an RWM, it is said that data is being read from memory. In this section, we will be concentrating on the semiconductor read/write memories shown previously in Figure 12-1. To begin with, let us discuss why read/write memories (RWMs) are more commonly referred to as **random-access memories (RAMs).**

12-2-1   *SAMs Versus RAMs*

The shift register discussed in Chapter 10 is an example of a **sequential-access memory (SAM),** or *serial memory.* To explain this, Figure 12-10 shows a 512-bit shift register which

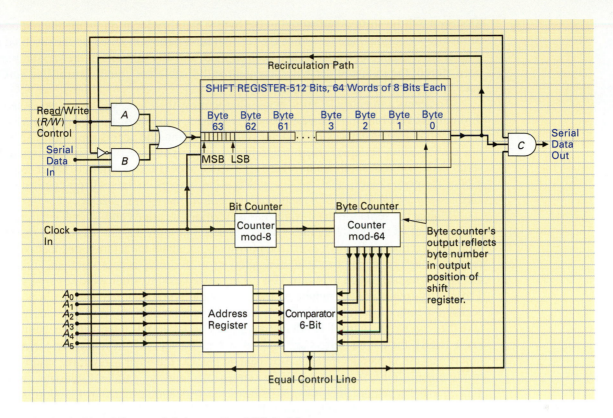

FIGURE 12-10 **A 64 × 8 Sequential-Access Read/Write Memory.**

can store sixty-four bytes (a 64 by 8 SAM). These sixty-four bytes of data are numbered byte 0 through byte 63 and are stored serially in 512 adjacent shift register storage elements. To store an 8-bit word in this serial memory, you must first apply a 6-bit address to specify the storage location for the data byte. This 6-bit address is loaded into the address register and then applied to the left input of the comparator.

For every clock pulse at the input, the data within the shift register is shifted one position to the right. The clock input is also applied to a mod-8 counter, which clocks the byte counter each time a complete word or eight bits have been shifted right. The byte counter, therefore, keeps track of the number of bytes shifted and which byte is presently being applied to the output.

To access memory location 9, for example, the address 001001 (decimal 9) is loaded into the address register and applied to the comparator. To read or retrieve the contents from memory location 9, the read/write control line is made HIGH, enabling AND gates A and C. The data bytes stored in the shift register are shifted right and recirculated back to the input via AND gate A until byte 9 is in the output position of the shift register. Since the byte counter's contents always reflect the number of the byte in the output position, the byte counter will also contain 9, and since both inputs to the comparator are equal, its output will go active or HIGH. This signal will enable AND gate C, allowing the 8-bit data contents in memory location 9 to be serially shifted out of the memory. After eight clock pulses have occurred and the entire contents of memory location 9 have been applied to the output, the byte counter will increment and the comparator's output will go inactive, completing the read-from-memory-9 operation.

To write or store an 8-bit data value into memory location 9, for example, we must once again load 9 into the address register and then shift and recirculate the contents of the shift register until byte 9 is in the output position. At this time, the byte counter will contain 9, the equal control line from the comparator will go HIGH, and this signal will be applied to AND gates B and C. Since the read/write control line is LOW in this instance, AND gates A and C will be disabled while AND gate B will be enabled. The 8-bit serial word to be stored will be shifted and stored in the byte 9 memory location, while the old contents of memory location 9 are lost since the recirculation path is blocked by the disabled AND gate *A*. After eight clock pulses have been applied, and all eight bits of the serial input have been stored, the byte counter will increment and disable the equal control line and AND gate *B,* completing the write operation.

To access a memory location in a serial memory, therefore, you must step or sequence through all of the bits until the desired word is located. This is why this memory type is called a sequential-access memory (SAM). As you can imagine, the access times for SAMs memory are very slow, and this is their major disadvantage. They are, however, still used today in serial data, low-cost applications where high-speed access is not required. Other sequential-access memory devices that are not widely used, except in specialized applications, are *charge-coupled devices (CCDs)* and *magnetic bubble memories (MBMs).*

Unlike the serial memory, which has its memory locations end-to-end, a *parallel memory* has all of its memory locations stacked, as seen in Figure 12-11. This means that there is no need to sequence through every memory location, since the address decoder will enable only one memory location and therefore allow us to randomly access any byte. For example, the 512-bit register block in Figure 12-11 is organized as sixty-four words of eight bits each. These sixty-four bytes are numbered byte 0 through byte 63 and are stacked in parallel. To access a memory location in this RAM, first apply a 6-bit address word to the decoder, then take the read/write control line HIGH (read) or LOW (write). If the read/write control line is made HIGH, the output buffers are enabled and the contents of the addressed memory location is read out in parallel onto the 8-bit data output. On the other hand, if the read/write control line is made LOW, the input buffers are enabled and the parallel 8-bit data word being applied to the data input lines is stored in the memory location enabled by the address decoder.

The RAM is more widely used than the SAM, because its parallel format results in faster memory access times. Before we leave this topic of random access, it is important to point out that ROMs are random access in nature, even though the term RAM is now only associated with read/write memories.

FIGURE 12-11   **A 64 × 8 Random-Access Read/Write Memory.**

12-2-2 RAM Types

There are two basic types of RAM—the static RAM (SRAM) and the dynamic RAM (DRAM). To begin with, the SRAM and DRAM are both **volatile,** which means that they will lose their data if power is interrupted. As far as storing data is concerned, both of these RAM types use a different technique. Static RAMs, for instance, latch bits of data in high-speed bistable flip-flops, whereas dynamic RAMs store each bit of data as a charge level in a capacitor. The DRAM capacitor circuit is smaller than the SRAM bistable flip-flop circuit; however, the DRAM needs additional refresh circuitry to replenish the charge lost from the capacitors due to capacitor leakage. Each RAM type, therefore, has its own advantage that makes it more suitable for certain applications. For example, the DRAM has a higher storage density but requires refresh circuitry, while the SRAM operates at a faster speed but is larger in size. Let us now examine these two types in more detail, beginning with the static RAM.

Static Random-Access Memories (SRAMs)

There are basically two types of **static RAMs** available. *Bipolar SRAMs* use TTL devices to construct bistable flip-flop memory circuits, which have a low storage density (small amount of storage per chip) but operate at a high speed. *MOS SRAMs* use either MOS or CMOS technology along with a capacitor to construct bistable flip-flop storage cells that have a high storage density but operate at a slower speed.

SRAM storage cell. Figure 12-12(a) shows the basic storage cell for a TTL static RAM. This storage cell, which could also be constructed using N-channel E-type MOSFETs, is capable of storing one single bit of data.

Write operation: To latch a binary 1 or binary 0 into the bistable flip-flop, the word-select control line must be pulsed HIGH to turn ON Q_3 and Q_4. If a HIGH is to be stored, it will appear on the data input line (Data Input = 1, $\overline{\text{Data Input}}$ = 0), and this positive voltage will be applied through Q_3 to the base of Q_2. This HIGH will force Q_2 to conduct and its dropping collector voltage will be applied to the base of Q_1, turning Q_1 OFF. With Q_1 OFF, its collector voltage will go HIGH, maintaining Q_2 ON. When the word-select control line goes inactive or LOW, the storage cell will remain in the same condition (Q_2 ON, Q_1 OFF) with a binary 1 latched in the flip-flop due to the cross coupling between Q_2 and Q_1. On the other hand, if we wanted to store or write a binary 0 into this storage cell, the data input line is made LOW (Data Input = 0, $\overline{\text{Data Input}}$ = 1), and then the word-select control line will again be pulsed HIGH.

When in the read mode, the data inputs are both floating and therefore no data is being applied to the flip-flop. To read or retrieve data out of the storage cell, the word-select line is pulsed HIGH in exactly the same way as in the write operation, turning ON Q_3 and Q_4 and switching the stored logic level to the output. For example, if the flip-flop has a logic 1 stored (Q_1 OFF, Q_2 ON), the left input of the sense amplifier (a comparator) will receive a HIGH voltage, while the right input of the sense amplifier will receive a LOW voltage. In this instance, the sense amplifier will produce a HIGH data output. On the other hand, if the flip-flop has a logic 0 stored (Q_1 ON, Q_2 OFF), the left input of the sense amplifier will receive a LOW voltage, while the right input of the sense amplifier will receive a HIGH voltage. In this instance, the sense amplifier will produce a LOW data output.

Figure 12-12(b) shows how sixty-four storage cells could be arranged to form a 16 by 4 (sixteen 4-bit words) static RAM. In this illustration, each of the squares represents one of the storage cells shown in Figure 12-12(a). For simplicity, only three of the sixteen 4-bit sets have been shown in Figure 12-12(b). Each of the data inputs is inverted to provide a *Data Input* and a *Data Input*. A sense amplifier is required for each of the four data output lines, so four sense comparator amplifiers are included at the bottom of the storage array.

SRAM storage array. Figure 12-12(c) shows all of the basic blocks in a 16 by 4 static RAM. The four-line to sixteen-line address decoder will decode the 4-bit address and make active one of the sixteen word-select control lines. The three-state input buffers are enabled by a HIGH signal on the write control line, while the three-state output buffers are enabled by a

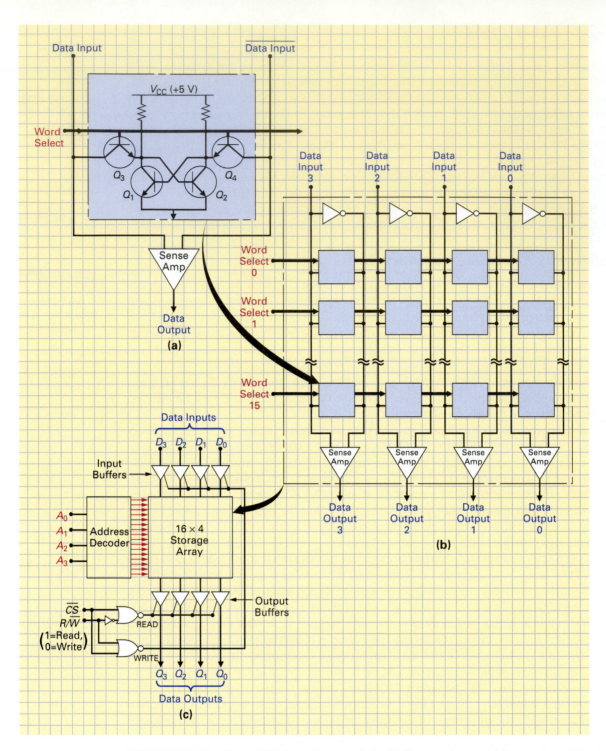

FIGURE 12-12 Static RAM. (a) Storage Cell. (b) Storage Array. (c) Internal Architecture.

HIGH on the read control line. These write and read control lines are controlled by the chip-select ($\overline{CS}$) and read/write ($R/\overline{W}$) inputs. When the $\overline{CS}$ and $R/\overline{W}$ control lines are both LOW, the input buffers are enabled and data is written into the addressed memory location. On the other hand, when the $\overline{CS}$ is LOW and the $R/\overline{W}$ control line is HIGH, the output buffers are enabled and the data at the addressed memory location is read out and applied to the Q outputs. Therefore, when the $\overline{CS}$ line is LOW (SRAM chip is selected or enabled), the logic level on the $R/\overline{W}$ control line will govern what operation is performed (1 = read, 0 = write). On the other hand, if the $\overline{CS}$ line is switched HIGH, both the input buffers and output buffers are disabled, and the SRAM chip is effectively disconnected from the data input and data output bus lines.

SRAM read and write timing. The logic symbol for the 16 by 4 SRAM is shown in Figure 12-13(a). Now that we have covered the operation of the SRAM cell and the operation of an SRAM's internal architecture, let us see how all of the control and input/output lines are used to perform a memory-read and a memory-write operation.

Figure 12-13(b) shows the typical timing waveforms for an SRAM read operation. Working from the top waveform to the bottom, you can see that the $R/\overline{W}$ control line input is first made HIGH (read operation). The desired address is then applied to the 4-bit address inputs, and then the chip-select control input is activated ($\overline{CS}$ = LOW). After a short delay,

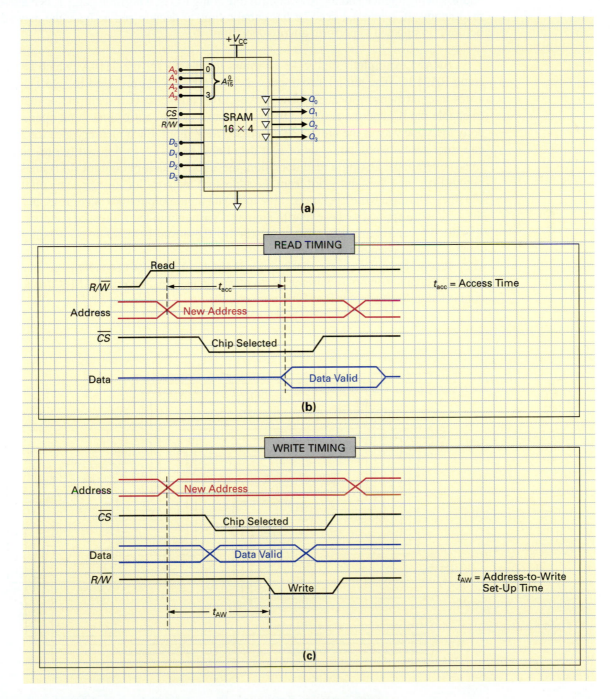

FIGURE 12-13 Static RAM (SRAM) Logic Symbol and Read/Write Timing.

the data stored at the addressed memory location will appear at the Q outputs. The *access time* (t_{acc}) of a memory IC is governed by the IC's internal address decoder Assuming no delays are introduced by the $R/\overline{W}$ or $\overline{CS}$ control lines, the read-access time of a memory IC is the time it takes for a new address to be decoded and the contents of the memory location to appear at the output.

Figure 12-13(c) shows the typical timing waveforms for an SRAM write operation. Working from the top waveform to the bottom, you can see that the desired address is first applied to the 4-bit address inputs, and then the chip-select control input is activated ($\overline{CS}$ = LOW). The data to be stored is then applied to the data inputs, and when the $R/\overline{W}$ control line input is brought LOW, the data is written into the addressed memory location. The *address-to-write set-up time* (t_{AW}) is a time delay that must be introduced between a new address being applied and the $R/\overline{W}$ control line going LOW. This time delay is needed so that the memory IC's internal decoder has enough time to decode and activate the new address before it is signaled to perform a write operation.

SRAM example. Figure 12-14 shows the details for the 2114A, which is a typical static RAM IC. The ten address input lines (A_0 through A_9) are used to select one of the 1,024 internal memory locations. The active-LOW chip-select ($\overline{CS}$) control line input is used to enable the IC, and the active-LOW write-enable ($\overline{WE}$) control line input operates in exactly the same way as the previously discussed $R/\overline{W}$ input. Each memory location in this IC can store a 4-bit word, and a set of four data lines is used for both input (write) and output (read) operations. During a read operation, these four data input/output (I/O) lines become output lines, and the 4-bit word from the addressed memory location appears on these four I/O lines. During a write operation, these four data input/output lines become input lines, and the 4-bit word applied to these four I/O inputs is written into the addressed memory location. It is the $\overline{WE}$ control line that determines whether these four I/O data lines are being used for input or output operations.

It is probably safe to say that a single RAM IC chip is not normally large enough for a digital system's memory requirements. In most applications, several RAM ICs will be interconnected to form a larger RAM memory circuit.

FIGURE 12-14   1,024 by 4-Bit Bipolar Static RAM IC.

Dynamic Random-Access Memories (DRAMs)

As mentioned previously, there are static RAMs and **dynamic RAMs.** As we saw in the previous section, memory systems using static RAMs appear to be ideal since they are fast, there are many types to choose from, and their circuitry is simple. The question, therefore, is, why bother having another type of RAM? There are two good reasons why dynamic RAMs are ideal in certain applications. The first is that you can package more storage cells into a dynamic RAM IC than you can into a static RAM IC. This can be easily seen by comparing the single static RAM cell shown in Figure 12-15(a), to the single dynamic RAM cell shown in Figure 12-15(b). A dynamic

FIGURE 12-15 **Dynamic RAMs. (a) Static Cell. (b) Dynamic Cell. (c) Internal Architecture. (d) Simplified DRAM Cell Operation.**

RAM storage cell has only one MOSFET and a capacitor compared to a static RAM storage cell, which normally contains six MOSFETs. Therefore, approximately three dynamic storage cells could fit into the same space as one static storage cell, giving the DRAM a large space-saving advantage. The second reason for using dynamic RAMs is power dissipation. Since power dissipation is dependent on the number of devices per storage cell, it is easy to imagine why DRAMs have a lower power dissipation since fewer MOSFETs are used for each storage cell.

DRAM storage cell. The DRAM storage cell shown in Figure 12-15(b) operates in the following way.

To write data into this storage cell, the data bit to be stored is first applied to the data-in line, and then the bit-select control line is pulsed HIGH. A HIGH on the bit-select control line will cause Q_1 to conduct (N-channel MOSFET), allowing either the HIGH or LOW data input to be applied directly to the capacitor, causing it to either charge (a 1 is stored) or discharge (a 0 is stored).

To read data out of this storage cell, the bit-select control line is switched HIGH to turn ON Q_1, allowing either the stored binary 1 (capacitor charged) or binary 0 (capacitor discharged) to pass through to the data-out line.

Now let us see how these dynamic storage cells are connected to form a complete dynamic RAM circuit.

DRAM storage array. Nearly all dynamic RAM ICs are organized as multiple 1-bit read/write memories. This means that they have multiple memory locations; however, each memory location has only a 1-bit word (for example, 16K by 1-bit, 64K by 1-bit, 256K by 1-bit, and so on).

Figure 12-15(c) shows the internal architecture of a typical dynamic RAM circuit. This dynamic RAM circuit has 16,384 1-bit memory locations, arranged into a 128 by 128 matrix. Each of the dynamic cells in this matrix has its own unique row and column position in the matrix. To address each of these 16,384 1-bit memory locations will require fourteen address inputs ($2^{14} = 16,384$). Looking at the dynamic RAM circuit in Figure 12-15(c), you will notice that this circuit only has seven address input pins. To reduce the number of pins

Address Multiplexing

A time-division technique in which two or more address codes are switched through a single input at different times.

on the IC package, manufacturers often use an **address multiplexing** technique, in which an address input pin is used to input two different address bits. With this technique, the 14-bit address is applied to the dynamic RAM address decoder in two 7-bit groups. First, the seven lower-order address bits (A_0 through A_6) are applied to the seven address inputs and are stored in the row address register and decoded by the row address decoder by activating the *row address strobe* ($\overline{RAS}$). Second, the seven higher-order address bits (A_7 through A_{13}) are applied to the seven address inputs and are stored in the column address register and decoded by the column address decoder by activating the *column address strobe* ($\overline{CAS}$). The row address strobe and column address strobe control-line inputs, therefore, allow us to input a 14-bit address on seven address input pins, which decreases the size of the IC package, decreasing the size of memory circuits and the final digital system. The outputs of the row and column address registers drive the row and column address decoders, and these decoders will bit-select one of the 16,384 dynamic RAM storage cells. Once activated, the read/write control-line input will determine whether a read operation (storage cell → data-out pin) or write operation (data-in pin → storage cell) is performed.

The dynamic RAM, therefore, uses an on-chip capacitor for each storage element. A charge is stored on the capacitor to indicate a binary 1, and no charge on the capacitor is used to indicate a binary 0. This technique simplifies the storage cell, permitting denser memory chips that dissipate less power. There is a problem, however, in that the charge leaks off the capacitor due to capacitor leakage, and therefore a binary 1 can become a binary 0 in just a few milliseconds. When a binary 1 is stored in a DRAM, therefore, it must be *refreshed* continually so that the data stored within the DRAM is accurate. Most DRAM ICs today have on-chip refresh circuits, as shown in Figure 12-15(c), to automatically rewrite the same data back into each memory location. This refresh action will typically take place every two ms. After a refresh, all of the binary 1s are restored to full charge and all of the binary 0s are left as no charge.

DRAM write, read, and refresh.   To explain the write, read, and refresh operations in more detail, Figure 12-15(d) shows a simplified DRAM cell and its input/output circuitry. In this diagram the circuit's three-state input buffer, three-state output buffer, and the internal cell MOSFETs have been represented as switches, which are under the control of the read/write input. The table associated with this illustration indicates which of the switches are open and closed for each of the three operations.

To write data to the cell, SW_1 and SW_2 are closed so that the data input can be applied directly to the capacitor. At the same time, SW_3, SW_4, and SW_5 are opened to disconnect the storage capacitor from the rest of the circuit.

To read data from the cell, SW_3 is closed to connect the stored capacitor voltage to the sense amplifier. The sense amplifier compares the stored capacitor voltage to a reference voltage in order to determine whether a logic 0 or logic 1 was stored, and then produces a valid 0 V or 5 V (binary 0 or 1) at its output. SW_5 switches the stored bit from the sense amplifier through to the data output. In this mode, SW_4 is also closed, along with SW_2, to refresh the capacitor voltage (recharge to reinforce the HIGH, or discharge to reinforce the LOW). During each read operation, therefore, a refresh action is performed.

To only refresh the cell, SW_2, SW_3, and SW_4 are closed, so that the logic level stored in the capacitor can be fed back to the capacitor to reinforce the stored HIGH or LOW.

DRAM read and write timing.   The logic symbol for the previously discussed 16,384 by 1-bit DRAM is shown in Figure 12-16(a). Now that we have covered the operation of the DRAM cell and the operation of a DRAM's internal architecture, let us see how all of the control and input/output lines are used to perform a memory-read and a memory-write operation.

Figure 12-16(b) shows the typical timing waveforms for a DRAM read operation. Working from the top waveform to the bottom, you can see that first the seven lower-order address bits (row address) are applied to the seven address input pins. This row address input is strobed, or latched, into the row address register when the row address strobe control line ($\overline{RAS}$) is brought LOW. Then the seven higher-order address bits (column address) are applied to the seven address input pins. This column address input is strobed, or latched, into the column address register when the column address strobe control line ($\overline{CAS}$) is brought LOW. The $R/\overline{W}$ control line is used to control the input and output buffers and therefore control whether a read or write operation is performed. In this example, the $R/\overline{W}$ control line goes HIGH after the $\overline{CAS}$ goes LOW, to enable a read operation. The three-state output buffers will therefore be taken out of their high impedance state and the stored binary 1 or binary 0 will appear at the data-out pin.

Figure 12-16(c) shows the typical timing waveforms for a DRAM write operation. Working from the top waveform to the bottom, you can see that the row address and column address are latched into the DRAM in exactly the same way as in the read operation. Once the 14-bit address has been inputted, the $R/\overline{W}$ control line will control whether a read or write operation is performed. In this example, the $R/\overline{W}$ control line goes LOW after the $\overline{CAS}$ goes LOW to enable a write operation. The input switch will therefore be enabled, and the binary 1 or binary 0 input being applied to the data-in pin will be stored or written into the addressed memory location.

As mentioned previously, nearly all dynamic RAM ICs are organized as multiple 1-bit read/write memories. This means that they have multiple memory locations; however, each memory location has only a 1-bit word (for example, 16K by 1-bit, 64K by 1-bit, 256K by 1-bit, and so on). Figure 12-17 shows how eight 16K by 1-bit DRAM ICs can be connected to form a 16K by 8-bit DRAM read/write memory circuit. An address multiplexer is used to switch the seven lower-order address bits through to the DRAMs just before $\overline{RAS}$ goes active and then switch the seven higher-order address bits through to the DRAMs just before the $\overline{CAS}$ goes active. The $\overline{RAS}$, $\overline{CAS}$, $R/\overline{W}$, and $\overline{CS}$ control line inputs are connected to all eight DRAM ICs so that they are all activated simultaneously.

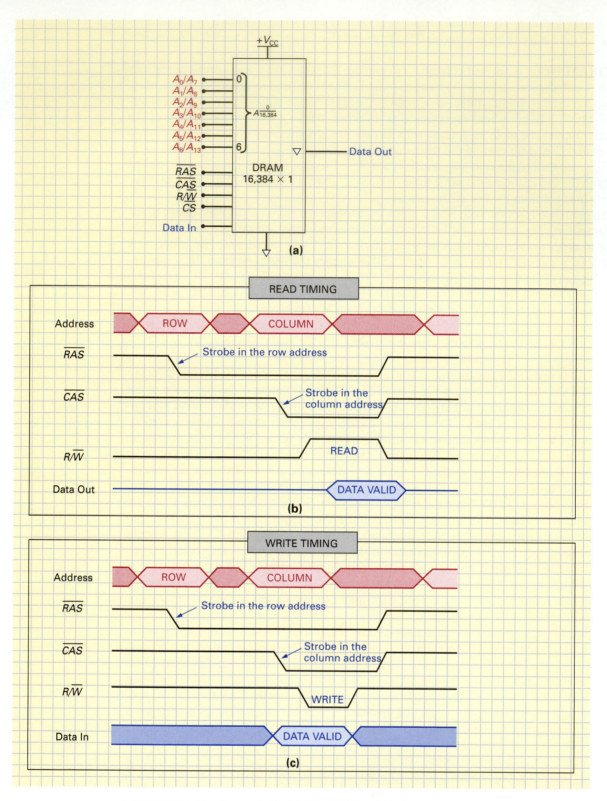

FIGURE 12-16 Dynamic RAM (DRAM) Logic Symbol and Read/Write Timing.

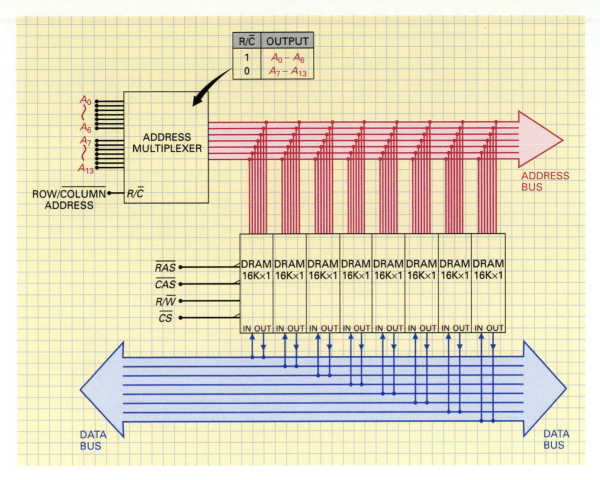

FIGURE 12-17 A 16K by 8-Bit DRAM Memory Circuit.

12-2-3   *RAM Applications*

As far as applications are concerned, dynamic RAMs are more widely used than static RAMs because they are smaller in size, lower in cost, and dissipate less power than an equivalent static RAM. Static RAMs, however, have faster access times and do not require refresh circuitry and the special timing requirements that go along with refreshing the stored data (you must refresh at least every 2 ms). Semiconductor memory manufacturers, however, are continually improving the DRAM, making its disadvantages easier to deal with.

SRAMs or DRAMs

In general, static RAMs are used when only a small amount of read/write memory is required (64K or less) or when high-speed access is needed. Dynamic RAMs, on the other hand, are used in applications where a large amount of read/write memory is needed (megabytes). Some systems employ a small amount of SRAM for high-speed read/write memory requirements and a large amount of DRAM for the bulk of the system's read/write memory needs.

A Typical RAM Memory Circuit

Whether the application is a personal computer or a digital oscilloscope, all digital computer systems use RAM. As discussed previously, a microprocessor unit (MPU) is at the heart of every digital computer system. The MPU is a complex integrated circuit that contains control logic circuits, instruction decoding circuits, temporary storage circuits, and arithmetic processing circuits. To be useful, the microprocessor IC must be able to exchange information with memory ICs (both ROM and RAM) input devices, such as a keyboard, and output devices, such as a display.

Figure 12-18(a) shows how a typical RAM memory circuit could be connected in a digital computer system. At this time, we will simply think of the MPU as a system controller that can select or address any memory location in RAM and then either read data from that memory location or write data to that memory location. It achieves this by controlling the ad-

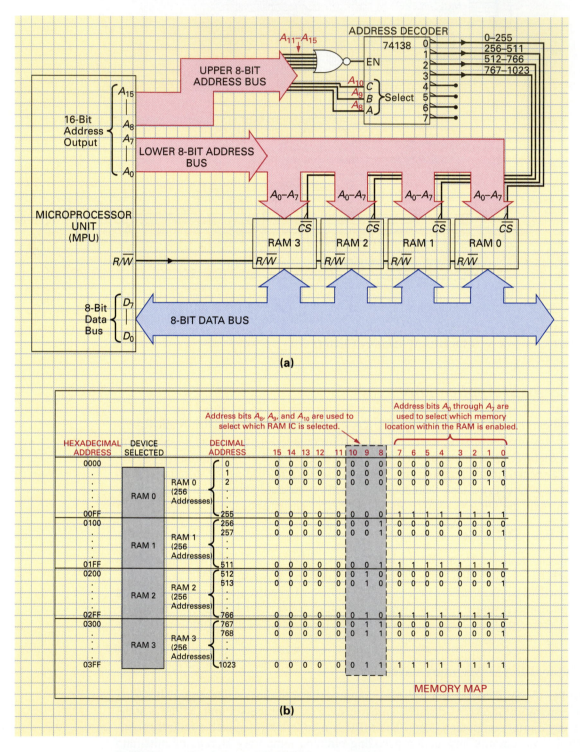

FIGURE 12-18   A Typical RAM Memory Circuit.

dress bus and the data bus and generating control signals. To review, the data bus is used for data transfers, and since all devices share this bus, the microprocessor will only activate one device at a time and then read data from that device via the data bus or write data to that device via the data bus (under the control of the $R/\overline{W}$ control line from the MPU). Before a data transfer can take place on the data bus, however, the microprocessor must output an address onto the address bus. This address specifies which memory location the microprocessor wishes to access.

In Figure 12-18(a), four 256 by 8-bit RAM ICs have been connected to form a 1,024 (4×256) by 8-bit RAM memory. Since most computer systems use more than one memory IC, the address from the MPU must indicate which memory IC should be selected and which word within that IC should be accessed. This is achieved by assigning a unique set of addresses to each memory IC.

The **memory map** shown in Figure 12-18(b) shows how each of the four RAM ICs has its own unique set of 256 addresses. For example, addresses 0_{10} through 255_{10} (0000_{16} through $00FF_{16}$) are assigned to RAM 0, addresses 256_{10} through 511_{10} (0100_{16} through $01FF_{16}$) are assigned to RAM 1, addresses 512_{10} through 766_{10} (0200_{16} through $02FF_{16}$) are assigned to RAM 2, and addresses 767_{10} through 1023_{10} (0300_{16} through $03FF_{16}$) are assigned to RAM 3.

Referring to the 16-bit address bus from the MPU in Figure 12-18(a), you can see that the upper eight bits of the address bus are applied to an **address decoder,** and the lower eight bits of the address bus are applied to all four RAMs. By decoding address bits A_8 through A_{15}, the 74138 address decoder can determine which one of the four RAM ICs should be enabled. The eight lower-order address bits from the MPU (A_0 through A_7) will then be decoded by the selected RAM IC to determine which word within the RAM IC is accessed. In the circuit in Figure 12-18(a), you can see that address bits A_8, A_9, and A_{10} are applied to the ABC select inputs of the 74138.

In the memory map in Figure 12-18(b), you can see how address bits A_8, A_9, and A_{10} will advance by one for every 256 addresses. For example, address bits A_8, A_9, and A_{10} will remain at 000 (decimal 0) for the first 256 addresses, advance to 001 (decimal 1) for the next 256 addresses, advance to 010 (decimal 2) for the next 256 addresses, and so on. Since these three bits are applied to the 74138s ABC select inputs, output 0 will go active (LOW) for the first 256 addresses, output 1 will go active (LOW) for the next 256 addresses, output 2 will go active (LOW) for the next 256 addresses, and so on. Therefore, by connecting each of the 74138's active-LOW outputs to each of the active-LOW chip-select inputs of the RAM ICs, as shown in Figure 12-18(a), we will ensure that RAM 0 is only enabled whenever any address from 0000_{16} to $00FF_{16}$ is generated by the MPU, RAM 1 is only enabled whenever any address from 0100_{16} to $01FF_{16}$ is generated by the MPU, RAM 2 is only enabled whenever any address from 0200_{16} to $02FF_{16}$ is generated by the MPU, and RAM 3 is only enabled whenever any address from 0300_{16} to $03FF_{16}$ is generated by the MPU. The rest of the higher-order address bits (A_{11} through A_{15}) are used to enable (EN) the decoder only when they are all LOW.

To summarize, with the address decoding technique, the lower-order address bits are sent directly to the memory's address line, and the higher-order address bits are decoded to generate the chip selects, with only one IC selected at any given time. In other circuit designs, you may see that the number of bits fed directly to the memory devices may change (depending on the number of words that need to be addressed, 2^n), and this will change which higher-order address bits are decoded to generate chip selects. In the following troubleshooting memory circuits section, you will see another example of how an address decoder is used to create 1K (1,024) memory map blocks instead of the 256 memory map blocks shown in Figure 12-18(b).

12-2-4   *RAM Testing*

When testing a ROM IC, all we had to do was read out the contents of each memory location and compare it to the ROM listing to see if a memory cell error was present. With a RAM IC, we have to test both the read and the write ability of all the memory cells, making the testing of a RAM a little more lengthy.

Memory Map
An address-listing diagram showing the boundaries of the address space.

Address Decoder
A circuit that decodes the applied addresses and activates certain outputs based on which zone the address is within.

Figure 12-19 shows a simple RAM test circuit. In this circuit, a 10-bit counter is triggered to sequence through the 1,024 addresses needed for a 2114 IC being tested. Four data switches can be used to generate any 4-bit word combination, which can then be applied to the 2114 and stored in any memory location (write operation). A 4-bit output register and LED display are included so that a 4-bit word retrieved from any memory location can be viewed (read operation). RAM ICs are normally tested by first storing 0s in every memory cell and then reading out all of these 0s and checking that no 1s are present. The second step is to store 1s in every memory cell and then read out all of these 1s and check that no 0s are present. These first two checks can be seen in the inset in Figure 12-19. These two checks, however, do not test for a memory cell short between two adjacent memory cells. For example, if one memory cell is shorted to an adjacent memory cell, storing all 1s or storing all 0s will not expose this fault since the short will cause both memory cells to be at the same logic level anyway.

To determine whether a memory cell short exists within a RAM IC, a **checkerboard test pattern** is normally written into the RAM IC, as shown in the inset in Figure 12-19. With

Checkerboard Test Pattern

A RAM testing method in which the computer first stores 0s in every memory location, then stores 1s, and then alternates the storage of 1s and 0s.

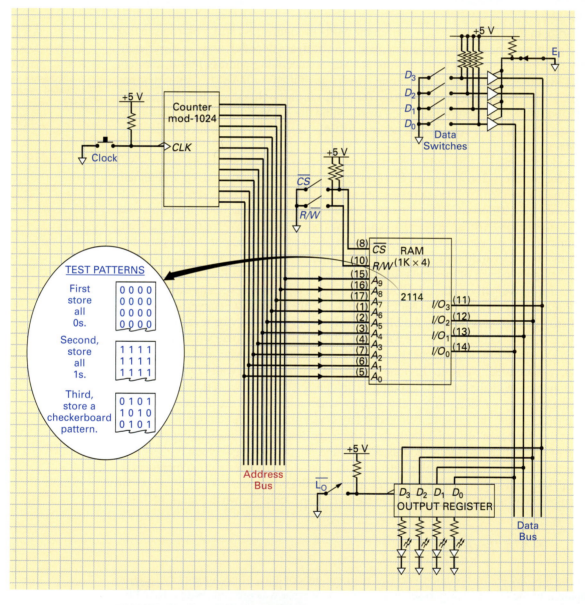

FIGURE 12-19   A Simple RAM Test Circuit.

the checkerboard test pattern, adjacent cells will have opposite bits stored, and if the logic levels do not alternate when the pattern is read out of the RAM, a memory cell problem exists. After the RAM is checked with this checkerboard pattern, the pattern is normally reversed to further test the RAM and then read out and checked again. Since there is no way to repair an internal RAM IC failure, the IC must be discarded and replaced if a problem is found.

The test circuit in Figure 12-19 would be ideal for testing a small 1K by 4 RAM; however, most digital system RAMs will be much larger, having possibly millions of memory locations, making a complete test with this circuit long and laborious. In these cases, a RAM tester can be used to check the RAM's read/write ability. This test instrument automatically writes several test patterns into the RAM and then reads back the data and compares it to check for any operation or storage error.

Most RAMs within digital electronic computer systems are tested by a built-in self-diagnostic program. This test program will generally use the checkerboard test pattern method to verify the RAM's operation.

SELF-TEST EVALUATION POINT FOR SECTION 12-2

Use the following questions to test your understanding of Section 12-2.

1. What is the difference between a read/write memory and a read-only memory?
2. What is the difference between a SAM and a RAM?
3. Are ROMs RAMs?
4. Is a RAM a volatile or nonvolatile device?
5. Describe the difference between a static RAM and a dynamic RAM.
6. What are the advantages and disadvantages of SRAMs and DRAMs?
7. What is a memory map?
8. How are RAMs usually tested?

REVIEW QUESTIONS

Multiple-Choice Questions

1. How many memory locations can be accessed by a ROM IC if it has ten address line inputs?
   a. 16
   b. 1,024
   c. 256
   d. 2,048

2. A memory is said to be _____ if it loses its data whenever power is removed.
   a. nonvolatile
   b. accessible
   c. volatile
   d. sequential

3. If a RAM IC has twelve address inputs and eight data input/outputs, what is its size and configuration?
   a. 2K by 8
   b. 4K by 8
   c. 1,024 × 8
   d. 8,192 by 8

4. The _____ of a memory is the time interval between the memory receiving a new address input and the data at that address being available at the data outputs.
   a. storage density
   b. programmable link
   c. access time
   d. memory configuration

5. The total storage capacity of an 8K by 8 RAM is _____ bits.
   a. 16,384
   b. 65,536
   c. 64,000
   d. 49,152

6. Which of the following ROM types is the cheapest when ordered in large quantities?
   a. MROM
   b. PROM
   c. EPROM
   d. EEPROM

7. Which of the following ROM types can easily be erased and reprogrammed?
   a. MROM
   b. PROM
   c. EPROM
   d. EEPROM

8. Which of the following ROM types can be erased and is low in cost?
   a. MROM
   b. PROM
   c. EPROM
   d. EEPROM

9. A read/write memory is normally always referred to as a
   _____.
   a. ROM
   b. EEPROM
   c. RAM
   d. MBM

10. The storage cell in a static RAM is a _____.
    a. bistable flip-flop
    b. FET and capacitor
    c. monostable flip-flop
    d. Both a. and c. are true.

11. Which of the RAM types has a higher storage density but operates slower?
    a. SRAM
    b. DRAM
    c. SAM
    d. All of the above

12. Which of the RAM types has a lower storage density but operates faster?
    a. SRAM
    b. DRAM
    c. SAM
    d. All of the above

13. Which of the following RAM types has a lower power consumption?
    a. SRAM
    b. DRAM
    c. SAM
    d. All of the above

14. Static RAMs and dynamic RAMs are both nonvolatile.
    a. True
    b. False

15. Dynamic RAM ICs will normally use an address multiplexing technique to _____.
    a. refresh
    b. reduce the number of data inputs
    c. enable an address decoder
    d. reduce the number of address inputs

16. The notebook computer used to develop the manuscript for this textbook has 32 megabytes of RAM. Which RAM type do you think is being used in this application?
    a. SRAM
    b. DRAM
    c. SAM
    d. ERAM

17. If address bits 5, 6, and 7 are applied to the *ABC* select inputs of an address decoder and address bits 0 through 4 are applied directly to the memory, what is the size of the memory map blocks?
    a. 16
    b. 32
    c. 8
    d. 64

18. If address bits 8, 9, and 10 are applied to the *ABC* select inputs of an address decoder, and address bits 0 through 7 are applied directly to the memory, what is the size of the memory map blocks?
    a. 256
    b. 1,024
    c. 512
    d. 128

19. If address bits 13, 14, and 15 are applied to the *ABC* select inputs of an address decoder, and address bits 0 through 12 are applied directly to the memory, what is the size of the memory map blocks?
    a. 4K
    b. 32K
    c. 8K
    d. 16K

20. The memory-mapped input/output technique assigns a block of addresses to memory and input and output ports.
    a. True
    b. False

Practice Problems

To practice your circuit recognition and operation skills, refer to Figure 12-20 and answer the following four questions.

21. What is the purpose of the $\overline{ROM}$ control signal input?

22. What is the purpose of the $\overline{READ}$ control signal input?

23. How is data written into this IC?

24. How is data read from this IC?

To practice your circuit recognition and operation skills, refer to the SAM schematic diagram on pages 419–422 and answer the following five questions.

25. Where is the 2316 ROM circuit shown in Figure 12-20 on the SAM schematic diagram?

26. The NRD (not read) control line input to the 2316 ROM IC comes from what IC?

27. The NROM (not ROM) control line input to the 2316 ROM IC comes from what IC?

28. Which two ICs output the 11-bit address that is applied to the 2316 IC?

29. The 8-bit data output of the 2316 ROM IC should connect to the data bus and then back to the MPU IC. Which IC do you think is the MPU?

To practice your circuit recognition and operation skills, refer to Figure 12-21 and answer the following four questions.

30. What is the purpose of the $\overline{RAM}$ control signal input?

31. What is the purpose of the $\overline{WRITE}$ control signal input?

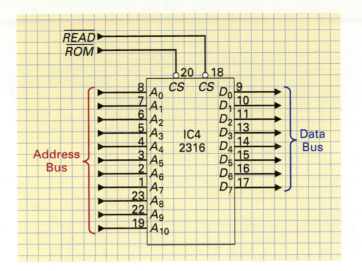

FIGURE 12-20 Application—The ROM Circuit from SAM (Simple All-Purpose Microcomputer).

FIGURE 12-21 Application—The RAM Circuit from SAM (Simple All-Purpose Microcomputer).

32. How is data written into this IC?

33. How is data read from this IC?

To practice your circuit recognition and operation skills, refer again to the SAM schematic diagram on pages 419–422 and answer the following three questions.

34. Where is the 2114 RAM circuit shown in Figure 12-21 on the SAM schematic diagram?

35. The NWR (not write) control line input to the 2114 RAMs comes from what IC?

36. The NRAM (not RAM) control line input to the 2114 RAMs comes from what IC?

To practice your circuit recognition and operation skills, refer to Figure 12-22 and answer the following three questions.

37. Which address bits are applied to the address decoder's *ABC* select inputs?

38. What is the size of each block in the memory map?

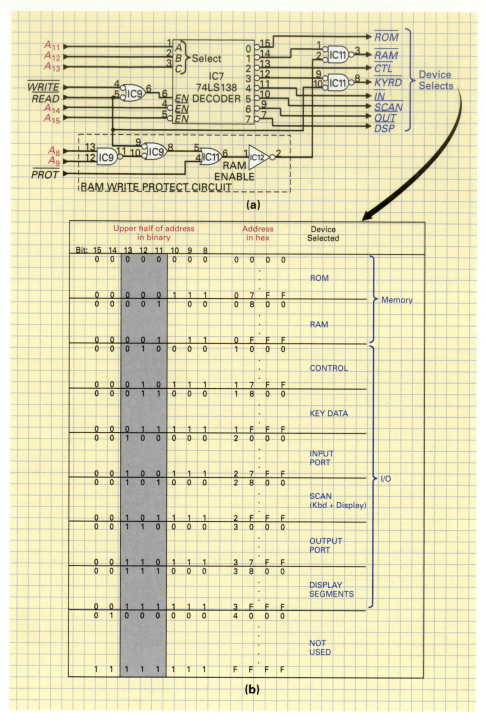

FIGURE 12-22   Application—The Address Decoder and Memory Map For a SAM (Simple All-Purpose Microcomputer).

39. What device is selected when A_{11}, A_{12}, and A_{13} equal:
  a. 000
  b. 001
  c. 010
  d. 011
  e. 100
  f. 101
  g. 110
  h. 111

To practice your circuit recognition and operation skills, refer again to the SAM schematic diagram on pages 419-422 and answer the following four questions.

40. Where is the address decoder shown in Figure 12-22 located on the SAM schematic diagram?

41. What is the purpose of the four-LED displays DS1, DS2, DS3, and DS4?

42. What is the purpose of the displays DS7 and DS8?

43. Is the SAM system using memory-mapped input/output?

Web Site Questions

Go to the web site http://www.prenhall.com/cook, select the textbook *Practical Digital Electronics,* select this chapter, and then follow the instructions when answering the multiple choice practice problems.

Analog and Digital Signal Converters

Go and Do "Something More Useful"

Vladimir Zworykin was born in Mourom, Russia, on July 30, 1889. He graduated in 1912 from the St. Petersburg Institute of Technology and then did his postgraduate work at the College de France in Paris. After World War I, he immigrated to the United States and joined the research staff of Westinghouse Electric Corporation in Pittsburgh. Here he invented the iconoscope (electronic camera) and the kinescope (CRT picture tube). In 1923 he proudly demonstrated to his employers what he considered his biggest achievement—a cloudy image of boats on the river outside his lab on a small screen. Management was not at all impressed, and Zworykin later said that he was told "to spend my time on something more useful."

In 1929 David Sarnoff, the founder of the Radio Corporation of America (RCA), asked Zworykin what it would take to develop television for commercial use. Zworykin, who at that time was not at all sure, answered confidently, "Eighteen months and $100,000." Sarnoff liked what he heard and hired Zworykin, putting him in charge of RCA's electronic research laboratory at Camden, New Jersey. Years later Sarnoff loved pointing out the slight difference in development cost and time, which finally worked out to be twenty years and $50 million.

While at RCA, Zworykin also developed the electron microscope (in only three months), the electron multiplier tube, and the first infrared sniper scopes, which were used in World War II.

In 1967 Zworykin received the National Medal of Science for his contributions to the instruments of science, engineering, and television and for his stimulation of the application of engineering to medicine. Recognized as the "father of television," he lived long enough to see all of his inventions flourish before he died in 1982. When asked in a television interview in 1980 what he thought of television today, he answered, "The technique is wonderful. It is beyond my expectation. But the programs! I would never let my children even close to this thing."

Introduction

Electronic information exists as either analog data or as digital data. With analog data, the signal voltage change is an analogy of the physical information it is representing. For example, if a person were to speak into a microphone, the pitch and loudness of the sound wave would determine the frequency and amplitude of the voltage signal generated by the microphone. With a digital signal, information is first converted into a group of HIGH and LOW voltage pulses that represent the binary digits 1 and 0. For example, the binary digit code 1101001 is generated whenever the lower case "i" key on an ASCII computer keyboard is pressed. In this case, therefore, the information "i" is encoded into a digital code, or a code that makes use of binary digits.

Whenever you have two different forms, there is always a need to convert back and forth between these two different forms. For example, in previous chapters we have seen how we have needed circuits to convert between binary and decimal, bipolar logic levels and MOS logic levels, serial data and parallel data, and so on. In this chapter, we will examine data conversion techniques and the circuits that are used to perform digital-to-analog (D/A) data conversion and analog-to-digital (A/D) data conversion. We will also discuss the application, testing, and troubleshooting of these data conversion circuits.

13-1 ANALOG AND DIGITAL SIGNAL CONVERSION

Today, microcomputers are used in almost every electronic system because of their ability to quickly process and store large amounts of data, make systems more versatile, and perform more functions. Digital processing within electronic systems has therefore become more and more predominant. Many of the input signals applied to a microcomputer for storage or processing are analog in nature, and therefore a data conversion circuit is needed to interface these analog inputs to a digital system. Similarly, a data conversion circuit may also be needed to interface the digital microcomputer system to an analog output device.

13-1-1 *Connecting Analog and Digital Devices to a Computer*

Figure 13-1 shows a simplified block diagram of a microcomputer. No matter what the application, a microcomputer has three basic blocks—a microprocessor unit, a memory unit,

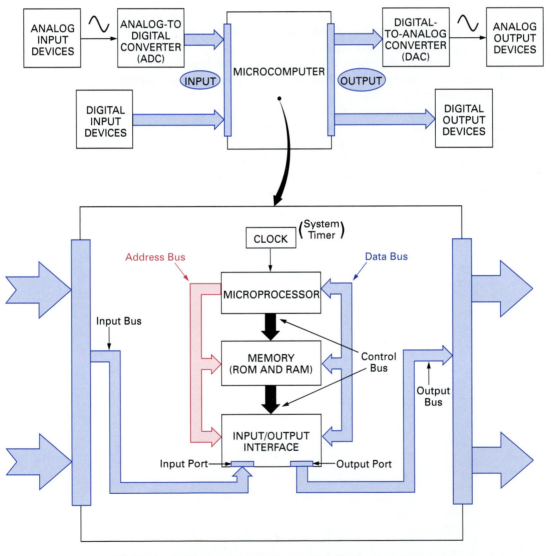

FIGURE 13-1 **Connecting Analog and Digital Devices to a Computer.**

and an input/output unit. Getting information into the computer is the job of input devices, and as you can see in Figure 13-1, there are two input paths into a computer—one path from analog input devices and the other path from digital input devices. Information in the form of light, sound, heat, pressure, or any other real-world quantity is analog in nature and has an infinite number of input levels. Sensors or transducers of this type, such as photoelectric cells for light or microphones for sound, will generate analog signals. Since the computer operates on information in only digital or binary form, a translation or conversion is needed if the computer is going to be able to understand or interpret an analog signal. This signal processing is achieved by an analog-to-digital converter (ADC), which transforms the varying analog input voltage into digital codes that the computer can understand. Once the analog information has been encoded into digital form, the information can then enter the computer through an electronic doorway called the **input port.** Some input devices, such as a computer keyboard, automatically generate a digital code for each key that is pressed (ASCII codes). The codes generated by these digital input devices can be connected directly to the computer without having to pass through a converter.

Similarly, the digital output information from a microcomputer's **output port** can be used to drive either an analog device or a digital device. If an analog output is desired, as in the case of a voice signal for a loudspeaker, the digital computer's output will have to be converted into a corresponding analog signal by a digital-to-analog converter (DAC). If only a digital output is desired, as in the case of a printer, which translates the binary codes into printed characters, the digital computer output can be connected directly to the output device without the need for a converter.

13-1-2 *Converting Information Signals*

Figure 13-2(a) shows how analog data can be converted into digital data, and Figure 13-2(b) shows how digital data can be converted into analog data. To give this process purpose, let us consider how the music information stored on a compact disc (CD) is first recorded and then played back. Like all sound, music is made up of waves of compressed air. When these waves strike the diaphragm of a microphone, an analog voltage signal is generated. In the past, this analog data was recorded on magnetic tapes or as a grooved track on a record. These data storage devices were susceptible to wear and tear, temperature, noise, and age. Using digital recording techniques, binary codes are stored on a compact disc to achieve near-perfect fidelity to live sound.

Figure 13-2(a) shows how an **analog-to-digital converter (ADC)** is used during the recording process to convert the analog music input into a series of digital output codes. These digital codes are used to control the light beam of a recording laser so that it will engrave the binary 0s and 1s onto a compact disc (CD) in the form of pits or spaces. The ADC is triggered into operation by a sampling pulse that causes it to measure the input voltage of the analog signal at that particular time and generate an equivalent digital output code. For example, on the active edge of sampling pulse 1, the analog input voltage is at 2 V so the binary code 010 (decimal 2) is generated at the output of the ADC. On the active edge of sampling pulse 2, the analog voltage has risen to 4 V, so the binary code 100 (decimal 4) is generated at the output of the ADC. On the active edge of sampling pulses 3, 4, and 5, the binary codes 110 (decimal 6), 111 (decimal 7), and 110 (decimal 6) are generated at the output, representing the analog voltages of 6 V, 7 V, and 6 V, respectively, and so on.

Figure 13-2(b) shows how a **digital-to-analog converter (DAC)** is used during the music playback process to convert the digital codes stored on a CD to an analog output signal. Another laser is used during playback to read the pits and spaces on the compact disc as 0s and 1s. These codes are then applied to the DAC, which converts the digital input codes into discrete voltages. The DAC is triggered into operation by a strobe pulse that causes it to convert the code currently being applied at the input into an equivalent output voltage.

Input Port

An electronic doorway used for transmitting information into a computer.

Output Port

An electronic doorway used for transmitting information out of a computer.

Analog-to-Digital Converter (ADC)

A circuit that converts analog input signals into equivalent digital output signals.

Digital-to-Analog Converter (DAC)

A circuit that converts digital input signals into equivalent analog output signals.

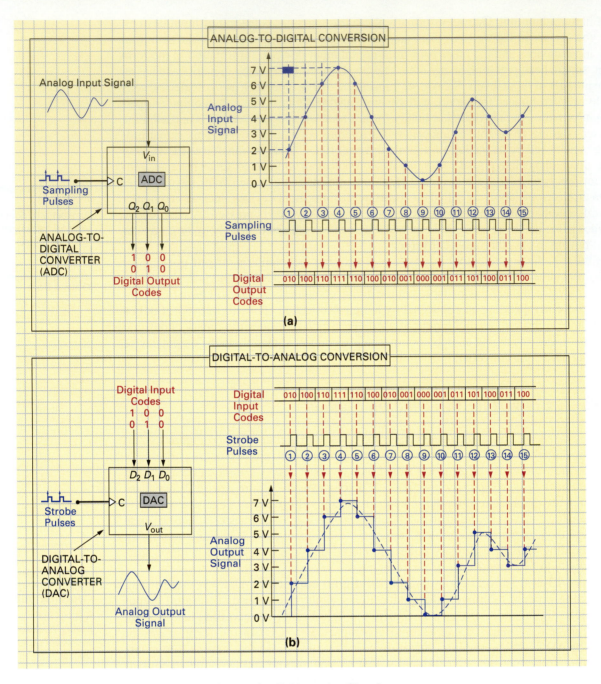

FIGURE 13-2   Converting Information Signals.

For example, on the active edge of strobe pulse 1, the digital code 010 (decimal 2) is being applied to the DAC, so it will generate 2 V at its output. On the active edge of strobe pulse 2, the digital code 100 (decimal 4) is being applied to the DAC, so it will generate 4 V at its output. On the active edge of strobe pulse 3, the digital code 110 (decimal 6) is being applied to the DAC, so it will generate 6 V at its output. On the active edge of strobe pulse 4, the digital code 111 (decimal 7) is being applied to the DAC, so it will generate 7 V at its output. On the active edge of strobe pulse 5, the digital code 110 (decimal 6) is being applied to the DAC, so it will generate 6 V at its output, and so on. If the output of a DAC is then applied to a lowpass filter, the discrete voltage steps can be blended into a smooth wave that closely approximates the original analog wave, as shown by the dashed line in Figure 13-2(b).

Use the following questions to test your understanding of Section 13-1.

1. Give the full names of the following abbreviations:
   a. ADC   b. DAC

2. To interface a microcomputer with analog devices, a _____ would be connected to the input port, and a _____ would be connected to the output port.

3. An ADC will convert a/an _____ input into an equivalent _____ output.

4. A DAC will convert a/an _____ input into an equivalent _____ output.

13-2   DIGITAL-TO-ANALOG CONVERTERS (DACs)

A DAC (pronounced "dak") is a circuit that generates an analog output voltage that is proportional to the value of the binary input code. There are a variety of digital-to-analog converter ICs available with varying characteristics. In this section, we will be discussing digital-to-analog conversion techniques and characteristics, a typical DAC data sheet, and DAC applications and testing.

13-2-1   *Binary-Weighted Resistor DAC*

One of the simplest digital-to-analog converter circuits is the **binary-weighted resistor DAC,** shown in Figure 13-3(a). In this circuit, the 4-bit digital input is used to control four HIGH-LOW input switches (SW_1 through SW_4). The setting of these input switches will determine the analog output voltage generated at the output, as listed in the table in Figure 13-3(b). A -10 V reference voltage is applied to a binary-weighted input resistor network, and an inverting op-amp is used to convert a current input into a positive analog output voltage.

    Let us now examine the operation of this circuit in more detail. In the example in Figure 13-3(a), a digital input of 1011 (decimal 11) is being applied to the DAC circuit, so SW_1 is set HIGH, SW_2 is set HIGH, SW_3 is set LOW, and SW_4 is set HIGH. This means that input resistors $8R$, $4R$, and R will contribute to the final output, whereas input resistor $2R$ will not contribute to the final output. A negative reference voltage has been used because the op-amp is connected as an inverting amplifier, so a negative reference voltage input will generate positive analog output voltages, as listed in the table in Figure 13-3(b). The final analog output voltage from this circuit is determined by the reference voltage and the ratio of the feedback resistor R_f to the contributing input resistors. The formula is as follows:

> **Binary-Weighted Resistor Circuit**
>
> A DAC circuit that uses a power-of-2 weighted resistor network to convert a digital signal input into an equivalent analog signal output.

$$V_{out} = -V_{ref}\left(\frac{R_f}{R} + \frac{R_f}{2R} + \frac{R_f}{4R} + \frac{R_f}{8R}\right)$$

For the 1101 (decimal 11) digital input example shown in Figure 13-3(a), the analog output voltage will be

$$V_{out} = -V_{ref}\left(\frac{R_f}{R} + \frac{R_f}{2R} + \frac{R_f}{4R} + \frac{R_f}{8R}\right)$$

$$= -(-10\text{ V})\left(\frac{8\text{ k}\Omega}{10\text{ k}\Omega} + 0 + \frac{8\text{ k}\Omega}{40\text{ k}\Omega} + \frac{8\text{ k}\Omega}{80\text{ k}\Omega}\right)$$

$$= 10\text{ V}(0.8 + 0 + 0.2 + 0.1)$$

$$= 10\text{ V} \times 1.1 = 11\text{ V}$$

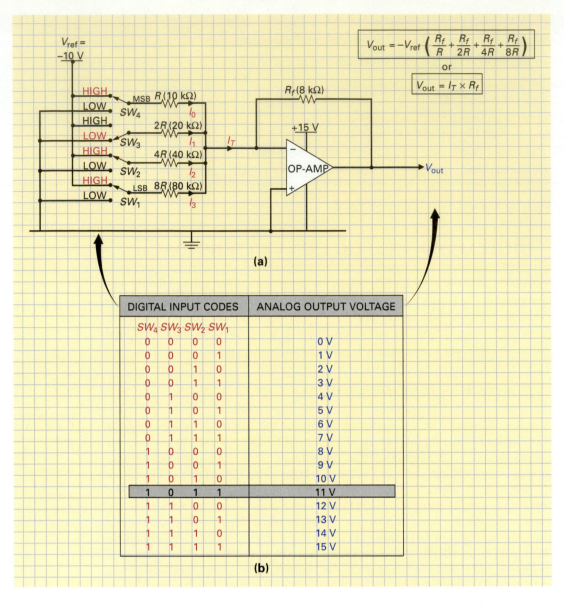

FIGURE 13-3 **Binary-Weighted Resistor Digital-to-Analog Converter (DAC).**

■ **EXAMPLE:**

What analog output voltage will be generated by the circuit in Figure 13-3(a) if a digital input of 1001 is applied?

■ *Solution:*

If a digital input of 1001 (decimal 9) is applied to the DAC circuit, SW_4 will be set HIGH, SW_3 will be set LOW, SW_2 will be set LOW, and SW_1 will be set HIGH. This means that input resistors $8R$ and R will contribute to the final output, whereas input resistors $4R$ and $2R$ will not contribute to the final output. The circuit will therefore generate the following analog output voltage:

$$V_{out} = -V_{ref}\left(\frac{R_f}{R} + \frac{R_f}{2R} + \frac{R_f}{4R} + \frac{R_f}{8R}\right)$$

$$= -(-10\text{ V})\left(\frac{8\text{ k}\Omega}{10\text{ k}\Omega} + 0 + 0 + \frac{8\text{ k}\Omega}{80\text{ k}\Omega}\right)$$

$$= 10 \text{ V}(0.8 + 0 + 0 + 0.1)$$
$$= 10 \text{ V} \times 0.9 = 9 \text{ V}$$

The analog output voltage generated by the circuit in Figure 13-3(a) can also be calculated using the formula

$$V_{\text{out}} = I_T \times R_f$$

For a 1101 (decimal 11) input, therefore, the circuit in Figure 13-3(a) will generate the following analog output voltage:

$$V_{\text{out}} = I_T \times R_f$$

$$I_0 = \frac{-V_{\text{ref}}}{R} = \frac{-(-10 \text{ V})}{10 \text{ k}\Omega} = \frac{10 \text{ V}}{10 \text{ k}\Omega} = 1 \text{ mA}$$

$$I_1 = 0 \text{ mA}$$

$$I_2 = \frac{-V_{\text{ref}}}{4R} = \frac{-(-10 \text{ V})}{40 \text{ k}\Omega} = \frac{10 \text{ V}}{40 \text{ k}\Omega} = 0.25 \text{ mA}$$

$$I_3 = \frac{-V_{\text{ref}}}{8R} = \frac{-(-10 \text{ V})}{80 \text{ k}\Omega} = \frac{10 \text{ V}}{80 \text{ k}\Omega} = 0.125 \text{ mA}$$

$$I_T = I_1 + I_2 + I_3 + I_4 = 1 \text{ mA} + 0 \text{ mA} + 0.25 \text{ mA} + 0.125 \text{ mA}$$
$$= 1.375 \text{ mA}$$

$$V_{\text{out}} = I_T \times R_f = 1.375 \text{ mA} \times 8 \text{ k}\Omega = 11 \text{ V}$$

EXAMPLE:

Using the $V_{\text{out}} = I_T \times R_f$ formula, calculate the analog output voltage generated by the circuit in Figure 13-3(a) if a digital input of 1001 were applied.

Solution:

$$V_{\text{out}} = I_T \times R_f$$

$$I_0 = \frac{-V_{\text{ref}}}{R} = \frac{-(-10 \text{ V})}{10 \text{ k}\Omega} = \frac{10 \text{ V}}{10 \text{ k}\Omega} = 1 \text{ mA}$$

$$I_1 = 0 \text{ mA}$$

$$I_2 = 0 \text{ mA}$$

$$I_3 = \frac{-V_{\text{ref}}}{8R} = \frac{-(-10 \text{ V})}{80 \text{ k}\Omega} = \frac{10 \text{ V}}{80 \text{ k}\Omega} = 0.125 \text{ mA}$$

$$I_T = I_1 + I_2 + I_3 + I_4 = 1 \text{ mA} + 0 \text{ mA} + 0 \text{ mA} + 0.125 \text{ mA} = 1.125 \text{ mA}$$

$$V_{\text{out}} = I_T \times R_f = 1.125 \text{ mA} \times 8 \text{ k}\Omega = 9 \text{ V}$$

13-2-2   *R/2R Ladder DAC*

One of the key disadvantages of the binary-weighted resistor DAC is that the weighted resistors create difficulties as the number of digital input bits increases. For example, an 8-bit DAC will need eight weighted resistors of R, $2R$, $4R$, $8R$, $16R$, $32R$, $64R$, and $128R$. A 16-bit DAC would need sixteen weighted resistors, with the largest resistance being 32,768 times larger than the smallest resistor. The large difference between the LSB resistor value and the MSB resistor value creates several problems. Another disadvantage

R/2R Ladder Circuit

A DAC circuit that uses a resistor network having only two different values— R and 2R.

is the extreme difference in current in each branch of the binary-weighted resistor network. For example, having a very large LSB resistor value will result in an extremely small current value, which is susceptible to noise. In addition, having an extremely small MSB resistor value will result in an extremely large current value, which will cause temperature and load changes.

The problems of the binary-weighted resistor DAC can be overcome by using the **R/2R ladder DAC,** shown in Figure 13-4(a). This R/2R resistor network does not have the resistance range problems of the weighted-resistor network because it only makes use of two resistance values, R and 2R. Like the binary-weighted resistor DAC, the R/2R ladder DAC circuit uses binary-controlled switches to switch binary-weighted currents through to the summing junction of the op-amp. The op-amp is included, as in the binary-weighted resistor DAC, to convert the input current into a proportional output voltage.

The R/2R ladder resistor network operates as a current divider. Figures 13-4(b) through (j) show how this circuit operates as a current divider. In Figure 13-4(b), you can see that below junction D we have a 2R resistor in parallel ($\|$) with a 2R resistor, which is equivalent to R, as redrawn in Figure 13-4(c). Moving on to the redrawn circuit in Figure 13-4(c), you can see that we now have R in series with R directly below junction C, which is equivalent to 2R, as redrawn in Figure 13-4(d). Moving on to the redrawn circuit in Figure 13-4(d), you can see that we now have 2R in parallel with 2R below junction C, which is equivalent to R, as redrawn in Figure 13-4(e). Moving on to the redrawn circuit in Figure 13-4(e), you can see that we now have R in series with R directly below junction B, which is equivalent to 2R, as redrawn in Figure 13-4(f). Moving on to the redrawn circuit in Figure 13-4(f), you can see that we now have 2R in parallel with 2R below junction B, which is equivalent to R, as redrawn in Figure 13-4(g). Moving on to the redrawn circuit in Figure 13-4(g), you can see that we now have R in series with R directly below junction A, which is equivalent to 2R as redrawn in Figure 13-4(h). Moving on to the redrawn circuit in Figure 13-4(h), you can see that we now have 2R in parallel with 2R below junction A, which is equivalent to R, as redrawn in Figure 13-4(i). The R/2R ladder resistor network is therefore equivalent to R, so the total current can be calculated with Ohm's law as follows:

$$I_T = \frac{V_{ref}}{R}$$

In the example circuit in Figure 13-4(a), $V_{ref} = -16$ V and $R = 1$ kΩ, so the total current will be

$$I_T = \frac{V_{ref}}{R} = \frac{-16 \text{ V}}{1 \text{ k}\Omega} = 16 \text{ mA}$$

The R/2R ladder circuit will divide this total current into binary levels, as shown in Figure 13-4(j). The 16 mA arriving at junction A will see a parallel path made up of 2R and 2R, as was shown in Figure 13-4(h), so the 16 mA will be halved, with 8 mA flowing through each branch, as seen in Figure 13-4(j). The 8 mA arriving at junction B will see a parallel path made up of 2R and 2R, as was shown in Figure 13-4(f), so the 8 mA will be halved, with 4 mA flowing through each branch, as seen in Figure 13-4(j). The 4 mA arriving at junction C will see a parallel path made up of 2R and 2R, as was shown in Figure 13-4(d), so the 4 mA will be halved, with 2 mA flowing through each branch, as seen in Figure 13-4(j). The 2 mA arriving at junction D will see a parallel path made up of 2R and 2R, as was shown in Figure 13-4(b), so the 2 mA will be halved, with 1 mA flowing through each branch, as seen in Figure 13-4(j). These binary-weighted current values are switched through to the summing junction of the op-amp based on the condition of the binary-controlled electronic switches. For example, if a 1011 digital input were applied to this R/2R DAC circuit, SW_1 would be in its HIGH position, SW_2 would be set LOW, SW_3 would be set HIGH, and SW_4 would be set HIGH, as shown in Figure 13-4(a).

FIGURE 13-4 *R/2R* Digital-to-Analog Converter (DAC).

This means that the 8 mA, 2 mA, and 1 mA current values will be switched through to the summing junction of the op-amp, so I_T will equal 11 mA (8 mA + 2 mA + 1 mA = 11 mA). Stated as a formula, I_T is equal to

$$I_{out} = I_{ref}\left(\frac{D_3}{2} + \frac{D_2}{4} + \frac{D_1}{8} + \frac{D_0}{16}\right)$$

$$I_{out} = I_{ref}\left(\frac{D_3}{2} + \frac{D_2}{4} + \frac{D_1}{8} + \frac{D_0}{16}\right)$$

$$= 16 \text{ mA}\left(\frac{1}{2} + \frac{0}{4} + \frac{1}{8} + \frac{1}{16}\right)$$

$$= 16 \text{ mA}\left(\frac{8 + 0 + 2 + 1}{16}\right)$$

$$= 16 \text{ mA} \times \left(\frac{11}{16}\right)$$

$$= 11 \text{ mA}$$

For a 1011 (decimal 11) digital input, therefore, this $R/2R$ ladder DAC will generate an analog output voltage of

$$V_{out} = I_T \times R_f = 11 \text{ mA} \times 1 \text{ k}\Omega = 11 \text{ V}$$

◼ EXAMPLE:

Calculate the analog output voltage generated by the circuit in Figure 13-4(a) if a digital input of 0110 were applied.

◼ *Solution:*

If a digital input of 0110 (decimal 6) were applied to the $R/2R$ ladder DAC circuit, SW_4 would be set LOW, SW_3 would be set HIGH, SW_2 would be set HIGH, and SW_1 would be LOW. The total current applied to the summing junction of the op-amp would be

$$I_T = I_0 + I_1 + I_2 + I_3 = 0 + 4 \text{ mA} + 2 \text{ mA} + 0 = 6 \text{ mA}$$

The analog output voltage will therefore be

$$V_{out} = I_T \times R_f = 6 \text{ mA} \times 1 \text{ k}\Omega = 6 \text{ V}$$

13-2-3   *DAC Characteristics*

The characteristics of a device determine its quality and its suitability for different applications. Digital-to-analog converters are normally classified as being either a current output or voltage output IC, based on whether a voltage-to-current op-amp is included within the IC's internal circuitry. The other key specifications to look for are the number of digital input bits and the output voltage range or current range. For example, Figure 13-5(a) shows a 0–5 V 8-bit DAC. In this test circuit, the DAC is being driven by an 8-bit free-running counter. This means that the DAC will receive a continually incrementing digital count input and will therefore generate a continually increasing analog output voltage waveform. This **staircase waveform** output will begin at 0 V when the digital input is 0000 0000 (decimal 0), will rise to a maximum analog output voltage when the digital input is 1111 1111 (decimal 255), and will then repeat the cycle. The table in Figure 13-5(b) explains the staircase output waveform in more detail. As you know, each of the digital input bits is weighted, and therefore:

> **Staircase Waveform**
>
> A waveform that continually steps up in value from zero to a maximum value and then repeats.

the LSB will have a weight of $^{5\text{ V}}/_{256} = 19.5$ mV,

the next bit will have a weight of $^{5\text{ V}}/_{128} = 39$ mV,

the next bit will have a weight of $^{5\text{ V}}/_{64} = 78$ mV,

the next bit will have a weight of $^{5\text{ V}}/_{32} = 156$ mV,

the next bit will have a weight of $^{5\text{ V}}/_{16} = 312$ mV,

the next bit will have a weight of $^{5\text{ V}}/_{8} = 625$ mV,

the next bit will have a weight of $^{5\text{ V}}/_{4} = 1.25$ V,

and the MSB will have a weight of $^{5\text{ V}}/_{2} = 2.5$ V.

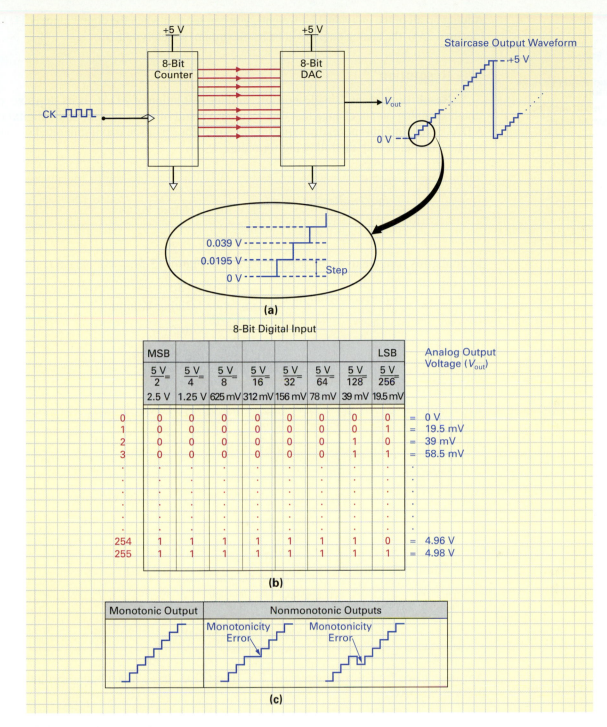

FIGURE 13-5 DAC Characteristics.

As the digital input count advances by 1, the analog output of the DAC will increase in 19.5 mV steps, as shown in the inset in Figure 13-5(a). The DAC's analog output, therefore, will generate 255 steps and have a **step size** of 19.5 mV.

$$Step\ Size = \frac{Max.\ Rated\ Output\ Voltage}{2^n}$$

$$Step\ Size = \frac{Max.\ Rated\ Output\ Voltage}{2^n} = \frac{5\ V}{2^8} = \frac{5\ V}{256} = 19.5\ mV$$

Adding up all of the individual bit weights, you can see that when the digital input is 1111 1111 (decimal 255), the analog output will be its maximum, which is 4.98 V.

Resolution

Resolution
A term used to describe the amount of detail present.

Resolution means "smallest detail," and the resolution of a DAC is its step size, or the smallest analog output change that can occur as a result of an increment in the digital input. A DAC's resolution is normally expressed as a percentage of the maximum rated output and can be calculated with the following formula:

$$Percent\ Resolution = \frac{Step\ Size}{Max.\ Rated\ Output\ Voltage} \times 100$$

To apply this formula to an example, the DAC in Figure 13-5(a) will have a resolution of

$$Percent\ Resolution = \frac{Step\ Size}{Max.\ Rated\ Output\ Voltage} \times 100$$
$$= \frac{0.0195\ V}{5\ V} \times 100$$
$$= 0.0039 \times 100 = 0.39\%$$

EXAMPLE:

Calculate the resolution of a 0−5 V 10-bit DAC.

Solution:

First calculate the step size, and then calculate the DAC's resolution.

$$Step\ Size = \frac{Max.\ Rated\ Output\ Voltage}{2^n} = \frac{5\ V}{2^{10}} = \frac{5\ V}{1,024} = 4.9\ mV$$
$$Percent\ Resolution = \frac{Step\ Size}{Max.\ Rated\ Output\ Voltage} \times 100$$
$$= \frac{4.9\ mV}{5\ V} \times 100$$
$$= 0.0009765 \times 100 = 0.098\%$$

As you can see from this example, a 0-5 V 10-bit DAC has a much better resolution than a 0-5 V 8-bit DAC.

Monotonicity

Monotonic
A term used to describe the condition of a DAC's output. A monotonic output from a DAC will have an increasing analog output for every increase in the digital input.

When a free-running counter is connected to the input of a DAC, its output voltage should continually step up, producing a staircase waveform. A DAC is said to be **monotonic** if its output increases for each successive increase in the digital input, as shown in the left portion of Figure 13-5(c). If an increase in the digital input does not cause the output of the DAC to step up, or if the output steps backwards, as shown in the right portion of Figure 13-5(c), the DAC is said to be nonmonotonic. Monotonicity errors can be caused by circuit malfunctions within the DAC IC, and they can also be caused by circuit problems external to the IC. As always, you will need to isolate the device to determine whether the problem is source or load. We will be discussing the testing of DACs and the troubleshooting of data conversion circuits in a later section in this chapter.

CHAPTER 13 / ANALOG AND DIGITAL SIGNAL CONVERTERS

Settling Time

The **settling time** of a DAC is the time it takes for the DAC's analog output to settle to 99.95% of its new value after a digital input has been applied. This DAC characteristic describes how fast a conversion can be made, and so it will determine the maximum frequency at which the DAC can operate. As an example, the MC1408 current DAC has a settling time of about 300 ns, whereas most voltage output DACs will have settling times ranging from 50 ns to 20 μs.

Settling Time
The time it takes for a DAC's analog output to settle to 99.95% of its new value.

Accuracy

The **relative accuracy** of a DAC describes how much the output level has deviated from its ideal theoretical output value. This accuracy specification is usually expressed as a percentage and describes the percentage of error rather than the percentage of accuracy. As an example, the MC1408 DAC has a ±0.19% error relative to the full-scale output.

Relative Acccuracy
The amount a DAC's output level has deviated relative to an ideal theoretical output value.

13-2-4 A DAC Application Circuit

Figure 13-6 shows the MC1408 8-bit current DAC. This IC contains an *R/2R* ladder circuit, a reference current circuit, and eight binary current steering switches. In Figure 13-6 this IC is connected to function as a 0 to +5 V 8-bit digital-to-analog converter. A +5 V reference voltage ($+V_{\text{ref}}$), in conjunction with rheostat R_1, sets up a reference current of 2 mA for the DAC's internal ladder circuit. When all of the digital inputs are HIGH, therefore, the analog output current (I_{out}) will be

$$I_{\text{out}} = I_{\text{ref}} \left(\frac{A_1}{2} + \frac{A_2}{4} + \frac{A_3}{8} + \frac{A_4}{16} + \frac{A_5}{32} + \frac{A_6}{64} + \frac{A_7}{128} + \frac{A_8}{256} \right)$$

$$= 2 \text{ mA} \left(\frac{1}{2} + \frac{1}{4} + \frac{1}{8} + \frac{1}{16} + \frac{1}{32} + \frac{1}{64} + \frac{1}{128} + \frac{1}{256} \right)$$

$$= 2 \text{ mA} \times \left(\frac{255}{256} \right)$$

$$= 1.99 \text{ mA}$$

FIGURE 13-6   Application—An 8-Bit Digital-to-Analog Converter Circuit.

The analog output current is applied to a 741 op-amp, which will develop an output voltage that is equal to

$$V_{out} = I_T \times R_f$$

■ EXAMPLE:

What will be the analog output voltage from the circuit in Figure 13-6 if a digital input of 1100 0001 is applied?

■ Solution:

The first step should be to calculate I_{out} and then calculate V_{out}.

$$I_{out} = I_{ref}\left(\frac{A_1}{2} + \frac{A_2}{4} + \frac{A_3}{8} + \frac{A_4}{16} + \frac{A_5}{32} + \frac{A_6}{64} + \frac{A_7}{128} + \frac{A_8}{256}\right)$$

$$= 2 \text{ mA}\left(\frac{1}{2} + \frac{1}{4} + \frac{0}{8} + \frac{0}{16} + \frac{0}{32} + \frac{0}{64} + \frac{0}{128} + \frac{1}{256}\right)$$

$$= 2 \text{ mA} \times \left(\frac{128 + 64 + 1}{256}\right)$$

$$= 2 \text{ mA} \times \left(\frac{193}{256}\right) = 1.51 \text{ mA}$$

$$V_{out} = I_T \times R_f = 1.51 \text{ mA} \times 2.5 \text{ k}\Omega = 3.8 \text{ V}$$

13-2-5   *Testing DACs*

One of the easiest ways to test a DAC is to have a free-running counter generate the digital input and have an oscilloscope monitor the analog output, as shown in Figure 13-7. Ideally, the output should be a straight-line monotonic staircase, as shown in Figure 13-7. If a fault exists, you may obtain an incorrect output similar to one of the waveform examples also shown in Figure 13-7. In this situation, you will need to isolate whether the problem is internal or external to the DAC IC and then replace the faulty device.

SELF-TEST EVALUATION POINT FOR SECTION 13-2

Use the following questions to test your understanding of Section 13-2:

1. What is the main disadvantage of the weighted-resistor DAC?
2. What is the purpose of the op-amp in a DAC circuit?
3. Define the following terms:
   a. staircase output
   b. resolution
   c. step size
   d. monotonicity
   e. settling time
   f. accuracy
4. How can a DAC be tested?

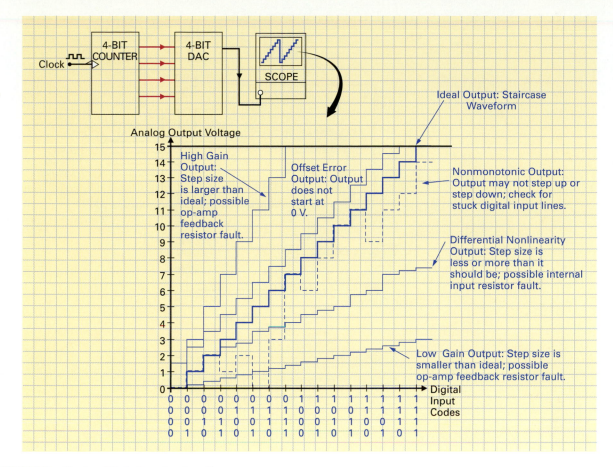

FIGURE 13-7 Testing Digital-to-Analog Converters.

13-3 ANALOG-TO-DIGITAL CONVERTERS (ADCs)

An ADC (pronounced "aye-dee-see") is a circuit that generates a binary output code that is proportional to an analog input voltage. A variety of analog-to-digital converter ICs are available with varying characteristics. In this section, we will be discussing analog-to-digital conversion techniques and characteristics, a typical ADC data sheet, and ADC applications and testing.

13-3-1 *Staircase ADC*

One of the simplest analog-to-digital converter circuits is the basic **staircase ADC,** shown in Figure 13-8(a). To convert an analog input into a digital output, this ADC circuit has a counter, comparator, AND gate, and a DAC. The analog input voltage is applied to the negative input of the comparator. The positive input of the comparator is a staircase voltage waveform from the DAC. The DAC generates a staircase voltage waveform because its digital input is connected to a counter incremented by a clock signal. This clock signal is either switched through the AND gate or blocked by the AND gate, based on the logic level output from the comparator.

To explain the operation of the circuit in Figure 13-8(a), let us step through the timing waveforms shown in Figure 13-8(b), which show how an analog input voltage of 5.9 V is converted to the equivalent 4-bit digital code 0110 (decimal 6). To begin with, the active-LOW start conversion ($\overline{SC}$) input is used to RESET the counter to 0 V. This zero count will be

Staircase ADC

An ADC circuit that uses a staircase waveform and comparator to determine the value of the analog input voltage and generate an equivalent digital output.

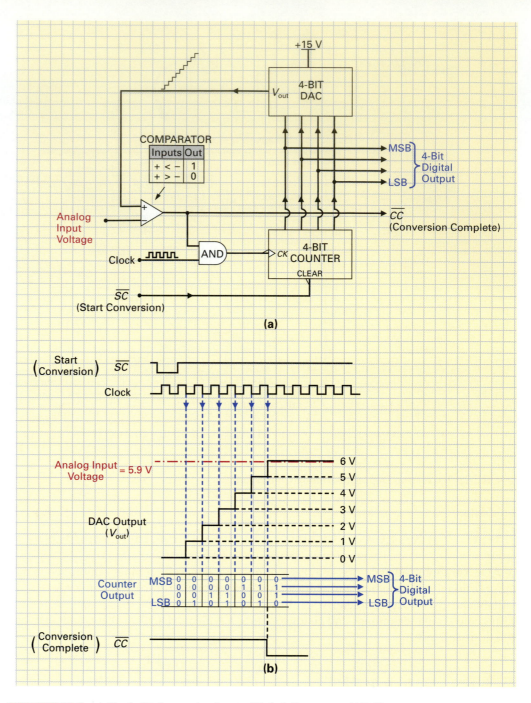

FIGURE 13-8 **A Basic Staircase Analog-to-Digital Converter (ADC).**

applied to the comparator's positive input. Since the comparator's positive input of 0 V is less than the negative analog input of 5.9 V at this time, the comparator will give a HIGH output.

The HIGH from the comparator will enable the AND gate, allowing the clock signal to pass through the AND gate and increment the counter to 0001 (decimal 1). The 0001 count will cause the DAC to generate a 1 V output, which is still less than the analog input voltage of 5.9 V, so the comparator will continue to enable the AND gate and allow the counter to be incremented further to 0010 (decimal 2). The 0010 count will cause the DAC to generate a 2 V output, which is still less than the analog input voltage of 5.9 V, so the comparator will continue to enable the AND gate and allow the counter to be incremented further to 0011 (decimal 3). This process will continue until the counter has been incremented to 0110 (decimal 6), and the DAC is therefore generating an output of 6 V. At this time, the comparator's positive input is greater than the analog input voltage of 5.9 V, so the comparator's output will go LOW.

The LOW output will disable the AND gate, preventing the clock signal from passing through to the counter, so the counter will hold its count of 0110. In addition to controlling the AND gate, the output of the comparator is also used to generate an active-LOW conversion complete ($\overline{CC}$) output. This $\overline{CC}$ output is used to indicate that the present 4-bit code appearing at the digital output is the binary equivalent of the analog input voltage.

EXAMPLE:

What digital output will be generated by the circuit in Figure 13-8(a) if an analog input voltage of 11.8 V were applied?

Solution:

If an analog input of 11.8 V were applied to the ADC circuit, the counter would be incremented until it reached a count of 1100 (decimal 12).

13-3-2   *Successive Approximation ADC*

One of the key disadvantages with the staircase ADC is that its conversion speed is dependent on the analog input voltage. For example, with the circuit in Figure 13-8, it would take fifteen clock pulses to convert an analog input voltage of 15 V to a digital output of 1111. For an 8-bit ADC, in the worst case, it would take 255 clock pulses to convert a maximum analog input voltage to 1111 1111. The worst-case conversion time, therefore, is equal to

$$t_{\text{conv(max)}} = (2^n - 1) \times t_{ck}$$

$$t_{\text{conv(max)}} = \text{maximum conversion time}$$
$$n = \text{number of bits}$$
$$t_{ck} = \text{clock pulse period}$$

EXAMPLE:

Calculate the maximum conversion time of a 12-bit staircase ADC converter with a 20 kHz clock pulse applied.

Solution:

First, we will need to determine the clock pulse input period, and then we can use the maximum conversion time formula.

$$t_{ck} = \frac{1}{f} = \frac{1}{20 \text{ kHz}} = 50 \text{ μs}$$

$$\begin{aligned} t_{\text{conv (max)}} &= (2^n - 1) \times t_{ck} \\ &= (2^{12} - 1) \times 50 \text{ μs} \\ &= (4{,}095) \times 50 \text{ μs} \\ &= 0.2 \text{ s} \end{aligned}$$

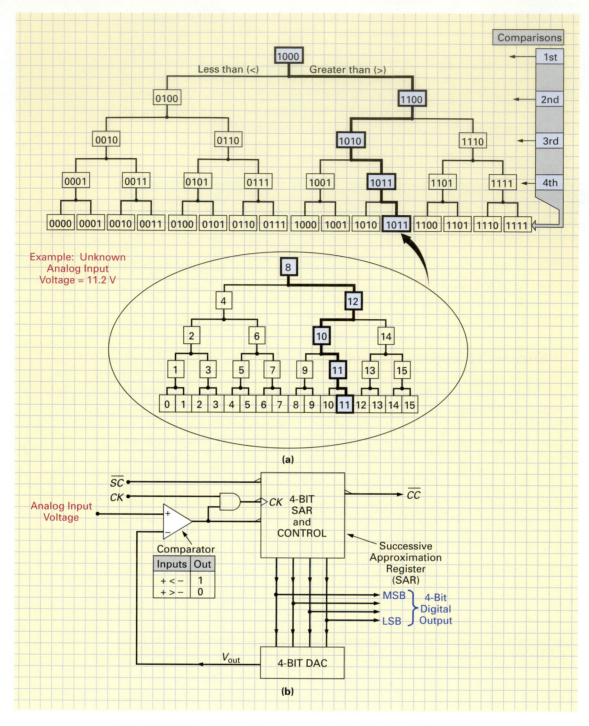

FIGURE 13-9 A Successive Approximation Analog-to-Digital Converter (ADC).

The slow conversion time of the staircase ADC can be overcome by using the **successive approximation ADC,** shown in Figure 13-9(b). Before we discuss the operation of this circuit, let us first examine the successive approximation (SA) method. As an example, Figure 13-9(a) shows how the SA method can be used to determine the digital equivalent of the analog input by making only four comparisons. In the example shown highlighted in Figure 13-9(a), an analog input voltage of 11.2 V has been applied. Following the comparisons shown in binary (and decimal in the inset), you can see that the SA method first sets the MSB HIGH (1000, or decimal 8) and then tests to see whether this value is greater than or less than the analog input voltage. Since the analog input voltage of 11.2 V is greater than the first SA value of 1000, the MSB is kept HIGH, the next MSB is set HIGH (1100, or decimal 12), and another comparison is made. Since the analog input voltage of 11.2 V is less than the second SA value of 1100, the second MSB is reset LOW, the next MSB is set HIGH (1010, or decimal 10), and another comparison is made. Since the analog input voltage of 11.2 V is greater than the third SA value of 1010, the third MSB is kept HIGH, the LSB is set HIGH (1011, or decimal 11), and a final comparison is made. Since the analog input voltage of 11.2 V is greater than the fourth SA value of 1011, the LSB is kept HIGH, and an equivalent digital code has been found.

Referring to the successive approximation ADC block diagram in Figure 13-9(b), you can see that this circuit also contains a DAC and a comparator. This circuit, however, does not contain a counter since the *successive approximation register (SAR)* and control circuits will generate the special sequence of binary codes needed to converge on the digital equivalent of the analog input voltage. The start conversion ($\overline{SC}$) input is used to reset the SAR, and initiate the series of four comparisons. The SA values at the output of the SAR register are applied to the DAC, and as with the staircase ADC, the DAC output controls a comparator, which in turn controls an AND gate, which in turn controls the clock input. After each of the bits has been set and tested, the conversion complete output ($\overline{CC}$) is used to indicate that the present 4-bit code appearing at the digital output is the binary equivalent of the analog input voltage.

The key advantage of the successive approximation ADC is its conversion speed. Unlike the staircase ADC, the SA ADC does not need to sequence through every possible binary combination to find a match. An SA ADC will only need to set and test each of the digital output bits, so a 4-bit ADC will only need to make four comparisons, an 8-bit ADC will only need to make eight comparisons, and so on. This means that an SA ADC's conversion time will be the same for any input value. Having a fixed conversion time for any input value will make system timing a lot easier.

13-3-3 *Flash ADC*

One of the simplest ways to convert an analog input voltage into an equivalent digital output is to use a **flash ADC.** To explain the operation of this ADC type, Figure 13-10(a) shows a 3-bit flash converter circuit. Flash ADCs use comparators to compare the analog input voltage to successively smaller equally spaced reference voltages provided by a resistive voltage-divider network. A comparator will generate a HIGH output whenever the analog input voltage is greater than the applied reference voltage. The outputs from the comparators are applied to a 3-bit priority encoder. The 3-bit binary output code generated by the priority encoder will be based on which of the highest-order inputs is HIGH, as shown in Figure 13-10(b). For example, when an analog input of 4.1 V is applied, the priority encoder inputs 4, 3, 2, and 1 will be HIGH, so the binary output code generated will be 100 (decimal 4) since 4 is the highest-order input.

The advantage of this analog-to-digital conversion method is its fast conversion time, which is typically in the nanoseconds (hence the name "flash"). The disadvantage of this ADC type is the large number of comparators that are needed. In most cases, $2^n - 1$ comparators are needed, so an 8-bit flash ADC will need 255 comparators, a 16-bit flash ADC will need 65,535 comparators, and so on.

Flash ADC
An ADC circuit that uses a resistive voltage divider and comparators for a fast conversion time.

13-3-4 *An ADC Application Circuit*

Figure 13-11 shows an ADC0803 8-bit successive approximation ADC that has an on-chip clock circuit connected to function as an 8-bit analog-to-digital converter. In this example circuit, the 8-bit digital output from the ADC0803 (pins 11 through 18) have been applied to an LED display. In most digital system applications, the parallel digital output will be under three-state control, based on the active-LOW $\overline{RD}$ control input. The active-LOW $\overline{WR}$ control input

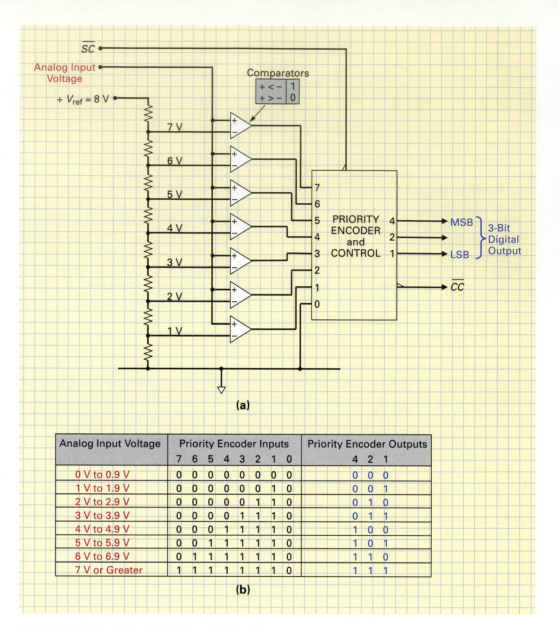

(a)

Analog Input Voltage	Priority Encoder Inputs								Priority Encoder Outputs		
	7	6	5	4	3	2	1	0	4	2	1
0 V to 0.9 V	0	0	0	0	0	0	0	0	0	0	0
1 V to 1.9 V	0	0	0	0	0	0	1	0	0	0	1
2 V to 2.9 V	0	0	0	0	0	1	1	0	0	1	0
3 V to 3.9 V	0	0	0	0	1	1	1	0	0	1	1
4 V to 4.9 V	0	0	0	1	1	1	1	0	1	0	0
5 V to 5.9 V	0	0	1	1	1	1	1	0	1	0	1
6 V to 6.9 V	0	1	1	1	1	1	1	0	1	1	0
7 V or Greater	1	1	1	1	1	1	1	0	1	1	1

(b)

FIGURE 13-10   A Flash Analog-to-Digital Converter.

is used to start a conversion, and the active-LOW $\overline{INTR}$ output is used to indicate when a conversion is complete. The $\overline{INTR}$ output is connected directly to the $\overline{WR}$ input to ensure that the conversion is continuous. The resistor, capacitor, and buffer circuit are included to ensure that the $\overline{WR}$ input is taken LOW when power is first applied to the circuit.

The RC circuit connected to the $CLK\text{-}R$ and $CLK\text{-}IN$ pins will determine the converter's internal clock circuit frequency based on the following formula:

$$f = \frac{1}{1.1\,RC}$$

In the example circuit in Figure 13–11, the converter clock frequency will be

$$f = \frac{1}{1.1\,RC} = \frac{1}{1.1 \times 10\ k\Omega \times 150\ pF} = 606\ kHz$$

FIGURE 13-11 Application—An 8-Bit Analog-to-Digital Converter Circuit.

The ADC0803 has both a positive analog input (pin 6) and a negative analog input (pin 7). This versatility means that you can ground pin 7 and have a positive input, ground pin 6 and have a negative input, or have the ADC generate an output code based on the difference in voltage between pins 6 and 7 (differential input).

The $+5$ V supply voltage is applied to pin 20, and both the analog and digital ground pins must be grounded. The $V_{ref}/2$ input to the ADC0803 is used to determine which analog input voltage will generate the maximum digital output of 1111 1111 (decimal 255). If this input is not connected, the supply voltage of $+5$ V will mean that the analog input range will be from 0 V to $+5$ V. When a voltage is applied to the $V_{ref}/2$ input, a different analog input range can be obtained. For example, to generate a maximum digital output code of 1111 1111 when an analog input of $+4$ V is applied, connect $+2$ V to the $V_{ref}/2$ input ($+4$ V$/2 = 2$ V).

The application circuit in Figure 13–11 includes an LF198 **sample-and-hold amplifier** at the analog input. This circuit has been included to prevent any error due to analog input voltage changes during conversion. Up until this time, we have assumed that the analog input voltage remains constant while the converter is making the conversion. This does not always happen, and if a change in the analog input voltage were to occur while

Sample-and-Hold Amplifier

A circuit that will sample the value of a changing analog input signal when triggered and hold this value constant at its output.

Input Voltage	Ideal Output		Incorrect Codes		Missing Codes		Offset Output	
	8 4 2 1	Display	8 4 2 1	Display	8 4 2 1	Display	8 4 2 1	Display
0 V to 0.9 V	0 0 0 0	0	0 0 0 0	0	0 0 0 0	0	0 0 1 1	3
1 V to 1.9 V	0 0 0 1	1	0 0 0 1	1	0 0 0 1	1	0 1 0 0	4
2 V to 2.9 V	0 0 1 0	2	0 0 0 1	1	0 0 1 0	2	0 1 0 1	5
3 V to 3.9 V	0 0 1 1	3	0 0 0 0	0	0 0 1 1	3	0 1 1 0	6
4 V to 4.9 V	0 1 0 0	4	0 1 0 0	4	0 1 0 0	4	0 1 1 1	7
5 V to 5.9 V	0 1 0 1	5	0 1 0 1	5	0 1 0 0	4	1 0 0 0	8
6 V to 6.9 V	0 1 1 0	6	0 1 0 1	5	0 1 1 0	6	1 0 0 1	9
7 V to 7.9 V	0 1 1 1	7	0 1 0 1	5	0 1 1 1	7	1 0 1 0	A
8 V to 8.9 V	1 0 0 0	8	0 1 1 0	6	1 0 0 0	8	1 0 1 1	b
9 V to 9.9 V	1 0 0 1	9	1 0 0 1	9	1 0 0 1	9	1 1 0 0	C
10 V to 10.9 V	1 0 1 0	A	1 0 1 0	A	1 0 1 0	A	1 1 0 1	d
11 V to 11.9 V	1 0 1 1	b	1 0 1 1	b	1 0 1 0	A	1 1 1 0	E
12 V to 12.9 V	1 1 0 0	C	1 0 1 1	b	1 1 0 0	C	1 1 1 1	F
13 V to 13.9 V	1 1 0 1	d	1 1 0 1	d	1 1 0 1	d	1 1 1 1	F
14 V to 14.9 V	1 1 1 0	E	1 1 1 0	E	1 1 1 0	E	1 1 1 1	F
15 V or Greater	1 1 1 1	F	1 1 1 0	E	1 1 1 1	F	1 1 1 1	F

FIGURE 13-12   **Testing Analog-to-Digital Converters.**

the converter was performing the conversion, the digital output would be inaccurate. To prevent this problem, a sample-and-hold (SAH) amplifier can be included at the input of an ADC to sample the analog input voltage and hold this sample constant while the ADC makes the conversion. The sample-and-hold amplifier contains two unity-gain noninverting amplifiers ($V_{out} = V_{in}$), a logic-controlled switch, and an external capacitor, as shown in the inset in Figure 13-11. When the SAH control input is made HIGH, the logic-controlled switch is closed, and the external capacitor charges quickly to equal the analog input voltage. When the logic-controlled switch is made LOW, the capacitor will retain the analog input voltage level, and this unchanging voltage will be applied to the ADC. Using a capacitor of 0.001 µF, the LF198 must have its switch closed for at least 4 µs in order to get an accurate sample (acquisition time), and when in the hold condition the output voltage will decrease due to leakage paths at a rate of 30 mV/s (droop rate).

13-3-5   *Testing ADCs*

One of the easiest ways to test an ADC is to apply a linear ramp to the analog input and have a display monitor the digital output, as shown in Figure 13-12. If operating correctly, the output should be an incrementing digital code, as listed in the table in Figure 13-12 in the ideal output column. If a fault exists, you may obtain incorrect output codes, missing codes, or an offset output, similar to the other columns listed in the table in Figure 13-12. In this situation, you will need to isolate whether the problem is internal or external to the ADC IC and then replace the faulty device.

Use the following questions to test your understanding of Section 13-3.

1. What is the main disadvantage of the staircase ADC?
2. What two types of ADCs make use of a DAC?
3. How does a successive approximation ADC operate?
4. What is the fastest type of ADC?
5. How can an ADC be tested?

REVIEW QUESTIONS

Multiple-Choice Questions

1. A _____ converts an analog input voltage into a proportional digital output code.
   - a. ROM
   - b. DAC
   - c. ADC
   - d. RWM

2. A _____ converts a digital input code into a proportional analog output voltage.
   - a. ROM
   - b. DAC
   - c. ADC
   - d. RWM

3. Which of the following signal converters is the most frequently used DAC?
   - a. Binary-weighted
   - b. Staircase
   - c. R/2R ladder
   - d. Successive approximation
   - e. Flash

4. Which of the following signal converters is the fastest ADC?
   - a. Binary-weighted
   - b. Staircase
   - c. R/2R ladder
   - d. Successive approximation
   - e. Flash

5. Most DACs give a/an _____ output.
   - a. parallel binary
   - b. analog current
   - c. serial binary
   - d. analog voltage

6. Most ADCs give a/an _____ output.
   - a. parallel binary
   - b. analog current
   - c. serial binary
   - d. analog voltage

7. The op-amp in a DAC circuit is used to convert a/an _____ input into a _____ output.
   - a. parallel binary, current
   - b. serial binary, voltage
   - c. current, voltage
   - d. analog voltage, current

8. Which of the following data converter types can be used to generate a staircase output waveform?
   - a. Staircase
   - b. Flash
   - c. R/2R
   - d. Successive approximation

9. The _____ of a DAC describes the smallest analog output change.
   - a. monotonicity
   - b. settling time
   - c. accuracy
   - d. resolution

10. What circuit can be used to hold the input to an ADC constant during the conversion cycle?
    - a. R/2R circuit
    - b. Flash circuit
    - c. V-to-I circuit
    - d. Sample-and-hold circuit

Practice Problems

11. If a 4-bit DAC has a 12 V reference applied, calculate each of the following:
    - a. step size
    - b. percent resolution
    - c. the weight of each binary column
    - d. the output voltage when 0010 is applied
    - e. the output voltage when 1011 is applied

12. If an 8-bit DAC has a 5 V reference applied, calculate each of the following:
    - a. step size
    - b. percent resolution
    - c. the weight of each binary column
    - d. the output voltage when 0000 0101 is applied
    - e. the output voltage when 1011 0111 is applied

13. If an MC1408 has a 5 mA reference current applied, what will be the output current for the following digital inputs?
    - a. 1010 0001
    - b. 0001 0101
    - c. 1111 1111
    - d. 0001 0001

14. Calculate the maximum conversion time of a 16-bit staircase ADC converter with a 200 kHz clock.

To practice your circuit recognition and operation skills, refer to Figure 13-13 and answer the following five questions.

15. Would the circuit indicated by the ? in this circuit be a DAC or an ADC?

16. Sketch the output you would expect from the ? circuit.

17. Sketch the output you would expect from the lowpass filter circuit.

18. Why was the lowpass filter included in this circuit?

19. If the converter in this circuit had a 3 V reference applied, calculate each of the following:
    - a. step size
    - b. percent resolution
    - c. the weight of each binary column
    - d. the output voltage when 0000 0101 is applied
    - e. the output voltage when 1011 0111 is applied

To practice your circuit recognition and operation skills, refer to Figure 13-14 and answer the following four questions.

20. What is the purpose of this circuit?

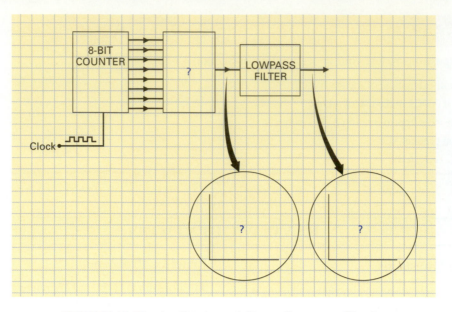

FIGURE 13-13   Application—A Ramp Generator Circuit.

21. Would the circuit indicated by the ? in this circuit be a DAC or an ADC?

22. What is the function of the PLD in this circuit?

23. Briefly describe the operation of this circuit.

To practice your circuit recognition and operation skills, refer to Figure 13-15 and answer the following four questions.

24. Is this circuit a DAC or an ADC?

25. What circuit name is given to this type of converter?

26. Indicate what logic level will appear at test points 1, 2, and 3 for each of the four input voltage ranges listed in the table in this figure.

27. Complete the table shown in this figure to show what logic levels will be present at the MSB and LSB outputs for the four input voltage ranges listed in the table.

To practice your circuit recognition and operation skills, refer to Figure 13-16 and answer the following three questions.

28. Which of the blocks in this system will perform the following conversions?
    a. Parallel digital input to a serial digital output
    b. Serial digital input to a parallel digital output
    c. Digital input to an analog output
    d. Analog input to a digital output

29. Sketch the output for the input shown in Figure 13-16(b).

30. Sketch the output for the input shown in Figure 13-16(c).

Web Site Questions

Go to the web site http://www.prenhall.com/cook, select the textbook *Practical Digital Electronics,* select this chapter, and then follow the instructions when answering the multiple choice practice problems.

FIGURE 13-14   Application—A Digital Thermometer.

FIGURE 13-15   Application—A Converter.

Input Voltage	TP1	TP2	TP3	MSB	LSB
0 V to 0.9 V					
1 V to 1.9 V					
2 V to 2.9 V					
3 V to 3.9 V					

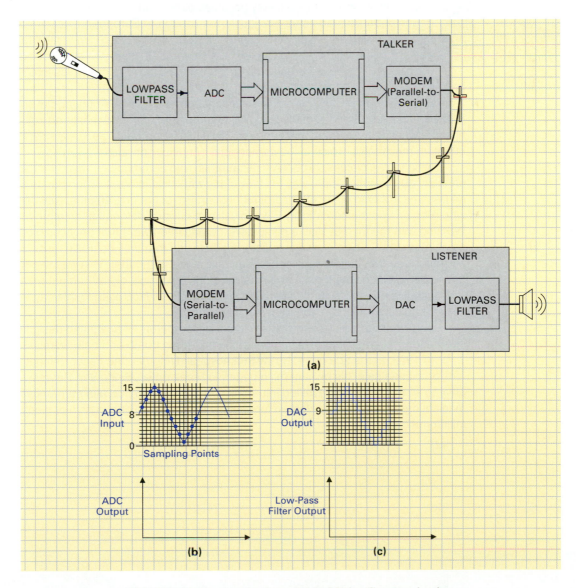

FIGURE 13-16   Application—Digital Voice Communication.

Computer Hardware and Software

Making an Impact

John Von Neuman, a mathematics professor at the Institute of Advanced Studies, delighted in amazing his students by performing complex computations in his head faster than they could with pencil, paper, and reference books. He possessed a photographic memory, and at his frequently held lavish parties in his home in Princeton, New Jersey, he gladly occupied center stage to recall from memory entire pages of books read years previously, the lineage of European royal families, and a store of controversial limericks. His memory, however, failed him in his search for basic items in a house he had lived in for seventeen years. On many occasions when traveling, he would become so completely absorbed in mathematics that he would have to call his office to find out where he was going and why.

Born in Hungary, he was quick to demonstrate his genius. At the age of six he would joke with his father in classical Greek. At the age of eight, he had mastered calculus, and in his mid-twenties he was teaching and making distinct contributions to the science of quantum mechanics, which is the cornerstone of nuclear physics.

Next to fine clothes, expensive restaurants, and automobiles, which he had to replace annually due to smash-ups, was his love of his work. His interest in computers began when he became involved in the top-secret Manhattan Project at Los Alamos, New Mexico, where he proved mathematically the implosive method of detonating an atom bomb. Working with the then-available computers, he became aware that they could become much more than a high-speed calculator. He believed that they could be an all-purpose scientific research tool, and he published these ideas in a paper. This was the first document to outline the logical organization of the electronic digital computer and was widely circulated to all scientists throughout the world. In fact, even to this day, scientists still refer to computers as "Von Neuman machines."

Von Neuman collaborated on a number of computers of advanced design for military applications, such as the development of the hydrogen bomb and ballistic missiles.

In 1957, at the age of 54, he lay in a hospital dying of bone cancer. Under the stress of excruciating pain, his brilliant mind began to break down. Since Von Neuman had been privy to so much highly classified information, the Pentagon had him surrounded with only medical orderlies specially cleared for security for fear he might, in pain or sleep, give out military secrets.

Introduction

The microprocessor is a large, complex integrated circuit containing all the computation and control circuitry for a small computer. The introduction of the microprocessor has caused a dramatic change in the design of digital electronic systems. Before microprocessors, random or hard-wired digital circuits were designed using individual logic blocks, such as gates, flip-flops, registers, counters, and so on. These building blocks were interconnected to achieve the desired end, as required by the application. Using random logic, each application required a unique design and there was little similarity between one system and another. This approach was very similar to analog circuit design, in that the structure of the circuit was governed by the function that needed to be performed. Once constructed, the function of the circuit was difficult to change.

The microprocessor, on the other hand, provides a general-purpose control system which can be adapted to a wide variety of applications with only slight circuit modification. The individuality of a microprocessor-based system is provided by a list of instructions (called the program) that controls the system's operation. A microprocessor-based system therefore has two main elements—the actual components, or hardware, and the programs, or software.

In the first section of this chapter, you will be introduced to the hardware that makes up a microprocessor system and the software that controls it. In the second section of this chapter, you will see how all of the circuits discussed in previous chapters come together to form a complete microprocessor-based system. The system is called "SAM," which is an acronym for "simple all-purpose microcomputer." We will be examining SAM's schematic diagram down to the component level to remove any mystery associated with microcomputers and to give you the confidence that comes from understanding a complete digital electronic microprocessor-based system. In addition, we will discuss general microprocessor system troubleshooting techniques and the detailed troubleshooting procedures for the SAM system.

FIGURE 14-1 A Modern Microcomputer.

14-1 MICROCOMPUTER BASICS

The earliest electronic computers were constructed using thousands of vacuum tubes. These machines were extremely large and unreliable and were mostly a laboratory curiosity. The *electronic numeric integrator and computer,* or ENIAC, was unveiled in 1946. The ENIAC weighed 38 tons, measured 18 feet wide and 88 feet long, and used 17,486 vacuum tubes. These vacuum tubes produced a great deal of heat and developed frequent faults requiring constant maintenance. The personal computer shown in Figure 14-1, on the other hand, makes use of semiconductor integrated circuits and is far more powerful, versatile, portable, and reliable than the ENIAC. Another advantage is cost: In 1946 the ENIAC calculator cost $400,000 to produce, whereas a present-day high-end personal computer can be purchased for about $1,200.

14-1-1 *Hardware*

Figure 14-2 shows the block diagram of a basic microcomputer system. The microprocessor (also called central processor unit, or CPU) is the "brains" of the system. It contains all of the logic circuitry needed to recognize and execute the program of instructions stored in

FIGURE 14-2 Basic Microcomputer System.

memory. The input port connects the processor to the keyboard so that data can be read from this input device. The output port connects the processor to the display so that we can write data to this output device. Combined in this way, the microprocessor unit, memory unit, and input/output unit form a **microcomputer.**

The blocks within the microcomputer are interconnected by three buses. The microprocessor uses the address bus to select which memory location, input port, or output port it wishes to put information into or take information out of. Once the microprocessor has selected the location using the address bus, data or information is transferred via the data bus. In most cases, data will travel either from the processor to memory, from memory to the processor, from the input port to the processor, or from the processor to the output port. The control bus is a group of control signal lines that are used by the processor to coordinate the transfer of data within the microcomputer.

A list of instructions is needed to direct a microcomputer system so that it will perform a desired task. For example, if we wanted the system in Figure 14-2 to display the number of any key pressed on the keyboard, the program would be as follows:

1. Read the data from the keyboard.
2. Write the data to the display.
3. Repeat (go to step 1).

For a microprocessor to perform a task from a list of instructions, the instructions must first be converted into codes that the microprocessor can understand. These instruction codes are stored, or programmed, into the system's memory. When the program is run, the microprocessor begins by reading the first coded instruction from memory, decoding its meaning, and then performing the indicated operation. The processor then reads the instruction from the next location in memory, decodes the meaning of this next instruction, and then performs the indicated operation. This process is then repeated, one memory location after another, until all of the instructions within the program have been fetched, decoded and then executed.

The input/output devices connected to the input and output ports (the keyboard and display, for example) are called the **peripherals.** Peripherals are the system's interface with the user, or the system's interface with other equipment such as printers or data storage devices. As an example, Figure 14-3 shows how a microcomputer could be made to operate as a microprocessor-based digital voltmeter. Its input peripherals are an analog-to-digital converter (ADC), and a range of function switches. The output peripheral is a digital display.

As you come across different microcomputer applications, you will see that the basic microcomputer system is always the same. The only differences between one system and another are in the program and peripherals. Whether the system is a personal computer or a digital voltmeter, the only differences are in the program of instructions that controls the microcomputer and the peripherals connected to the microcomputer.

Microcomputer

Complete system, including CPU, memory, and I/O interfaces.

Peripheral

Any interface device connected to a computer. Also, a mass storage or communications device connected to a computer.

FIGURE 14-3 A Microprocessor-Based Digital Voltmeter.

14-1-2 *Software*

Microcomputer programs are first written in a way that is convenient for the person writing the program, or *programmer*. Once written, the program must then be converted and stored as a code that can be understood by the microprocessor. When writing a computer program, the programmer must tell the computer what to do down to the most minute detail. Computers can act with tremendous precision and speed for long periods of time, but they must be told exactly what to do. A computer can respond to a change in conditions, but only if it contains a program that says "if this condition occurs, do this."

The Microcomputer as a Logic Device

To understand how a program controls a microcomputer, let us see how we could make a microcomputer function as a simple two-input AND gate. As can be seen in Figure 14-4, the two AND gate inputs will be applied to the input port, and the single AND gate output will appear at the output port. To make the microcomputer operate as an AND gate, a program of instructions will first have to be loaded into the system's memory. This list of instructions will be as follows:

1. Read the input port.
2. Go to step 5 if all inputs are HIGH; otherwise, continue.
3. Set output LOW.
4. Go to step 1.
5. Set output HIGH.
6. Go to step 1.

Studying the steps in this program, you can see that first the input port is read. Then the inputs are examined to see if they are all HIGH, since that is the function of an AND gate. If the inputs are all HIGH, the output is set HIGH. If the inputs are not all HIGH, the output is set LOW. Once the procedure has been completed, the program jumps back to step 1 and repeats indefinitely, so that the output will continuously follow any changes at the input.

You may be wondering why we would use this complex system to perform a simple AND gate function. If this were the only function that we wanted the microcomputer to perform, then it would definitely be easier to use a simple AND logic gate; however, the microcomputer provides tremendous flexibility. It allows the function of the gate to be arbitrarily redefined just by changing the program. You could easily add more inputs, and the gate function could be extremely complex. For example, we could have the microcomputer function as an eight-input electronic lock. The output of this lock will only go HIGH if the inputs are turned ON in a specific order. Using traditional logic, this would require a complex circuit; however, this function could be easily implemented using the same microprocessor system

FIGURE 14-4   **A Microprocessor-Based AND Gate.**

FIGURE 14-5   Flowchart Symbols.

used for the simple AND gate. Of course, a new program would be needed, more complex than the simple AND gate program, but the hardware will not change. Additionally, the combination of this electronic lock could easily be changed by simply modifying the program.

Flowcharts

To develop a program of instructions to solve a problem, a programmer would generally follow three steps:

1. Define the problem.
2. Flowchart the solution.
3. Write the program.

To understand these steps, let us begin by examining the elements of the flowchart in more detail.

   Flowcharts provide a graphic way of describing the operation of a program. They are composed of different types of blocks interconnected with lines, as shown in Figure 14-5. A rectangular block describes each action that the program takes. A diamond-shaped block is used for each decision, such as testing the value of a variable. An oval block is used at the beginning of the flowchart with the name of the program placed inside it. This oval block can also be used to mark the end of the flowchart.

   As an example, Figure 14-6 shows a flowchart for the AND gate program discussed previously. For each line of the program there is a block, except for the two "go to" instructions. These are represented as lines indicating how the program will flow from one block to

Flowchart or Flow Diagram

Graphical representation of program logic. Flowcharts enable the designer to visualize the procedure necessary for each item in the program. A complete flowchart leads directly to the final code.

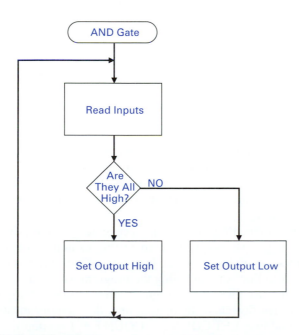

FIGURE 14-6   AND Gate Flowchart.

another. The flowchart contains the same information as the program list but in a graphic form. When you first set out to write a program, a flowchart is a good way to organize your thoughts and document what the program must do. By going through the flowchart manually, you can check the logic of the flowchart and then write the actual program from it. Flowcharts are also useful for understanding a program that has been written in the past.

Programming Languages

Writing programs in English is convenient, since it is the language most people understand. Unfortunately, English is meaningless to a microprocessor. The language understood by the microprocessor is called **machine language,** or machine code. Because microprocessors deal directly with digital signals, machine language instructions are binary codes, such as 00111100, and 11100111.

Machine language is not easy for people to use, since 00111100 has no obvious meaning. It can be made easier to work with by using the hexadecimal representation of 0011 1100, which is 3C; however, this still does not provide the user with any clue as to the meaning of this instruction. To counteract this problem, microprocessor manufacturers replace each instruction code with a short name called a **mnemonic,** or memory aid. As an example, Intel's 8085 microprocessor uses the mnemonic "INR A" for the code 3C, since this code instructs the 8085 microprocessor to "increment its internal A register." The mnemonics are much easier to remember than the machine codes. By assigning a mnemonic to each instruction code, you can write programs using the mnemonics instead of machine codes. Once the program is written, the mnemonics can easily be converted to their equivalent machine codes. Programs written using mnemonics are called **assembly language** programs.

The machine language is generally determined by the design of the microprocessor chip and cannot be modified. The assembly language mnemonics are made up by the microprocessor's manufacturer as a convenience for programmers and are not set by the microprocessor's design. For example, you could write INC A instead of INR A, as long as both were translated to the machine code 3C. A microprocessor is designed to recognize a specific list or group of codes called the **instruction set,** and each microprocessor type has its own set of instructions.

Although assembly language is a vast improvement over machine language, it is still difficult to use for complex programs. To make programming easier, **high-level languages** have been developed. These are similar to English and are generally independent of any particular microprocessor. For example, a typical instruction might be "LET COUNT = 10" or "PRINT COUNT." These instructions give a more complicated command than those that the microprocessor can understand. Therefore, microcomputers using high-level languages also contain long, complex programs (permanently stored in their memory) that translate the high-level language program into a machine language program. A single high-level instruction may translate into dozens of machine language instructions. Such translator programs are called **compilers.**

A Programming Example

To show the difference between machine language, assembly language, and a high-level language, let us use a simple programming example. Figure 14-7(a) shows the flowchart for a program that counts to ten. There are no input or output operations in this program, since all we will be doing is having the contents of a designated memory location count from zero to ten, and then repeat.

The translation from the flowchart to a high-level language is fairly simple, as seen in Figure 14-7(b). The high-level language used in this example is called **BASIC,** which stands for Beginner's All-purpose Symbolic Instruction Code and has the advantage of being simple and similar to English. Following the program listing shown in Figure 14-7(b), you can see that the first two lines of the program correspond exactly to the first two action blocks of the count-to-ten flowchart. In the first line, the memory location called COUNT is set to 0. In the second line, LET COUNT = COUNT + 1 is simply a way of saying "increment the count." Lines 3 and 4 perform the function of the decision block in the count-

Machine Language

Binary language (often represented in hex) that is directly understood by the processor. All other programming languages must be translated into binary code before they can be entered into the processor.

Mnemonic Code

To assist the human memory, the binary numbered codes are assigned groups of letters (or mnemonic symbols) that suggest the definition of the instruction.

Assembly Language

A program written as a series of statements using mnemonic symbols that suggest the definition of the instruction. It is then translated into machine language by an assembler program.

Instruction Set

Total group of instructions that can be executed by a given microprocessor.

High-Level Language

A language closer to the needs of the problem to be handled than to the language of the machine on which it is to be implemented.

Compiler

Translation program that converts high-level instructions into a set of binary instructions (machine code) for execution.

BASIC

An easy-to-learn and easy-to-use language available on most microcomputer systems.

Line No.	Instruction	Description
1	LET COUNT = 0	Set Count to 0
2	LET COUNT = COUNT + 1	Increment Count
3	IF COUNT = 10 THEN 1	Go to 1 if Count = 10
4	GO TO 2	Otherwise go to 2

(b)

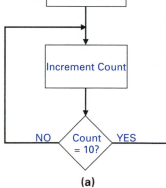

(a)

Label	Instruction	Comments
START:	MVI A,0	;Set A register to 0
LOOP:	INR A	;Increment A register
	CPI 10	;Compare A register to 10
	JZ START	;Go to beginning if A = 10
	JMP LOOP	;Repeat

(c)

Memory Address (Hex)	Memory Contents (Hex)	(Binary)
07F0	3E	00111110
07F1	00	00000000
07F2	3C	00111100
07F3	FE	11111110
07F4	0A	00001010
07F5	CA	11001010
07F6	F0	11110000
07F7	07	00000111
07F8	C3	11000011
07F9	F2	11110010
07FA	07	00000111

(d)

BASIC Language		8085 Assembly Language			8085 Machine Language		
Line No.	Instruction	Label	Instruction		Address	Contents	
1	LET COUNT = 0	START:	MVI	A,0	07F0	3E	Opcode
					07F1	00	Data
2	LET COUNT = COUNT + 1	LOOP:	INR	A	07F2	3C	Opcode
3	IF COUNT = 10 THEN 1		CPI	10_{10}	07F3	FE	Opcode
					07F4	0A	Data
			JZ	START	07F5	CA	Opcode
					07F6	F0	Address
					07F7	07	
4	GO TO 2		JMP	LOOP	07F8	C3	Opcode
					07F9	F2	Address
					07FA	07	

(e)

FIGURE 14-7 A Count-to-Ten Programming Example. (a) Flowchart. (b) High-Level Language Program. (c) Assembly Language Program. (d) Machine Language Program. (e) Summary of Program Languages.

to-ten flow chart. Line 3 specifies that if COUNT = 10, then the next instruction executed should be line 1. If the count is not equal to ten, the COUNT = 10 instruction has no effect, and the program continues with line 4, which says "go to line 2." To test whether this BASIC program will count to 10 and then repeat, try following the program step by step to see if it works, starting with a value of zero.

Assembly language is not one specific language but a class of languages. Each microprocessor has its own machine language and therefore its own assembly language, which is defined by the manufacturer. Figure 14-7(c) shows the assembly language listing for the count-to-10 program. This program is certainly more cryptic than the BASIC language program, but it performs exactly the same function.

The three columns in Figure 14-7(c) are for labels, instructions, and comments. The label provides the same function as the line number in the BASIC program. Instead of numbering every line, you simply make up a name (called a label) for each line to which you need to refer. A colon (:) is used to identify the label. A line needs a label only if there is another instruction in the program that refers to that line. The comments are an aid to understanding the program, and a semicolon (;) is used to identify the beginning of a comment. High-level language programs do not always need many comments because the instructions themselves are more descriptive. For assembly language programs, however, comments are an invaluable aid for people other than the programmer, or when the programmer returns to a program after some time.

The first instruction is MVI A, 0, which means "move immediately to the accumulator the data zero." The **accumulator** is also called the A register and is a storage location inside the microprocessor. This assembly language instruction is equivalent to the BASIC instruction LET COUNT = 0, except instead of making up a name for the variable (COUNT), we used a preassigned name (A) for a register inside the microprocessor. This MVI A, 0 instruction will therefore load the data 0 into the microprocessor's internal A register. The next instruction, INR A, means "increment the value in the accumulator," and this assembly language instruction is equivalent to the BASIC instruction LET COUNT = COUNT + 1. The next three instructions implement the decision function. The instruction CPI 10 means "compare the value in the accumulator with the value 10." The result of this comparison will determine whether a special flip-flop within the microprocessor called the *zero flag* is SET or RESET. If the value in the accumulator is not equal to 10, the zero flag is RESET LOW, whereas if the value in the accumulator is equal to 10, the zero flag is SET HIGH. The next instruction, JZ START, means "jump if the zero flag is SET to the line with the label START." This instruction tests the zero flag, and if it is SET (accumulator = 10), it will cause the program to jump to the line with the label START. Together, these two instructions (CPI 10 and JZ START) perform the function of the BASIC statement IF COUNT = 0 THEN 1. The last instruction, JMP LOOP, means "jump to the line with the label LOOP." This instruction simply causes the program to jump to the line with the label LOOP and is equivalent to the BASIC statement GO TO 2.

Figure 14-7(d) shows the machine language listing for the count-to-ten program. Although this language looks the most alien to us, its sequence of 1s and 0s is the only language that the microprocessor understands. In this example, each memory location holds eight bits of data. To program the microcomputer, we will have to store these 8-bit codes in the microcomputer's memory. Each instruction begins with an **opcode,** or operation code, that specifies the operation to be performed, and since all 8085 opcodes are eight bits, each opcode will occupy one memory location. An 8085 opcode may be followed by one byte of data, two bytes of data, or no bytes, depending upon the instruction type used.

Stepping through the machine language program shown in Figure 14-7(d), you can see that the first byte ($3E_{16}$) at address $07F0_{16}$ is the opcode for the instruction MVI A. The MVI A instruction is made up of two bytes. The first byte is the opcode specifying that you want to move some data into the accumulator, and the second byte stored in the very next memory location (address $07F1_{16}$) contains the data 00_{16} to be stored in the accumulator. The third memory location (address 07F2) contains the opcode for the second instruction, INR A. This opcode (3C) tells the microprocessor to increment the accumulator. The INR A instruction has no additional data, and therefore the instruction only occupies one memory location. The next code, FE, is the opcode for the compare instruction, CPI. Like the MVI A, 0 instruction, the memory location following the opcode contains the data required by the instruction. Since the machine language program is shown in hexadecimal notation, the data 10 (decimal) appears as 0A (hex). This instruction compares the accumulator with the value 10 and sets the zero flag if they are equal, as described earlier. The next instruction, JZ, has the opcode CA and appears at address 07F5. This opcode tells the microprocessor to jump if the zero flag is set. The next two memory locations contain the address to jump to. Because addresses in an 8085 system are sixteen bits long, it takes two memory locations (eight bits each) to store an address. The two parts of the address are stored in an

Accumulator

One or more registers associated with the arithmetic and logic unit (ALU), which temporarily store sums and other arithmetical and logical results of the ALU.

Opcode

The first part of a machine language instruction that specifies the operation to be performed. The other parts specify the data, address, or port. For the 8085, the first byte of each instruction is the opcode.

order that is the reverse of what you might expect. The least significant half of the address is stored first and then the most significant half of the address is stored next. The address 07F0 is therefore stored as F0 07. The assembly language instruction JZ START means that the processor should jump to the instruction labeled START. The machine code must therefore use the actual address that corresponds to the label START, which in this case is address 07F0. The last instruction, JMP LOOP, is coded in the same way as the previous jump instruction. The only difference is that this jump instruction is independent of any flag condition, and therefore no flags will be tested to determine whether this jump should occur. When the program reaches the C3 jump opcode, it will always jump to the address specified in the two bytes following the opcode, which in this example is address 07F2. The machine language program, therefore, contains a series of bytes, some of which are opcodes, some are data, and some are addresses.

You must know the size and format of an instruction if you want to be able to interpret the operation of the program correctly. To compare the differences and show the equivalents, Figure 14-7(e) combines the high-level language, assembly language, and machine language listings for the count-to-ten program.

SELF-TEST EVALUATION POINT FOR SECTION 14-1

Use the following questions to test your understanding of Section 14-1.

1. Briefly describe the three following languages:
   a. machine language   b. assembly language   c. high-level language

2. What is a flowchart?

3. Define the following terms:
   a. instruction set   c. programmer   e. compiler   g. flag register
   b. peripheral   d. mnemonic   f. accumulator h. opcode

14-2   A MICROCOMPUTER SYSTEM

In this section, you will see how all of the circuits discussed in the previous chapters come together to form a complete microprocessor-based system. The system is called "SAM," which is an acronym for "simple all-purpose microcomputer." We will be examining SAM's schematic diagram down to the component level to remove any mystery associated with microcomputers and give you the confidence that comes from understanding a complete digital electronic microprocessor-based system. In addition, we will discuss general microprocessor system troubleshooting techniques and the detailed troubleshooting procedures for the SAM system.

This section of the textbook will read very similar to a manufacturer's technical service manual. This has been done intentionally to introduce you to the format and style of a typical tech manual and so you can see how this resource will provide you with the detailed data that you need to have before attempting to troubleshoot any electronic system.

14-2-1   *Theory of Operation*

On pages 419 through 422 you will find the schematic diagram for the SAM system. You may wish to copy and combine this schematic, since we will be referring to it constantly throughout the circuit description. You may also want to color code your copy of the schematic to help you more easily see the circuit interconnections. For example, using highlighters, you could make all address lines yellow, all data lines blue, and all control signal lines green.

Introduction

The simple all-purpose microcomputer system was designed specifically for educational use. It uses Intel's 8085 microprocessor and has 2kB (kilobytes) of ROM and 1kB of RAM. The system contains two input ports that are connected to a keyboard from which you can enter programs, store data, and give commands to control the operation of the microcomputer, and a set of slide switches. The output port display allows you to view the contents of the memory and registers. In addition, there is an output port using LEDs and a speaker that is controlled by the processor. There are also LEDs on the address bus, the data bus, and the major control lines so you can monitor their activity. The ROM in the SAM contains programs to read the keyboard, execute keyboard commands, and send data to the display. The entire operation of the system is governed by a built-in software program called the *monitor*. Using the keyboard and the display, a program can be loaded into RAM, and then it can be run at full speed or executed one step at a time to allow you to follow the operation of the program in detail. Special switches are also provided to establish loop operations for test purposes, and moveable jumpers are included to insert faults in the circuits for training in troubleshooting microprocessor systems.

Block Diagram Description

A simplified block diagram of the SAM system is shown in Figure 14-8. The three main functional blocks of this microcomputer are the microprocessor, the memory, and the input/output ports. These blocks within the microcomputer are interconnected by three groups of signals—address, data, and control.

Address bus. The microprocessor is the only address bus "talker," which means that it is the only device that ever sends out addresses onto the 16-bit address bus. Every other device connected to the address bus, therefore, is a "listener" since these devices only receive addresses from the microprocessor. This one-direction flow of address information on this bus accounts for why the address bus is called a **unidirectional bus.** Before any data transfer can take place via the data bus, the microprocessor must first output an address onto the address bus to specify which memory location or input/output port the processor wishes to access. In this way, the processor can select any part of the system it wishes to communicate with.

Unidirectional Bus

Wire or group of wires in which data flows in only one direction. Each device connected to a unidirectional bus is either a transmitter, or a receiver, but not both.

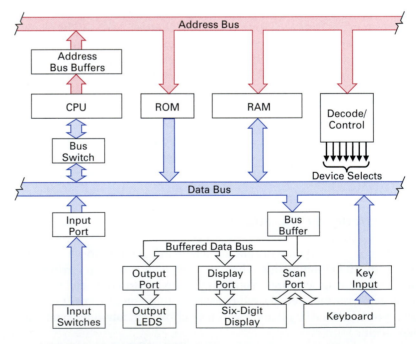

FIGURE 14-8   The SAM System Block Diagram.

Data bus. The 8-bit data bus is a **bidirectional bus** (two-direction), and therefore any device that talks on this bus must have eight driver outputs, and any device that listens must have eight data inputs. The microprocessor and the RAM are data bus talkers and listeners, while the ROM is only a data bus talker. The input ports are talkers since they input data from outside the system and put it on the data bus, whereas the output ports are listeners since they take data off the data bus and send it outside the system. The arrows on the data bus lines in Figure 14-8 summarize how each of the devices communicates with the data bus.

The microprocessor, RAM, ROM, and input ports contain three-state drivers on their data outputs. The select input to the RAM, ROM, and input ports enables the three-state drivers and causes the data from the selected device to appear on the data bus. The microprocessor is the controller of the system, so it will ensure that no more than one device is trying to use the data bus at any given time. For example, if the microprocessor wants to read data from the ROM, it would three-state its own data lines and generate the control signals needed to cause the ROM's select input to be true. The ROM's outputs would then appear on the data bus, and the microprocessor would read the data off the bus. Reading the RAM or an input port is done in a similar manner. On the other hand, if the microprocessor wanted to write data to the RAM or an output port, the microprocessor would first place the data to be written on its data lines. It would then generate the control signals needed to cause a write pulse to be sent to the appropriate device, which would then input the data from the bus.

Control bus. The address bus is used to select a particular memory location or input/output port. The data bus is then used to carry the data, and the control bus is used to control this process. The control bus consists of a number of control signals, most of which are generated by the microprocessor.

The three main control signals generated by the microprocessor are $\overline{RD}$ (not read, or NRD), $\overline{WR}$ (not write, or NWR), and $IO/\overline{M}$ (input/output/not memory, or IO/NM). If NRD is LOW (active), it indicates that a read is in progress and the microprocessor will be expecting the device that is being addressed to put data on the data bus. If NWR is LOW (active), a write is in progress and the microprocessor will be putting data on the data bus, expecting the addressed device to store this data. If IO/NM is LOW, the operation in progress (which may be a read or write) is a memory operation. If IO/NM is HIGH, the operation in progress (which may be an output or input) is an input/output port operation.

Inside the 8085

At the heart of the SAM system is an 8085 microprocessor (U3 on the SAM schematic). In most microcomputer circuit descriptions, the microprocessor is treated as a device with known characteristics but whose internal structure is of no concern. Since a knowledge of the microprocessor's internal operation will help you gain a clearer understanding of the operation of the SAM system, we will take a look at the internal structure and operation of the 8085.

Figure 14-9 shows the basic signals that connect to a microprocessor. There are sixteen address outputs that drive the address bus, and eight data lines connected to the data bus. The $\overline{RD}$ and $\overline{WR}$ output control signals coordinate the movement of data on the data bus, and the RESET input is used to initialize the microprocessor's internal circuitry. The two connections at the top of the microprocessor are for an external crystal, which is used as the frequency-determining device for the microprocessor's internal clock oscillator. The output from the microprocessor's internal clock oscillator is called the *system clock*, and it is used to synchronize all devices in the system and set the speed at which instructions are executed.

The inset in Figure 14-9 shows a simplified internal block diagram of the 8085 microprocessor. The accumulator, or A register, connects to the data bus and the arithmetic and logic unit (ALU). The ALU performs all data manipulation, such as incrementing a value or adding two numbers. The temporary register is automatically controlled by the microprocessor's control circuitry, and this register feeds the ALU's other input. The flags in the flag register are a collection of flip-flops that indicate certain characteristics about the result of the most recent operation performed by the ALU. For example, the zero flag is set if the

Bidirectional Bus

Indicates that signal flow may be in either direction. Common bidirectional buses are three-state or open collector TTL.

FIGURE 14-9   The Microprocessor.

result from an operation is 0, and as discussed earlier, this flag is tested by the JZ (jump if 0) instruction. The instruction register, instruction decoder, program counter, and control and timing logic are used for fetching instructions from memory and directing their execution.

Let us now take a closer look at the microprocessor's *fetch-execute cycle.* As an example, let us imagine that we need to read an MVI A instruction from ROM at address 0200. The first step is to retrieve the opcode for this instruction. This process is called the *instruction fetch* and is shown in Figure 14-10(a). As its name implies, the program counter is used to keep count by pointing to the current address in the program. In this example, the program counter will contain the address 0200, and this will be placed on the address bus and cause memory location 0200 to be selected. The ROM will then place the contents of memory location 0200 on the data bus, and the microprocessor will store this data, which is the opcode for MVI A (3E) in the instruction register.

Once the opcode is stored, the *instruction execute* process begins, as shown in Figure 14-10(b). The instruction register feeds the instruction decoder, which recognizes the opcode and provides control signals to the control and timing circuitry. For example, for an MVI A instruction, the control and timing logic circuit first reads the opcode 3E and then increments the address in the program counter. The instruction decoder determines that this opcode is always followed by a byte of data, so the contents of the memory location pointed to by the program counter is read, and the second byte of this instruction is placed into the accumulator. The program counter is again incremented, and the next byte of the main program (which will be the next instruction's opcode) is read into the instruction register, and then the execution of this instruction begins.

The real work is done in the execute phase of the instruction. Four basic types of operations can be performed by the 8085 microprocessor:

1. Read data from memory or an input port.
2. Write date to memory or an output port.
3. Perform an operation internal to the microprocessor.
4. Transfer control to another memory location.

The first two operation types are self-explanatory. The third involves manipulating the 8085's internal registers (such as the accumulator) without accessing the external memory or

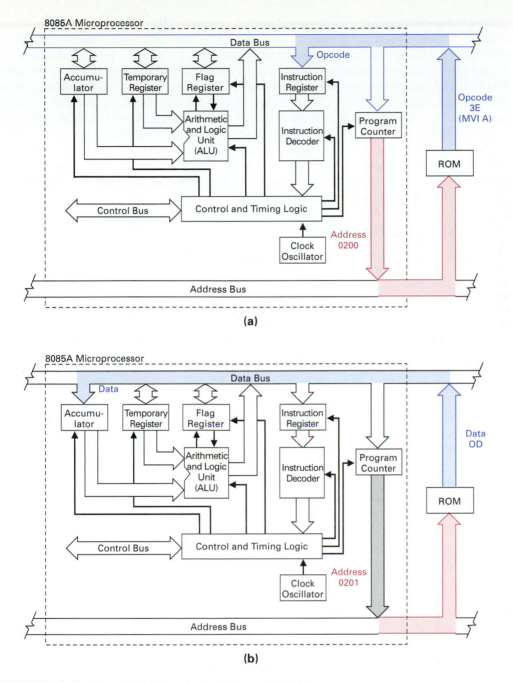

(a)

(b)

FIGURE 14-10   The Fetch-Execute Cycle for an MVI A Instruction.

input/output ports. For example, the contents of one register may be incremented or decremented. The fourth operation type includes instructions such as JMP, which causes the program counter and therefore the program to shift to point to a new memory location.

To summarize the operation of the microprocessor, we can say that the microprocessor keeps reading sequentially through the memory, one location after another, performing the indicated operations. Exceptions to this occur when a jump instruction is executed or at the occurrence of an **interrupt.** The 8085 has a few interrupt input pins, which are used to tell the microprocessor to stop its current task and perform another task immediately. Jump instructions or interrupts will cause the microprocessor to stop its sequential flow through a program and begin executing instructions from another address.

Remember that opcodes and data are intermixed in memory. One address might contain an opcode, the next two addresses a jump address, the next address an opcode, and the next address a piece of data. It is the programmer's responsibility to be sure that the memory contains a valid sequence of opcodes and data. The microprocessor cannot distinguish among opcodes, jump addresses, and data since they are all simply bit patterns stored in the memory. All such information is read in exactly the same way and it travels over the same data bus. The microprocessor must always keep track of whether it is reading an opcode or data and treat each appropriately. The processor always assumes that the first location it reads contains an opcode and goes from there. If the opcode requires a second byte of data, the microprocessor is aware of this (because of the instruction decoder) and treats the next byte accordingly. It then assumes that the byte following the last byte in an instruction is the next opcode. The microprocessor will not know if a programmer has placed data in a position that should have an opcode. Such errors will cause the system to "crash" since the microprocessor would decode mispositioned data as opcodes.

SAM Circuit Description

As we proceed through this section, you should have your SAM schematic open, since we will be referring to devices on the schematic along with simplified SAM circuit diagrams shown in this textbook.

Decoding. The SAM system does not use I/O mapping for its I/O ports, so the IO/NM output from the microprocessor is not used for decoding. The SAM system uses memory-mapped input/output, which means that the I/O ports are treated as addressable devices similar to memory space.

Figure 14-11(a) shows the SAM address decoder (U7 on the SAM schematic), and Figure 14-11(b) shows the memory map or address map for the SAM system. Only the first quarter of the address space is used, so address bits 14 and 15 are always 0. The address space used in the SAM system is divided into eight equal sections, each of which has 2K, or 2,048, locations. The ROM occupies the first 2K addresses (0000 to 07FF hex), and the RAM has been assigned the next 2K addresses (0800 to 0FFF). The six address sections following the RAM are used for I/O ports. The control port is used by the monitor program (system control program in ROM) to provide some special functions, which will be described later. The key data, scan, and display segment ports control the keyboard and display, and the input and output ports are used for the switches and LEDs.

To simplify the address decoding hardware, 2,048 addresses have been assigned to each memory type and each of the input and output ports. The SAM system makes use of 16K addresses for 2K of ROM, 1K of RAM, and six I/O ports. The SAM system could have used only $2,048 + 1,024 + 6 = 3,078$ addresses, which would result in more unused address space and a much more complicated address decoding circuit. Referring to the simplified SAM address decoder circuit shown in Figure 14-11(a) and U7 on the SAM schematic, let us now see how this circuit operates. There are two things to note in this decoding arrangement:

1. Some of the read/write control is mixed with the decoding.
2. There is a special circuit for RAM write protect.

The binary addresses listed in the memory map in Figure 14-11(b) show that the A_{11}, A_{12}, and A_{13} lines specify which section is to be addressed. Therefore, these lines are used to provide the binary-select inputs to U7 (a 74138 binary to one-of-eight decoder IC). This provides eight separate active-LOW outputs, one for each of the 2K byte blocks in the SAM memory map. Address decoder U7 has three enable inputs, two that are active-LOW and one that is active-HIGH. All three enable inputs must be true to allow any of the outputs to be true. This is achieved by connecting the A_{14} and A_{15} lines to the two active-LOW enables. This will prevent any of the outputs from being true unless both A_{14} and A_{15} are LOW.

The read/write control gating is distributed throughout the circuitry, using the enables of the decoder, the memory devices, and the I/O ports to reduce the number of gates required. The decoder's third enable input is connected to a gate that generates the OR of *NRD* and *NWR*. This has the effect of allowing the device-select output of U7 to be true only when either a read or a write is in progress. This is necessary because the address bus will not contain meaningful in-

Upper Half of Address in Binary

Bit:	15	14	13	12	11	10	9	8	Address in Hex	Device
	0	0	0	0	0	0	0	0	0 0 0 0	
									.	ROM
	0	0	0	0	0	1	1	1	0 7 F F	
	0	0	0	0	1	0	0	0	0 8 0 0	
									.	RAM
	0	0	0	0	1	1	1	1	0 F F F	
	0	0	0	1	0	0	0	0	1 0 0 0	
									.	CONTROL
	0	0	0	1	0	1	1	1	1 7 F F	
	0	0	0	1	1	0	0	0	1 8 0 0	
									.	KEY DATA
	0	0	0	1	1	1	1	1	1 F F F	
	0	0	1	0	0	0	0	0	2 0 0 0	
									.	INPUT PORT
	0	0	1	0	0	1	1	1	2 7 F F	
	0	0	1	0	1	0	0	0	2 8 0 0	
									.	SCAN (Kbd + Display)
	0	0	1	0	1	1	1	1	2 F F F	
	0	0	1	1	0	0	0	0	3 0 0 0	
									.	OUTPUT PORT
	0	0	1	1	0	1	1	1	3 7 F F	
	0	0	1	1	1	0	0	0	3 8 0 0	
									.	DISPLAY SEGMENTS
	0	0	1	1	1	1	1	1	3 F F F	
	0	1	0	0	0	0	0	0	4 0 0 0	
									.	NOT USED
	1	1	1	1	1	1	1	1	F F F F	

(Memory: ROM, RAM; I/O: CONTROL, KEY DATA, INPUT PORT, SCAN, OUTPUT PORT, DISPLAY SEGMENTS, NOT USED)

(b)

FIGURE 14-11 SAM's Address Decoder and Address Map.

formation if either *NRD* or *NWR* are true. In addition, the ROM and the input ports are to be selected only if a read is being performed. If they respond to both a read or a write, a bus conflict could occur. For example, if a write to the ROM is attempted, the microprocessor would put data on the data bus to be written to the ROM. If the ROM were allowed to be enabled by a write, then it would also put data on the data bus, which is an unacceptable situation.

To solve this problem, U11C ANDs the *NRD* signal with the device select. This is shown for the KYRD port, but not for the ROM and IN ports. The ROM and IN port chips each have two enables, so one is used for the device select and one for *NRD*. This effectively ANDs the *NRD* signal with the device select. For the output ports, the situation is slightly different. In an attempt to read an output port (which is not a valid operation), a write will be performed instead, and the port will be loaded with invalid data. This is acceptable, since the software should know not to do this and even if it does, no real damage will be caused. This is in contrast to the situation of writing to an input port, which causes a hardware conflict and must not be allowed. Therefore, it is not necessary to AND the *NWR* signal with the device select for the output port U15.

The RAM's device select should be true when a read or write to the RAM's address space is in progress. The gate (U11A) on the RAM's device-select line is for the write protect circuit. The write protect circuitry helps prevent the RAM's contents from being accidentally lost. A programming error may result in the microprocessor running wild (usually by interpreting data as an opcode), and often this will result in incorrect data being written into the RAM. To prevent this, the SAM system contains control latch U8, which provides an *NPROT* input. When this latch is SET, the RAM will be protected. The monitor program will set this protect latch whenever a program is running and reset the protect latch at all other times so that data can be stored in the RAM so SAM can be programmed. Since the program may want to use the RAM to store data during program execution, only the first three-quarters of the RAM is protected. Address lines A_8 and A_9 indicate which quarter of the RAM is being addressed, and if they are both HIGH, then the last quarter is being addressed and the memory will not be protected. To achieve this, A_8 and A_9 are ANDed together in U9D and the result is then ORed with *NRD* and *NPRT* in U11B. This produces the RAM enable signal, which will be true if A_8 and A_9 are HIGH if a read is in progress or if the protect latch is not set. If the protect latch is set, then the RAM will be disabled unless a read is in progress or if A_8 and A_9 are HIGH.

RAMs. Figure 14-12 shows the SAM system's RAM circuit (U5 and U6 on the SAM schematic). The SAM system uses two 1K by 4-bit SRAMs connected to form a 1K

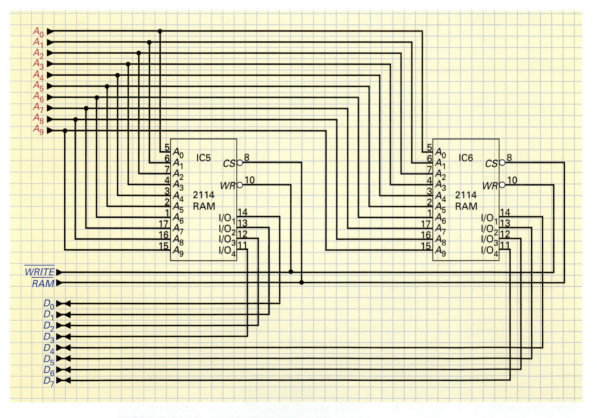

FIGURE 14-12 **SAM's RAM Circuit.**

FIGURE 14-13 SAM's ROM Circuit.

by 8-bit RAM memory circuit. The address and control pins for both chips are parallel connected together, while U5 connects to data lines 0 through 3 and U6 connects to data lines 4 through 7.

ROM. Figure 14-13 shows the SAM system's ROM circuit (U4 on the SAM schematic). The eleven low-order address lines (A_0 through A_{10}) supply an address to this 2K by 8-bit masked-program ROM. The 8-bit data output from this IC is connected to the system's data bus. The two selects for this IC must be true for the ROM's three-state output drivers to be enabled. The ROM will therefore only drive the data bus if the ROM select is true and the operation is a read.

Speaker. Figure 14-14 shows the SAM system's serial output circuit, which will have to be traced carefully on the SAM schematic. The speaker is driven by the microprocessor's serial output, so this is in effect a 1-bit output port. The U3 (8085) SOD output is buffered by U18A and sent to the edge connector for use by external hardware, as desired. It is then buffered again before driving the speaker. The speaker draws so much current that the edge connector would not have valid logic levels if the speaker buffer were not included. A 100 Ω resistor in series with the speaker limits the current to levels that will not damage the buffer. The other connection to the speaker is tied to +5 V because the TTL buffer can sink more current than it can source (it can pull more current through the speaker than it can push through the speaker). All of the actual tone generation is controlled by the software. A beep program within the ROM turns the serial output ON and OFF several hundred times a second. This several-hundred-hertz square wave is applied to the speaker for a few seconds.

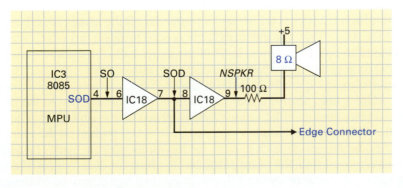

FIGURE 14-14 SAM's Serial Output Circuit.

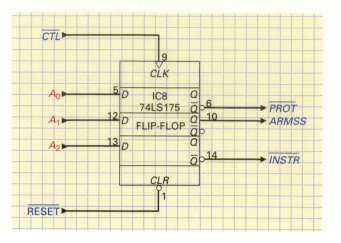

FIGURE 14-15 SAM's Control Port Register.

The control port. Figure 14-15 shows the SAM system's control port register (U8 on the SAM schematic). The control port is used by the microprocessor to send signals to special circuits. The control port IC (U8) is a 4-bit register clocked by the control port select signal, which is generated by address decoder U7. This is similar to the other output ports; however, it is unusual because the data inputs are connected to the address bus instead of the data bus. Therefore, the data written to the port is independent of the data on the data bus. The control port will be selected by any address from 1000 to 17FF. This allows the eleven low-order address lines to contain any value and still select the port. Note that A_0, A_1, and A_2 provide the data inputs to this register, and the data sent to the port is determined by the address used to select the port. For example, an address of 1000 would RESET or clear all of the bits, whereas an address of 1001 would SET the Q output and therefore RESET the $\overline{PROT}$ bit. As another example, an address of 1004 would SET the Q output and therefore RESET the $\overline{INSTR}$ bit. This technique simplifies the software. Since it does not matter what data is sent to the port (only the address matters), the software does not need to set up a value within the microprocessor before it writes the value to the port. The $\overline{PROT}$ bit of this port is used to control the memory protect circuit, and if active, the first three-fourths of the RAM will be write protected as described earlier. The other two control bits from this port are used to control the single-step circuitry.

The multiplexed bus. Figure 14-16(a) shows the SAM system demultiplexing circuit (U1 and U2 on the SAM circuit diagram). To keep the number of IC pins to a minimum, the 8085 microprocessor chip (U3) multiplexes the data bus with the lower half of the address bus. An 8-bit address bus from the 8085 carries the upper half of the address bus, and an 8-bit address/data bus carries the data and the lower half of the address. The address latch enable (*ALE*) signal is generated by the microprocessor to indicate when the address/data bus contains address information. This signal is used to latch the address/data bus contents into the address latch IC (U2) when the lower half of the address is being generated by the 8085. The 8-bit latch U2 latches the address information off the address/data bus on the negative edge of the *ALE* signal (the inverter U12 is needed to select this edge). The three-state buffer U1 is used to boost the current level of the upper half of the address bus for the address bus LED display and is not part of the demultiplexing circuit.

The waveforms in Figure 14-16(b) show the 8085 microprocessor's bus timing. The A_8 through A_{15} address lines always contain the high-order address byte. At the beginning of each memory cycle, the low-order address byte is placed on the address/data bus. The trailing edge of the *ALE* indicates that a low-order address byte is present on the address/data bus and causes the demultiplexing latch to store the low-order address byte. The address information is then removed from the address/data bus to allow for data transfer. If a read operation is in progress, the microprocessor will generate a read signal, and the addressed memory

FIGURE 14-16 SAM's Multiplexed Bus.

or I/O device will place the data on the address/data bus. On the trailing edge of the *NRD* signal, the microprocessor will read the data off the bus.

The write cycle is similar, except that the direction of the data transfer is reversed. At the beginning of the cycle, the low-order address byte is placed on the address data bus and the *ALE* line is pulsed. The microprocessor then issues a write pulse and places the data on the address/data bus. On the trailing edge of the *NWR*, the addressed memory device or I/O device will store the data from the bus.

FIGURE 14-17 SAM's Interrupt and Reset Circuitry.

Interrupts and reset. Figure 14-17 shows the SAM interrupt and reset circuit. Interrupts provide a means for hardware external to the microprocessor to request immediate action by the processor. They allow the usual program flow to be interrupted and cause control to be transferred to a special program routine. There are two groups of interrupts available on the 8085 microprocessor: TRAP, RST 5.5, 6.5, and 7.5, which are controlled by the individual pins on the microprocessor; and RST 1, 2, 3, 4, 5, 6, and 7, which are controlled by INTR (interrupt request) and INTA (interrupt acknowledge). Only the TRAP, RST 7.5, and RST 6.5 interrupts are used in the SAM system. All that is required to initiate one of the interrupts is to apply a signal to the corresponding pin on the microprocessor. The RST 5.5 and 6.5 interrupt inputs will respond to a HIGH logic level, whereas the RST 7.5 interrupt input will only respond to a positive edge. The SAM system uses the *TRAP* input for the keyboard's RESET key, the RST 6.5 input for the INTRPT key on the keyboard, and the RST 7.5 input for the SA (signature analysis) switch. (Signature analysis is a troubleshooting method that will be described later in the SAM troubleshooting section.) The NAND gate U9A (which is equivalent to a bubbled-input OR) is used to allow the single-step circuitry to access the *TRAP* input. The 100 Ω resistor and the capacitor are used to debounce the RESET switch, which is necessary to cause only one interrupt. With the other interrupts, switch debouncing is not necessary because the software disables the interrupt as soon as it is acknowledged and so prevents a second interrupt. The *TRAP* input cannot be disabled, so it must be debounced using hardware.

Some allowances must be made for the fact that more than one interrupt may occur at the same time. To accommodate this, each interrupt is assigned a priority, and the interrupt with the highest priority will be acknowledged first. The 8085 *TRAP* interrupt has the highest priority, followed by the RST 7.5, 6.5, and 5.5, in that order, with INTR having the lowest priority.

The reset-in pin on U3 (pin 36) is used for power-up initialization. When a LOW level is applied to this pin, the microprocessor's internal circuitry is cleared. The program counter is set to 0000 (hex), so program execution begins from this address. This address is in ROM and contains a start-up program that initializes the complete SAM system. The slow-start circuit connected to the reset-in line will provide an automatic power-ON pulse.

FIGURE 14-18   SAM's Single-Step Circuit (Advances Microprocessor One Machine Cycle).

Ready.   Figure 14-18 shows the SAM system's single-step circuit and, as before, the IC numbers on this circuit correspond to the "U" numbers on the SAM schematic. When the ready input to the microprocessor (pin 35) is brought LOW, the microprocessor enters a wait state. The buses are not put in a three-state position, but remain at their current status until the ready input is brought HIGH again. This input is used for the hardware (HDWR) single-step mode, which allows the observation of the address bus, data bus, and status LEDs for each step of a program.

Data bus buffer.   The 8085 microprocessor will not generate enough current at its address and data bus outputs to drive all of the devices connected to the buses plus a set of address bus monitoring LEDs (DS1 through DS4) and data bus monitoring LEDs (DS5 and DS6). Bus buffers are therefore included to boost current. Figure 14-19 shows how a data bus buffer IC14 (U14 on the SAM schematic) is included in this circuit. Since data flow can only travel in one direction (from the data bus to the buffered data bus), only output devices

FIGURE 14-19   SAM's Data Bus Buffer.

can be connected to the buffered data bus. Any input devices must be connected to the data bus since they will be sending data back to the microprocessor.

Timing. In order for any microprocessor-based system to operate correctly, many timing relationships must be satisfied. To give you an example of the timing involved, Figure 14-20 shows a write-and-read timing operation.

Referring to the write timing waveforms shown in Figure 14-20(a), you can see that the address must be stable for some period of time (called the access time) before any operation can be performed. This allows the memory's internal address decoders to select the specified

FIGURE 14-20   **Timing. (a) Write Timing. (b) Read Timing.**

CHAPTER 14 / COMPUTER HARDWARE AND SOFTWARE

memory location. The data must then be stable for some time before the write pulse (called set-up time) and after the write pulse (called hold time).

Referring to the read timing waveforms shown in Figure 14-20(b), you can see that, as with the write operation, the address must be stable for some time to allow the memory's internal decoders to settle. A read pulse is then generated, and after a short time (equal to the memory IC's access time), the accessed data will be placed on the data bus. This data must be allowed to stabilize (set-up time) before the data is read into the microprocessor on the rising edge of the NRD pulse, and the data must also be stable for some time after the read pulse (hold time).

Figure 14-21(a) shows the microprocessor timing for a typical instruction. In microprocessor-based systems, the basic unit of time is called the *state*, or timing state

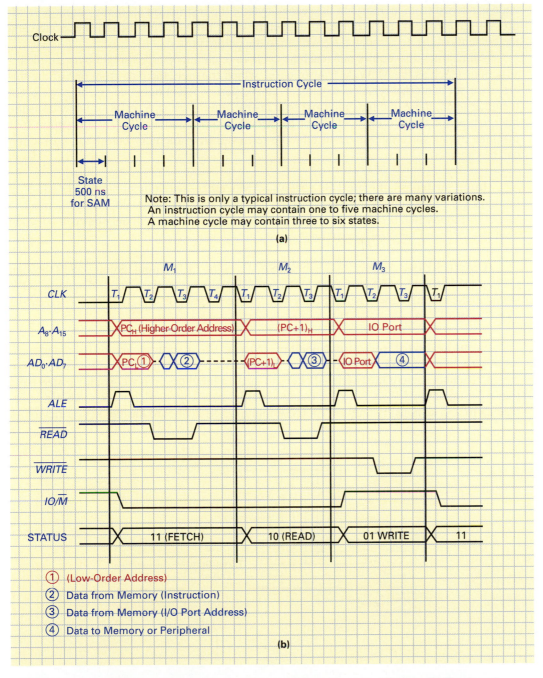

FIGURE 14-21 Instruction Cycle. (a) Summary of Microprocessor Timing. (b) Fetch and Execution of an OUT Instruction.

Machine Cycle

Basic period of time
required to manipulate
data in a system.

Instruction Cycle

All of the machine cycles
necessary to fully execute
an instruction.

(T-state), and is equal to one clock period. The 8085 microprocessor in SAM generates a 2 MHz system clock signal, and so each T-state is equal to 500 ns (1/2 MHz = 500 ns). A **machine cycle** consists of between three and six T-states, and it is the time it takes the machine or microprocessor to complete an operation. An **instruction cycle** is the time it takes a microprocessor to fetch and execute a complete instruction. Most simple instructions have an instruction cycle that has only one machine cycle, whereas more complex instructions may consist of five machine cycles.

As an example, Figure 14-21(b) shows the complete system timing for the two-byte OUT instruction. Timing states in this diagram use the prefix T (T_1, T_2, and so on), and machine cycles use the prefix M (M_1, M_2, and so on). The first byte of this two-byte instruction (the opcode) tells the microprocessor to send the contents of the accumulator to the output port specified in the second byte of the instruction. In the first machine cycle (M_1), the opcode is fetched from memory. In the second byte of the instruction (M_2), the output port address is read from memory. In the third machine cycle (M_3), the instruction is executed and the contents of the accumulator are written to the output port.

All instructions require one machine cycle to fetch the opcode. Single-byte simple instructions (such as the MOV A, B instruction, which transfers the data from one of the microprocessor's internal registers to another) can be performed in only one machine cycle. On the other hand, more complex multiple-byte instructions (such as STA 0837, which tells the microprocessor to store the contents of the accumulator at address 0837) will require four machine cycles since three bytes need to be retrieved from memory before the instruction can be executed.

Input switches. Figure 14-22 shows SAM's eight-switch input circuit (S3 and U13 on the SAM schematic). The switches at the inputs of the input port will cause the input to be LOW if the switch is closed. The input port's three-state buffer will only be enabled when both the *NRD* and the *NIN* control lines are active.

Output LEDs. Figure 14-23 shows SAM's eight-LED output circuit (DS12 through DS19—U15 on the SAM schematic). Since the LEDs form a common-anode display, a LOW stored in the output register will turn ON a LED. The output port's register is clocked by the *NOUT* control line.

Input keyboard and output display. Figure 14-24 shows a simplified diagram of SAM's keyboard circuit (S4 through S29, U17, and U18B on the SAM schematic). If we were to connect each of the twenty-six keys to its own input port bit, we would need to have four 8-bit input ports (4 × 8 = 32). Using a multiplexed keyboard, however, we can interface up to 256 keys using only two ports. A multiplexed keyboard has the keys arranged in

FIGURE 14-22   **SAM's Eight-Switch Input Port.**

FIGURE 14-23   **SAM's Eight-LED Output Port.**

a matrix that has intersecting column lines and row lines. An output port (U17, address 2800) drives the rows, and an input port (U18B, address 1800) reads the columns. The SAM monitor program scans the keyboard by making LOW only one row at a time and then reading in the contents of the input port. If any of the keys are pressed in the row that is LOW, then the column line which that key is on will be forced LOW, and so that input port bit will be LOW.

FIGURE 14-24   **SAM's Keyboard Interface.**

If no keys are pressed, the input port lines will all be pulled up to a HIGH level. The monitor program knows which row the activated key is in based on which output port bit is set LOW. It also knows which column the activated key is in based on which port bit is LOW (as seen in the example in Figure 14-24 in which key 2 has been pressed). To check all the keys, each output port bit must be set LOW in turn or scanned. At each step, only four keys are checked, and the process is so fast that the entire keyboard can be checked in less time than it takes a person to do the fastest possible press of a key.

Figure 14-25 shows SAM's display circuit (DS9 through DS11, U19, U17, and U20 on the SAM schematic). Like the keyboard, the seven-segment display also uses a scanning technique. The SAM system has six seven-segment display digits; however, only one display

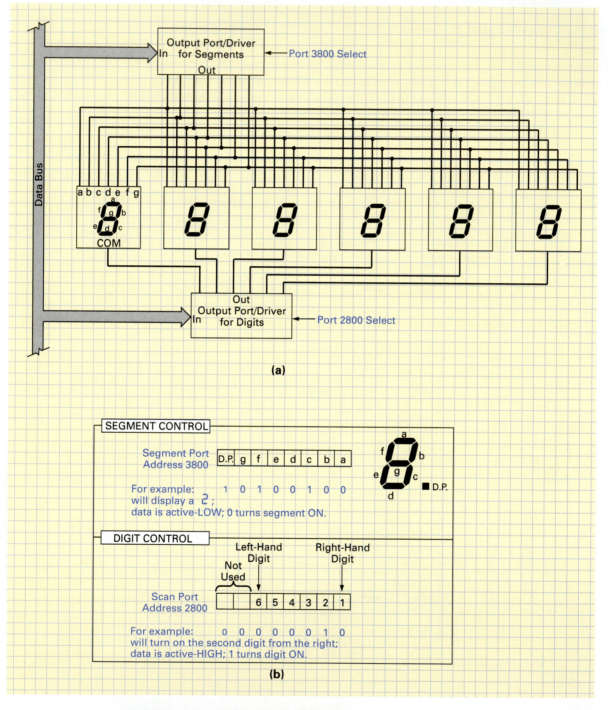

FIGURE 14-25 **SAM's Display Interface.**

is on at any instant. They are each turned on in sequence, but this happens so fast that they appear to all be on at the same time. A character is displayed by putting a HIGH level on the common connection and a LOW on the segment inputs based on which of the segments we wish to turn ON. In this multiplexed display, one 8-bit output port (U17 and U20, address 2800) is used to select one of the six digits, and another output port (U19, address 3800) supplies the segment information to the selected digit. A control program within the SAM monitor program can be run to control the display and display data stored in a set of six preselected RAM addresses. The control program operates in the following way. First, the segment information for digit 1 is sent to the segment port and the digit port is loaded with a value that will activate only digit 1. Then, the segment information for digit 2 is sent to the segment port and the digit port is loaded with a value that will activate only digit 2. Then, the segment information for digit 3 is sent to the segment port and the digit port is loaded with a value that will activate only digit 3. This continues until all six digits have been driven with segment data. The program then jumps back to its start point and the process is repeated.

SELF-TEST EVALUATION POINT FOR SECTION 14-2

Use the following questions to test your understanding of Section 14-2.

1. What is the difference between a unidirectional and bidirectional bus?
2. Define the function of the following 8085 microprocessor pins.
   - a. NWR
   - b. NRD
   - c. IO/NM
   - d. RESET
   - e. READY
   - f. ALE
   - g. HOLD
   - h. TRAP
3. Briefly describe a microprocessor's fetch-execute cycle.
4. What is a microprocessor
   - a. interrupt?
   - b. T-state?
   - c. machine cycle?
   - d. instruction cycle?

REVIEW QUESTIONS

Multiple-Choice Questions

1. Microprocessor-based systems are more flexible than hard-wired logic designs because _____.
   - a. they are faster
   - b. they use LSI devices
   - c. their operation is controlled by software
   - d. the hardware is specialized

2. The peripherals of a microcomputer system are _____.
   - a. the memory devices
   - b. the microprocessor
   - c. the software
   - d. the I/O devices

3. The personality of a microprocessor-based system is determined primarily by _____.
   - a. the microprocessor used
   - b. the program and peripherals
   - c. the number of data bus lines
   - d. the type of memory ICs used

4. The language that the microprocessor understands directly is called _____.
   - a. assembly language
   - b. high-level language
   - c. English language
   - d. machine language

5. The main purpose of a microprocessor's accumulator is _____.
   - a. keeping track of the next instruction to be executed
   - b. temporary data storage
   - c. selecting which interrupts should be enabled
   - d. storing opcodes

6. The microprocessor's program counter is used for _____.
   - a. keeping track of the next instruction to be executed
   - b. temporary data storage
   - c. selecting which interrupts should be enabled
   - d. storing opcodes

7. Interrupts are used primarily for _____.
   - a. breaking a program into modular segments
   - b. responding quickly to unpredictable events
   - c. speeding up program execution
   - d. halting the system

8. When an interrupt occurs, the microprocessor will _____.
   - a. jump to the interrupt service routine
   - b. halt until another request is made
   - c. continue executing the main program
   - d. complete the current program and then stop

9. The microprocessor knows which bytes to interpret as opcodes because _____.
   a. every byte is an opcode
   b. every third byte is an opcode
   c. each opcode indicates the number of information bytes that follow
   d. the programmer has highlighted the opcodes

10. The purpose of the ALU in a microprocessor is to _____.
    a. interpret the opcodes
    b. perform arithmetic and logic operations
    c. control the address bus
    d. calculate the number of machine cycles required

11. In a microprocessor system, the data bus is _____.
    a. unidirectional and three-state
    b. bidirectional and three-state
    c. unidirectional and bidirectional
    d. All of the above are true.

12. In a system using several 1 kB memory devices, which address lines are connected to the memory chips?
    a. A0–A9
    b. A10–A15
    c. A0–A15
    d. A0–A7

13. In a system with a 16-bit address bus, what is the maximum number of 2 kByte memory devices it could contain?
    a. 16
    b. 32
    c. 64
    d. 128

14. Memory-mapped input/output _____.
    a. treats I/O ports as memory locations
    b. tends to be more wasteful of address space than I/O-mapped decoding
    c. is used in the SAM system
    d. is all of the above

15. A bus conflict will occur when _____.
    a. an output port is enabled by a *NRD* signal
    b. more than one device is reading from the data bus
    c. more than one device is writing to the data bus
    d. MOS and TTL devices are both connected to a data bus

Practice Problems

16. Describe the function of the SAM system program shown in Figure 14-26.

To practice your circuit recognition and operation skills, refer to U4 on the SAM schematic and answer the following three questions.

17. What is the size and organization of this IC?

18. How is this IC enabled, and where do the enables originate from?

19. Is this IC a data bus talker or listener or both?

To practice your circuit recognition and operation skills, refer to U5 and U6 on the SAM schematic and answer the following four questions.

20. What type of ICs are these?

21. What is the size and organization of these ICs?

22. How are these ICs enabled, and where do the enables originate from?

23. Are these ICs data bus talkers or listeners or both?

To practice your circuit recognition and operation skills, refer to U3 on the SAM schematic and answer the following seven questions.

24. Where does the signal input to pin 8 originate from?

25. What is connected between pins 1 and 2, and how is this related to the output on pin 37?

26. How are the address and data information separated if both appear on pins 12 through 19?

27. Where does the signal output on pin 4 go to?

28. How does the *RC* connected to pin 36 input operate?

29. What is the function of the signal outputs on pins 30, 31, and 32?

30. Are pins 21 through 28 multiplexed?

To practice your circuit recognition and operation skills, refer to the SAM schematic and answer the following ten questions.

31. What is the function of DS1 through DS4?

32. Describe the operation of U7 and how it governs the system memory map.

33. What is the function of DS5 and DS6?

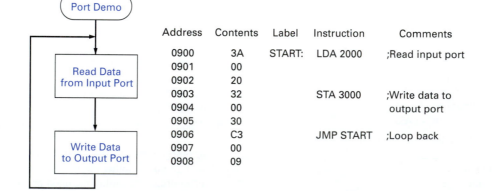

Address	Contents	Label	Instruction	Comments
0900	3A	START:	LDA 2000	;Read input port
0901	00			
0902	20			
0903	32		STA 3000	;Write data to
0904	00			output port
0905	30			
0906	C3		JMP START	;Loop back
0907	00			
0908	09			

FIGURE 14-26 **An Echo Program for the SAM System.**

REFERENCE
DESIGNATIONS

A1
C1-C16
CR1, CR2
DS1-DS19
LS1
R1-R6
S1-S29
U1-U20
W1-W14
XU4
Y1
A2
C1-C5
CR1, CR2
S1, S2
U1, U2

TABLE OF ACTIVE ELEMENTS

A1		
Reference Designation	HP Part Number	Mfg or Industry Part Number
CR1	1901-0518	1901-0518
CR2	1901-0731	1901-0731
DS1-DS6, DS8	1990-0652	HLMP-6220 1X4
DS7A, DS7B	1990-0685	1990-0685
DS9-DS11	1990-0667	1990-0667
DS12, DS16	1990-0673	5082-4690
DS13, DS15, DS17, DS19	1990-0675	5082-4590
DS14, DS18	1990-0674	5082-4990
U1, U13, U14	1820-1794	DM81LS95N
U2	1820-1997	SN74LS374PC
U3	1820-2074	P8085
U4	1818-0773	1818-0773
U5, U6	1818-0438	P2114
U7	1820-1216	SN74LS138N
U8	1820-1195	SN74LS175N
U9	1820-1197	SN74LS00N
U10	1820-1112	SN74LS74N
U11	1820-1208	SN74LS32N
U12	1820-1416	SN74LS14N
U15, U16, U17	1820-1730	SN74LS273N
U18	1820-1759	DM81LS97N
U19	1820-2138	DS8871N
U20	1820-1231	SN75492N
Y1	0410-1142	0410-1142
A2		
CR1, CR2	1901-0662	MR751
U1, U2	1826-0122	7805UC

NOTES

1. THE TRACE BETWEEN THE TERMINALS AT J1,J2,J3 AND J4 MUST BE CUT IF THESE LINES ARE TO BE USED WITH PERIPHERALS.

2. S3 SWITCHES ARE SHOWN IN CLOSED (LOGIC Ø) POSITION.

3. S1 SWITCHES ARE SHOWN IN CLOSED (NORM) POSITION. ALL SECTIONS OF S1 MUST BE OPEN FOR FREERUN MODE.

4. SIGNAL MNEMONICS PRECEDED BY THE LETTER "N" INDICATE LOW POLARITY.

SAM Schematic Diagram (through p. 422).

34. Describe the operation of the eight-switch input port.

35. Describe the operation of the eight-LED output port.

36. How are the scanning of the keyboard circuit and the scanning of the seven-segment display circuit related?

37. Describe the operation of the keyboard circuit.

38. Describe the operation of the seven-segment display circuit.

39. What is the function of DS7 and DS8?

40. Why does this circuit have a buffered data bus?

Web Site Questions

Go to the web site http://www.prenhall.com/cook, select the textbook *Practical Digital Electronics,* select this chapter, and then follow the instructions when answering the multiple choice practice problems.

420

422

Answers to Self-Test Evaluation Points

STEP 1-1
1. Section 1-1-1
2. Section 1-1-1
3. Section 1-1-2
4. Circuit simplicity and accuracy
5. 2
6. Section 1-1-2

STEP 1-2
1. CPU, memory, I/O
2. Analog to Digital Converter
   Digital to Analog Converter
3. Analog, digital
4. Digital, analog

STEP 2-1
1. No difference
2. Base 10
3. $2 = 1000, 6 = 100, 3 = 10, 9 = 1$
4. Reset and carry

STEP 2-2
1. 26_{10}
2. 10111_2
3. LSB = Least Significant Bit, MSB = Most Significant Bit
4. 1101110_2

STEP 2-3
1. (a) Base 10   (b) Base 2   (c) Base 16
2. $21_{10}\ 15_{16}$
3. BF7A
4. Binary equivalent = 100001
   Hexadecimal equivalent = 21

STEP 2-4
1. 8
2. 232_B
3. 4
4. 4267_B

STEP 3-1
1. OR, AND
2. Binary 1
3. AND gate
4. OR
5. Binary 0
6. AND

STEP 3-2
1. NOT gate
2. Binary 0
3. NAND
4. NOR
5. Binary 1
6. NAND

STEP 3-3
1. OR
2. XOR
3. XNOR
4. XNOR

STEP 3-4
1. IEEE/ANSI
2. Instead of using distinctive shapes to represent logic gates, it uses a special dependency notation system to indicate how the output is dependent on the input.

STEP 4-1
1. **a.** AND
   b. XOR
   c. OR
   d. NAND
   e. NOR
   f. NAND
2. $\overline{A \cdot B} = \overline{A} + \overline{B}$
3. **a.** $\overline{A} \cdot \overline{B}$ **b.** $(\overline{A} \cdot \overline{B}) \cdot C$
4. **a.**

STEP 4-2
1. **a.** 1
   b. 1
   c. 0
   d. A
2. $A(B + C) = (AB) + (AC)$
3. $(AB)C$ or $(AC)B$
4. $A(B + C + D)$

STEP 4-3

1. Product, sum
2. **a.** $A\,\overline{B}\,C\,D$
   b. $\overline{A}\,B\,C\,\overline{D}$
3. SOP
4. Fundamental products are ORed and then the circuit is developed.

STEP 4-4

1. 4
2. $y = B(A + \overline{A}) = B(1) = B$
3. Larger
4. With 1s for larger grouping

STEP 5-1

1. Transistor to transistor logic
2. Ten
3. HIGH
4. LOW, HIGH, and high impedance
5. High-speed applications
6. Higher packing density

STEP 5-2

1. True
2. Voltage control, high input impedance, and low power dissipation
3. P channel and N channel
4. Low power

STEP 5-3

1. **(a)** Transistor outline
   (b) Dual-in-line package
   (c) Surface-mount technology
2. SMT
3. (a) SSI
4. E-MOSFET, MOS

STEP 5-4

1. Lower cost
2. CMOS
3. Signal voltage and current incompatibility
4. Provides interfacing by shifting signal voltage for compatibility
5. BiCMOS family
6. TTL series

STEP 6-1

1. Section 6-1, margin definitions
2. Section 6-1
3. Section 6-1
4. Section 6-1

STEP 6-2

1. Section 6-2, margin definitions
2. Section 6-2
3. Section 6-2
4. Section 6-2

STEP 7-1

1. Voltmeter
2. Oscilloscope
3. Pulser
4. Current tracer, logic probe

STEP 7-2

1. New chip substitution
2. Open, short
3. Midpoint start during isolation
4. 2-state timing signal
5. Logic pulser and current tracer, logic probe

STEP 7-3

1. **a.** Record orientation of IC
   b. Desolder using solder wick or vacuum bulb (grounded iron is required with MOS devices)
   c. Avoid overheating board or component
2. A final test is the application of power and clock to ensure all aspects of the system operate properly.

STEP 8-1

1. All HIGH except line 9 which would be LOW
2. **a.** 10
   b. 16
3. Output "b" would be the only HIGH
4. Matching low power consumption

STEP 8-2

1. Ten to four line
2. Inverter gates wired to each 74147 output
3. 0111 (7 has priority)
4. **a.** 1101
   b. 0001
   c. HLHH
   d. OFF, ON, OFF, ON

STEP 8-3

1. (b) One of several, one
2. 4
3. False
4. Parallel
5. Parallel
6. 74157

STEP 8-4

1. (a) One, one of several
2. Decoder
3. $C = 1, B = 0, A = 0$
4. True

STEP 8-5

1. XNOR
2. The three outputs are labeled $A < B$, $A > B$, and $A = B$
3. For additional chips to handle larger word size
4. $A < B = 1, A > B = 0, A = B$ is 0

STEP 8-6

1.
Decimal	Gray Code	Odd-Parity Bit
0	0000	1
1	0001	0
2	0011	1
3	0010	0
4	0110	1
5	0111	0

2. True
3. Both, as a parity generator and a parity checker
4. 8-bits (7-bit ASCII code plus 1-bit for parity)

STEP 8-7
1. Diagnose, isolate, and repair
2. A single pulse condition with no clock
3. Midpoint selection to begin isolation
4. Both displays would be OFF at the same time and ON at the same time. Displays would therefore show same data.

STEP 9-1
1. Latch
2. Level-triggered, edge-triggered, and pulse-triggered
3. No change, race
4. **a.** Edge triggering
   b. Pulse triggering
5. A latch to eliminate switch noise
6. Input delay is accomplished by capacitor charge time

STEP 9-2
1. Data
2. No
3. D flip-flop has a single bit input.
4. (d) All of the above

STEP 9-3
1. (d)
2. HIGH, toggle
3. (b) A 63 kHz square wave
4. (c) *J-K, D*-type, *S-R*

STEP 9-4
1. Monostable
2. Bistable
3. Pulse generation, pulse delay, and timing
4. An internal 5Ω voltage divider

STEP 10-1
1. (c) PIPO
2. Data storage

STEP 10-2
1. 8 pulses
2. SIPO, PISO
3. 1 pulse
4. Bidirectional

STEP 10-3
1. High impedance (floating)
2. BUS
3. Bus conflict occurs when more than one signal seeks control of the bus due to lack of signal isolation. It is avoided through the use of tristate devices.
4. HIGH, LOW, and control inputs

STEP 10-4
1. LOAD = bus data enters register. ENABLE = data read (output) is allowed
2. **a.** A bus is a group of parallel wires designed to carry binary data.
   b. A bus organized digital system occurs when one set of wires is connected to all devices—as long as those that transmit data onto the bus have three-state output control.
   c. Bus conflict occurs when more than one signal seeks control of the bus due to lack of signal isolation.
   d. A 16-bit bus utilizes 16 wires.

3. Universal Asynchronous Receiver and Transmitter. It can be used to interface the parallel data format within a digital electronics system to the serial data format of external equipment.
4. 64 times larger
5. The feedback path is twisted with the johnson counter.

STEP 10-5
1. Are not, do not
2. Modulus = 256, maximum count = 255
3. 30 ns $\times$ 8 = 240 ns, $1 \times 10^9 \div 240 = 4.2$ MHz
4. Natural modulus = 256, preset modulus = 226

STEP 10-6
1. False
2. Asynchronous has fewer gates, synchronous counter is faster
3. Carry, borrow
4. Count

STEP 10-7
1. Binary counter, frequency divider
2. They function as counters.
3. They function as frequency dividers.
4. They function as counters.

STEP 10-8
1. Diagnose, isolate, and repair
2. Count window 74LS160s function as frequency dividers. Mod-10 74LS160s function as counters.
3. No problem exists within dividers.
4. Display should read 0000 Hz.

STEP 11-1
1. **(a)** 1011 + 11101 = 101000
   (b) 1010 − 1011 = 1111 0001
   (c) 101 $\times$ 10 = 1010
   (d) 10111 $\div$ 10 = 1011
2. **(a)** 0101
   (b) 1001 0101
   (c) 1101
   (d) 0111 1001
3. **(a)** −5
   (b) +6
   (c) −5
   (d) +6
4. Adder circuit

STEP 11-2
1. 2, 3 bits
2. (c) Parallel adder
3. The 7483 is an 8-bit, parallel, 2-word adder or a 4-bit binary full-adder with fast carry.
4. SIGMA

STEP 11-3
1. ICs are the two 7486 XOR chips.
2. A 1 is added via the carry-in input of the 7483.
3. **(a)** $F = \overline{B}$
   (b) $F = A$ minus B
4. 1001 ($F = A$ plus B)

STEP 12-1
1. Nonvolatile
2. A memory's address input is a gating system designed to access a certain cell. The number of address lines is related to these memory locations as function of 2^n.

3. (a) Storage density is a comparison of one memory content to another.
   (b) Memory configuration describes how the circuit is organized into groups of bits or words.
   (c) Burning-in is the process of entering data into a ROM.
   (d) Access time is the interval between a new address input and the data stored at that address being available at the output.
4. (a) MROM = Masked Read-Only Memory
   (b) PROM = Programmable Read-Only Memory
   (c) EPROM = Erasable Programmable Read-Only Memory
   (d) EEPROM = Electrically Erasable PROM
5. Input code is applied to the address inputs of a ROM where it is decoded internally to select that memory location, which, in turn, supplies the desired output code.
6. A ROM listing is a sheet that lists the data values that should be stored in each memory location within the ROM. ROM tests may be made by accessing each word location to check stored content with the listing.

STEP 12-2

1. Once programmed, the contents of a ROM cannot be changed.
2. With a SAM, one must sequence through all bits until the desired word is located. This is accomplished directly with a RAM due to its random-access capabilities.
3. No. Although access is the same, a ROM has no "write" capability.
4. A volatile device
5. A dynamic RAM MOS cell capacitor requires refreshing, whereas a static RAM has bipolar cells.
6. A SRAM has fast, simple circuitry, whereas the DRAM has more packing density and lower power consumption due to fewer MOSFETs per cell.
7. A memory map is an address-listing diagram showing which address blocks are assigned to each part of the system.
8. RAMs are tested by writing a test pattern into every memory location and then reading out and testing the stored contents. The stored patterns could be all 0s, then all 1s and then a checkerboard pattern.

STEP 13-1

1. (a) ADC = Analog-to-Digital Converter
   (b) DAC = Digital-to-Analog Converter
2. ADC, DAC
3. Analog, digital
4. Digital, analog

STEP 13-2

1. A weighted resistor DAC disadvantage is that as bits increase, large branch current differences may introduce noise or cause output load change.
2. The op-amp within a DAC is used to convert a current input into an equivalent voltage output.
3. (a) A staircase is a continually increasing analog output.
   (b) Resolution of a DAC is the smallest analog output change that may occur as a result of an increment in the digital input.
   (c) Step size is the DAC resolution.
   (d) Monotonicity describes correct staircase waveform behavior.
   (e) Settling time is the time it takes for a change in a DAC's analog output to settle to 99.95% of the new value.
   (f) The relative accuracy of a DAC describes how much the output level has deviated from its ideal theoretical output value.

4. DACs may be tested with an input counter to produce an output that may be viewed on an oscilloscope to determine high gain, offset, nonmonotonicity, and other problems.

STEP 13-3

1. The main disadvantage of the staircase ADC is that its conversion speed is dependent on the analog input voltage.
2. The staircase and successive approximation ADC circuits make use of a DAC.
3. A successive approximation ADC operates by first setting and then comparing bits to obtain an equivalent digital output.
4. The flash converter is the fastest type of ADC.
5. An ADC may be tested by applying a linear ramp to the analog input while monitoring the digital output with a display.

STEP 14-1

1. a. Machine language—the digital code understood by the MPU
   b. Assembly language—programs written using mnemonics, which are easier to remember
   c. High-level language—an independent language designed to make programming easier that later is compiled into machine language for the MPU
2. A flowchart is a graphic form that describes the operation of a program using oval, rectangular, and diamond-shaped blocks.
3. a. Instruction set—contains a list of the machine codes that are recognized by a specific type of microprocessor
   b. Peripherals—input and output devices connected to microprocessor system ports
   c. Programmer—a person who writes a program in any language that will be understood by the MPU
   d. Mnemonic—a short name, or memory aid, used in assembly language programming
   e. Compiler—a program that translates high-level language into machine code
   f. Accumulator—an MPU internal register (register A), used in conjunction with the ALU; a storage location within an MPU chip
   g. Flag register—an internal register within an MPU chip that is a collection of flip-flops used to indicate the results of certain instructions
   h. Op-code—the operation code or "do" portion of an instruction

STEP 14-2

1. A unidirectional bus receives data from only one transmitter chip, or "talker." Data transfers are therefore one-way only. A bidirectional bus has several "talkers"; therefore, data flow may be reversed. However, those transmitters must have three-state output control to avoid bus conflict.
2. 8085 pin-out signals are:
   a. WR—indicates data bus data is to be written into the selected memory or I/O location
   b. RD—indicates selected memory or I/O location is to be read. Data bus is ready for transfer.
   c. IO/NM—HIGH indicates I/O read/write is to I/O; LOW indicates read/write is to memory
   d. RESET—resets CPU. Once the program counter, interrupt enable, and HLDA flip-flops are reset, it allows a peripheral read.
   e. READY—HIGH allows completion of read/write cycle, while a LOW creates a wait state
   f. ALE—the address-latch enable signal clocks the lower-order address bits for storage

g. *HOLD*—causes CPU to relinquish the use of buses at the end of the present machine cycle for use by another master, usually a peripheral

h. *TRAP*—highest-priority signal of any interrupt request; recognized as INTR occurs

3. During FETCH, the op-code is read from memory and transferred to the instruction register. During EXECUTE, control/timing logic decodes the op-code, increments the program register, and retrieves the bytes necessary to complete the task.

4. a. Interrupt—provides a means for devices external to the microprocessor to request immediate action by the processor

b. T-state—the basic unit of time for a microprocessor system equal to one clock period

c. Machine cycle—the time it takes an operation to be completed; from three to six states

d. Instruction cycle—the time it takes to fetch and execute a complete instruction; usually from one to five machine cycles.

APPENDIX
B Answers to Odd-Numbered Problems

Chapter 1
1. b 3. d 5. c 7. b

Chapter 2
1. b
3. c
5. d
7. a
9. d
11. 1001, Decimal 9
13. Decimal equivalents are:
   a. 57   **b.** 39   **c.** 06
15. BCD equivalents are:
   a. 0010 0011 0110 0101   **b.** 0010 0100
17. Codes in order are: BCD, Gray, Excess-3, and ASCII.
19. Stx practice makes perfect etx.

Chapter 3
1. a   11. b
3. b   13. d
5. b   15. c
7. b   17. b
9. b   19. a
21. OR gate
23. Light, $Y = 1$; dark, $Y = 0$
25. OR
27. Truth table outputs (Y) should be:

a. 0 (AND)	c. 0 (XOR)	e. 1 (NOR)
0	1	0
0	1	0
1	0	0

b. 0 (OR)	d. 1 (XNOR)	f. 0 (NOT)
1	0	1
1	0	
1	1	

29. C = HIGH
   Y_0 0
   Y_1 1
   Y_2 1
   Y_3 0

Chapter 4
1. c   11. c
3. c   13. a
5. b   15. b
7. c   17. c
9. a   19. c
21. **a.** $Y = (\overline{A} \cdot B) + (B \cdot C) + D$
   b. $Y = (A \cdot B) + (\overline{A} \cdot \overline{B})$
   c. $Y = (\overline{\overline{A} \cdot \overline{B} \cdot \overline{C}}) + (\overline{A \cdot B \cdot \overline{C}})$
   d. $Y = (\text{LIT} + \text{NLT}) \times (\text{SLW} + \text{FST})$

23. **a.**

   b.

25.
$$Y_2 = \overline{B}\,\overline{C} + \overline{A}BC$$

27. $Y_1 = \overline{A}\,\overline{B}\,\overline{C}\,\overline{D} + \overline{A}\,\overline{B}C\overline{D} + \overline{A}B\overline{C}\,\overline{D} + \overline{A}BC\overline{D} +$
   $A\overline{B}\,\overline{C}D + A\overline{B}CD + AB\overline{C}\,\overline{D} + AB\overline{C}D +$
   $ABC\overline{D} + ABCD$

$$Y_1 = \overline{A}\,\overline{B} + AB + AD$$
(50 inputs to 9 inputs)

29.

(Question 25, Y_1)

428

(Question 27, Y_2)

Chapter 5

1. c	**9.** d
3. c	**11.** a
5. a	**13.** d
7. a	**15.** a

17. NOR logic, any HIGH in = LOW out
19. Symbol means device will have open collector output
21. Yes, 24 mA needed, 33 mA is maximum
23. 74H05
25. Truth table output is:
    a. 0
      1
      Float
      Float
27. The 74LS125 has a LOW activated control input while the 72LS126 has a HIGH activated control input.
29. Circuit clips, or rectifies, an ac signal, which is then formed into a positive pulse waveshape within the trigger circuit.
31. N-type E-MOSFET
33. NOR Function

Chapter 6

1. c	**11.** c
3. c	**13.** b
5. a	**15.** d
7. a	**17.** b
9. e	

19. PLA
21. Input/output function table
23. **a.** 0000110   **c.** 0100100
    b. 0100100   **d.** 0001100
25. An intact fuse at the NOR inputs will generate a LOW output.

Chapter 7

1. b
3. a
5. b
7. b
9. b
11. Pulse or rectangular wave (positive going)
13. **a.** P_w = 50 ns
    b. Period = 500 μs
    c. f = 2 kHz
    d. Duty Cycle = 10%
15. No
17. Replace 7427
19. 7408 has open output

Chapter 8

1. c	**11.** b
3. c	**13.** b
5. a	**15.** a
7. d	**17.** d
9. a	**19.** d

21. 4-Line BCD to 10-Line Decimal Decoder
23. Common-Anode
25.

D	C	B	A	Output
1	0	1	0	INVALID
0	0	1	0	LED 2
1	0	0	1	LED 9
3	2	1	0	

    Input
    Switches

27. BCD to seven-segment decoder/driver circuit
29. Active-HIGH
31. Continuous counting
33. Encoder
35.

	A	B	C
(1)	1	1	1
(2)	1	1	0
(3)	1	0	1
(4)	0	0	0
(5)	0	0	1
(6)	0	1	0
(7)	0	1	1

37. D_0 D_7
    1 0 1 1 0 0 1 1
39. 4(100), block, OFF-ON-ON

Chapter 9

1. c	**9.** d
3. d	**11.** b
5. c	**13.** c
7. b	**15.** b

17. All zeros
19. Negative edge triggering
21. Reset all Q outputs LOW when $\overline{CLR}$ is taken LOW.
23. OFF; $X = 0$ ON; $X = 1$
25. Positive edge-triggered D-type flip-flop
27.

    200 kHz CLK
    100 kHz Q_A
    50 kHz Q_B

29. Positive edge triggering
31. When D input and clock are in phase, Q will go HIGH, and LED will turn OFF.
33. NAND S-R flip-flop
35. Edge triggered
39.

Clock

J   0   1   0   1   1   1   0   0

K   0   0   1   1   1   0   0   1

Q

$\overline{Q}$

41. Astable
43. Non-retriggerable
45. Negative-edge triggered
47. 24–34(a)

Chapter 10

1. b	**11.** b	**19.** c
3. a	**13.** a	**21.** a
5. a	**15.** d	**23.** d
7. b	**17.** c	**25.** a
9. d		

27. D-type
29. SIPO
31. **a.** 15
    b. 63
    c. 16,777,199
33. $f = \dfrac{1 \times 10^9}{n \times t_p}$
    a. 100 ns, 10 MHz
    b. 150 ns, 6.6 MHz
    c. 600 ns, 1.7 MHz
35. 32
37. ÷32
39. 9.1 MHz
41. **a.** Depressing SW1 momentarily will FLOAT the CLR input HIGH, clearing the counter. The clock NAND gate is controlled by the selected output from 74154.
    b. The 74154 decoder output lines produce a LOW decoded output.
    c. After being cleared, the counter will count from 0000 to 1100. Line 12 (pin 14) will go LOW when the counter applies 1100 to the decoder's input, stopping the clock and turning the LED ON.

Chapter 11

1. b	**9.** d
3. d	**11.** d
5. a	**13.** b
7. d	**15.** c

17. **a.** 1001 0000
    b. 0001 0001
    c. 0110 0000
19. **a.** 1011 (R 110)
    b. 0010 0010
    c. 0001 0101 (R 001)
21. **a.** 1001 0110
    b. 0101 0101
    c. 0000 1111
23. **a.** +119 = 0111 0111
    b. −119 = 1000 1001
    c. −34 = 1101 1110
25. −32 to +31
27. To store and apply the control "s" inputs to the 74181.
29. **a.** Set input switches for A PLUS B "s" code (HLLH), pulse E_I HIGH, pulse $\overline{L_C}$ LOW. Set input switches for A word value, pulse E_I HIGH, $\overline{L_A}$ LOW. Set input switches for B word value, pulse E_I HIGH, pulse $\overline{L_B}$ LOW. Set Logic/Arith input LOW, pulse E_F HIGH, pulse $\overline{L_0}$ LOW.
    b. Input words as in **a.**, except code for A or B will be LLLH (M = L) or HHHL (M = H). Set Logic/Arith input to match "s" control code, pulse E_F HIGH, pulse $\overline{L_0}$ LOW.
    c. Input words as in **a.**, except code for A and B will be HLHH. Set Logic/Arith input HIGH, pulse E_F HIGH, pulse $\overline{L_0}$ LOW.

Chapter 12

1. b	**11.** b
3. b	**13.** b
5. b	**15.** d
7. d	**17.** a
9. c	**19.** a

21. Enable ROM IC (Chip Select)
23. When installed in a circuit, this MROM is a read-only device.
25. 2316 is designated as U4 on schematic.
27. U7, pin 15
29. U3, 8085
31. Control read/$\overline{\text{write}}$ control
33. $\overline{CS}$ should be active, $\overline{WE}$ should be HIGH
35. U18 pin 3
37. A_{11}, A_{12}, and A_{13}
39. **a.** ROM   **e.** INPUT
    b. RAM   **f.** SCAN
    c. CONTROL   **g.** OUTPUT
    d. KEY DATA   **h.** DISPLAY
41. Display address bus logic levels
43. Yes

Chapter 13

1. c
3. c
5. b
7. b
9. d
11. **a.** Step size $= \dfrac{V_0}{2^n} = \dfrac{12\text{ V}}{16} = 0.75$ V
    b. % Res. $= \dfrac{\text{s/size}}{V_0} = \dfrac{0.75\text{ V}}{12\text{ V}} = 6.25\%$
    c. LSB = 12 V/16 = 0.75 V
           1 = 12V/8 = 1.5 V
           2 = 12V/4 = 3.0 V
       MSB = 12V/2 = 6.0 V
    d. 0010 = 1.5 V
    e. 1011 = 8.25 V
13. **a.** $5\left(\dfrac{64 + 32 + 1}{256}\right) = 1.9$ mA
    b. $5\left(\dfrac{16 + 5}{256}\right) = 0.4$ mA
    c. $5\left(\dfrac{255}{256}\right) = 4.9$ mA
    d. $5\left(\dfrac{16 + 1}{256}\right) = 0.33$ mA
15. DAC
17.

19. **a.** 0.012 mV
    b. 0.0004%
    c. LSB = 12 mV, 24 mV, 48 mV, 96 mV, 384 mV, 768 mV, MSB = 1.54 V
    d. 60 mV
    e. 2.2 V
21. ADC
23. Temperature change will vary dc input voltage to ADC. Equivalent digital output is decoded by PLD and displayed in decimal on 7-segment displays.
25. Flash converter

27.

MSB	LSB
0	0
0	1
1	0
1	1

29. 1010, 1100, 1110, 1111, 1110, 1100, 1001, 0111, 0101, 0011, 0001, 0011, 0101, 0111.

Chapter 14

1. c		**9.** c	
3. b		**11.** b	
5. b		**13.** b	
7. b		**15.** c	

17. U4 is a 2 k $\times$ 8 ROM.

19. A talker (read-only)

21. Together U5 and U6 form a 1 k $\times$ 8 RAM.

23. Both talkers (read) and listeners (write)

25. External crystal, determines clock signal frequency output (pin 37)

27. The serial output data goes to U18A (6), and then on to the speaker.

29. Pin 30 (address latch enable, ALE): when active it indicates an address is present at the microprocessor's address and address/data output pins (active-LOW).
Pin 31 (write control output): indicates to external devices that data on data bus should be written into addressed location (active-LOW).
Pin 32 (read control output): indicates to external devices that addressed device should place addressed data on data bus (active-LOW).

31. Displays contents of address bus.

33. Displays contents of buffered data bus

35. When U15 is enabled by NOUT, the contents on the data bus will be latched and displayed on the eight common-anode LED display.

37. See section 29-2-1

39. These LEDs show which device has been enabled by the address decoder.

Index